Techniques and Methodological Approaches in Breast Cancer Research

Jose Russo • Irma H. Russo

Techniques and Methodological Approaches in Breast Cancer Research

Springer

Jose Russo
Fox Chase Cancer Center
Philadelphia, PA, USA

Irma H. Russo
Fox Chase Cancer Center
Philadelphia, PA, USA

ISBN 978-1-4939-0717-5 ISBN 978-1-4939-0718-2 (eBook)
DOI 10.1007/978-1-4939-0718-2
Springer New York Heidelberg Dordrecht London

Library of Congress Control Number: 2014934436

Printed on acid-free paper

Springer is part of Springer Science+Business Media (www.springer.com)

We dedicate this book to the memory of Irma H. Russo, M.D., who passed away on June 25, 2013.

I met Irma in Medical School in 1962. Our first encounter was in the lobby of the hospital where we were both doing our training. We noticed each other as we crossed paths, and when I smiled at her she smiled back. But it wasn't until several years later, when we began working in the Laboratory of Development and Cell biology together, that we really started talking and getting to know each other.

We married on February 9 of 1968 and from that point forward our life together was sealed by a continuous flow of communication of ideas and dreams.

Irma was a natural in science. She was not afraid of asking big questions and was able to see the problem of cancer in different ways and from multiple angles. In 1973 we consolidated our laboratories and, fittingly, called it the Breast Cancer Research Laboratory. Since then we have dedicated our energy towards forming a coherent research program that would also act as a training center in breast cancer.

Irma's compassion and magnetism drew people from around the globe to our lab. She always looked beyond initial appearances and was able to see the true potential in those who approached her seeking opportunities to work in breast cancer research.

Irma was also a thinking machine; she introduced new ideas about the role of the hypothalamus in the breast cancer, circadian rhythm, the developmental concept in breast cancer, the idea of windows of susceptibility to cancer and developed the idea that hormones like hCG could be used in cancer prevention.

Through the years, we became fused and it was difficult to know whether an idea was hers or mine. But what made Irma unique was that she never stopped asking questions about the meaning of what we were doing and the final repercussions of our work. One of her main desires was to leave behind a legacy; a Breast Cancer Research Laboratory that could continue beyond our lifetimes. On December 18, 2013 Fox Chase Cancer Center honored her memory and scientific contributions by formally dedicating the lab to which she gave so much of herself in her name. May the "Irma H Russo, M.D., Breast Cancer Research Laboratory at Fox Chase Cancer Center" continue to strive towards the complete understanding and eradication of this disease.

Jose Russo, M.D.

Preface

This book provides descriptions of the most common techniques used in breast cancer research with a special emphasis on the study of the normal breast as the basis for understanding the pathological processes in this organ.

The authors have poured their 37 years of experience in the field of breast cancer research into this book and described the process of handling and interpreting the complex organ that is the human breast using original data and unique material collected over the years.

Chapter 1 discusses the normal architecture of the human breast using the whole mount preparations and provides the parameters of how to identify each single structure in the tissues of nulliparous and parous women.

Chapter 2 describes possible sources of normal breast tissue, from autopsy specimens to core needle biopsies, and the morphological characteristics of each source when the normal tissue is studied. The pros and cons of each source, depending on the study planned in this organ, are also outlined.

Chapter 3 is an introduction to laser capture microdissection (LCM) in both fresh tissue and paraffin-embedded material. A detailed analysis of the extraction of DNA and RNA and the quality required for molecular studies is described and analyzed.

In Chap. 4, the technique of tissue microarray (TMA) is described and guidelines for the preparation of TMA for breast cancer as well as normal breast tissue are provided. There is original information on the limitations and quality expected for TMA of normal breast tissue arrays.

Chapter 5 discusses the role of tissue culture in the understanding of breast cancer with a historical perspective, followed by a description of the cell lines frequently used in any breast cancer research laboratory. Space is dedicated to analysis of the phenotypes commonly used for detecting malignant transformation in vitro and the use of cell lines for testing epigenetic targeting processes.

Chapter 6 provides novel information on detection of stem cells in paraffin-embedded tissue and data on how the use of immunocytochemical markers of stemness can provide prognostic indication of tumor behavior.

Chapter 7 provides a comparative analysis of the use of flow cytometry and immunofluorescence techniques compared with immunocytochemistry. The value of in situ hybridization is also explored with a critical approach to the use of these techniques in breast cancer laboratories.

Chapter 8 is dedicated to the use of the in vivo model for studying breast cancer with extensive description of the xerograph model and the use of the tail vein and intra-auricular injection of neoplastic cells for studying metastasis.

Chapter 9 describes the use of RNA sequencing in breast tissue and the methodology for purification of small amounts of RNA when the tissue obtained is from normal core biopsies.

Finally, in Chap. 10, the latest techniques for the study of epigenetic changes in the human breast are described. This chapter provides a new path for studying epigenetic changes induced by physiological processes in the normal breast, providing a methodology on how to analyze the data and their interpretation.

Altogether, the authors have established the methodological basis for studying the human breast in its normal and pathological conditions, providing a valuable reference for those researchers who are pursuing a deeper understanding of the human breast.

Philadelphia, PA, USA

Jose Russo, M.D., F.A.C.P.
Irma H. Russo, M.D., F.A.C.P.

Acknowledgments

Our special acknowledgment and thanks to *Ms. Patricia A. Russo* for her insightful editorial suggestions, critiques, and the delightful moments spent discussing the manuscript and its ideas. Our thanks also go to *Ms. Rose Sonlin* for verifying the accuracy of the references and, lastly, to *Pathology Consultation Services* in Rydal, PA, that financed the writing and editing of this book.

Jose Russo, M.D., F.A.C.P.
Irma H. Russo, M.D., F.A.C.P.

Contents

Contributors

Maria Barton, M.S. Irma H Russo MD Breast Cancer Research Laboratory, Fox Chase Cancer Center, Philadelphia, PA, USA

Steven G. Doll Drexel University at the Irma H Russo Breast Cancer Research Laboratory, Fox Chase Cancer Center, Philadelphia, PA, USA

Janine Gomes Irma H Russo MD Breast Cancer Research Laboratory, Fox Chase Cancer Center, Philadelphia, PA, USA

Pedro J. Gutiérrez Diez, Ph.D. School of Economics, University of Valladolid, Valladolid, Spain

Nathan Hopfinger, B.S. Irma H Russo MD Breast Cancer Research Laboratory, Fox Chase Cancer Center, Philadelphia, PA, USA

Ricardo Lopez de Cicco, Ph.D. Irma H Russo MD Breast Cancer Research Laboratory, Fox Chase Cancer Center, Philadelphia, PA, USA

Justin Newton Irma H Russo MD Breast Cancer Research Laboratory, Fox Chase Cancer Center, Philadelphia, PA, USA

Theresa Nguyen, B.S. Irma H Russo MD Breast Cancer Research Laboratory, Fox Chase Cancer Center, Philadelphia, PA, USA

Colleen O'Malley Drexel University at the Irma H Russo Breast Cancer Research Laboratory, Fox Chase Cancer Center, Philadelphia, PA, USA

Julia Santucci-Pereira, Ph.D. Irma H Russo MD Breast Cancer Research Laboratory, Fox Chase Cancer Center, Philadelphia, PA, USA

Thomas J. Pogash, B.S. Irma H Russo MD Breast Cancer Research Laboratory, Fox Chase Cancer Center, Philadelphia, PA, USA

Jose Russo, M.D., F.A.C.P. Irma H Russo MD Breast Cancer Research Laboratory, Fox Chase Cancer Center, Philadelphia, PA, USA

Irma H. Russo, M.D., F.A.C.P. Irma H Russo MD Breast Cancer Research Laboratory, Fox Chase Cancer Center, Philadelphia, PA, USA

Fathima Sheriff, M.D. Irma H Russo MD Breast Cancer Research Laboratory, Fox Chase Cancer Center, Philadelphia, PA, USA

Ryan R. Smalley Drexel University at the Irma H Russo Breast Cancer Research Laboratory, Fox Chase Cancer Center, Philadelphia, PA, USA

Yanrong Su, Ph.D. Irma H Russo MD Breast Cancer Research Laboratory, Fox Chase Cancer Center, Philadelphia, PA, USA

Chapter 1
The Use of Whole Mounts for Studying the Architecture of the Human Breast

Jose Russo and Irma H. Russo

Keywords Whole mounts • Architecture of the human breast • Lobular structures • Tridimensional reconstruction • Parenchyma–stroma relationship • Familial breast cancer • Subgross anatomy

1.1 Role of Age in the Architecture of the Normal Human Breast

1.1.1 Introduction

Age is an important risk factor for breast cancer because this type of cancer, practically nonexistent before age 24, exhibits maximal incidence during the postmenopausal years [1–5]. The majority of breast cancer patients are women in their sixth and seventh decades of life, and the mortality for breast cancer continues to rise after menopause. The age-specific incidence, that is, the number of cases per year per 100,000 women in each age group, climbs rapidly after the age of 30, reaching a peak of maximal incidence of 500 cases per 100,000 women in the 60- to 70-year-old group [1].

No other organ presents such dramatic changes in size, shape, and function as does the breast during growth, puberty, pregnancy, and lactation [6, 7]. It is agreed that the developmental phase of the human breast starts as early as the stage of nipple epithelium during embryonic development, continuing steadily with body

J. Russo, M.D. (✉) • I.H. Russo, M.D.
Irma H Russo MD Breast Cancer Research Laboratory, Fox Chase Cancer Center, Cottman Avenue 333, Philadelphia, PA 19111, USA
e-mail: jose.russo@fccc.edu; Irma.Russo@fccc.edu

J. Russo and I.H. Russo, *Techniques and Methodological Approaches in Breast Cancer Research*, DOI 10.1007/978-1-4939-0718-2_1,

growth, and undergoing a spurt of growth with lobule formation at puberty [6–8]. Lobular formation is accomplished by interaction of two processes, development and differentiation, both of which take place almost simultaneously [7–9].

1.1.2 Lobular Structures in Whole Mount Preparations

Using whole mount preparations we have been able to identify terminal structures (Figs. 1.1a–d and 1.2a–c) and lobular structures, each one representing sequential developmental stages [7] (Tables 1.1 and 1.2). Lobules type 1 are the most undifferentiated lobular structures because they are present in the immature female breast before menarche (Figs. 1.2b–d, 1.3a–d, 1.4a–f, and 1.5a–d). They are composed of clusters of 6–11 ductules per lobule. Lobules type 2 evolve from the lobules type 1 and have a more complex morphology, being composed of a higher number of ductular structures per lobule (Fig. 1.5a–d). These progress to lobules type 3, which are characterized by having an average of 80 ductules or alveoli per lobule (Table 1.2); they are frequently seen in the breasts of women under hormonal stimulation or during pregnancy [6] (Figs. 1.6a–d, 1.7a–d, and Table 1.2). A fourth type of lobule, lobule type 4, has been described during the lactational period of the mammary gland but is not found in the breast of nulliparous postpubertal women. It is considered to be the maximal expression of development and differentiation [7–9]. Each of these structures has different proliferative activity, the highest being in the structures of the terminal end bud and the lowest in the lobule type 3 (Fig. 1.8).

The identification and quantization of these lobular structures serve as an important parameter for evaluating the effect of age and reproductive history on the normal development of the gland, as well as the pathological response of the mammary parenchyma to either hormonal stimulation or unknown carcinogenic stimuli. The fact that the breast is the source of the most frequent malignancy in the female population, and the knowledge that breast cancer is heavily influenced by the reproductive history of the individual, requires a thorough understanding of the developmental pattern of the breast during the life-span of a woman. It is known that women with a history of early pregnancy are at a lower risk of developing breast cancer than nulliparous women [10–15]. Based upon studies in an experimental animal model, we have postulated that the protective effect of pregnancy is due to differences in the degree of differentiation of the breast [16], since in rodents the initiation of the neoplastic process is inversely related to the degree of differentiation of the mammary gland, which in turn is a function of age and reproductive history [17–23]. In the study of the pathogenesis of human breast cancer, it has been reported that the terminal ductal lobular unit (TDLU) [10, 24], which is equivalent to lobules type 1 [8, 10], is the site of origin of pre-neoplastic lesions such as atypical ductal hyperplasias, which evolve into ductal carcinoma in situ and progress to invasive carcinoma, development of atypical lobular hyperplasia,

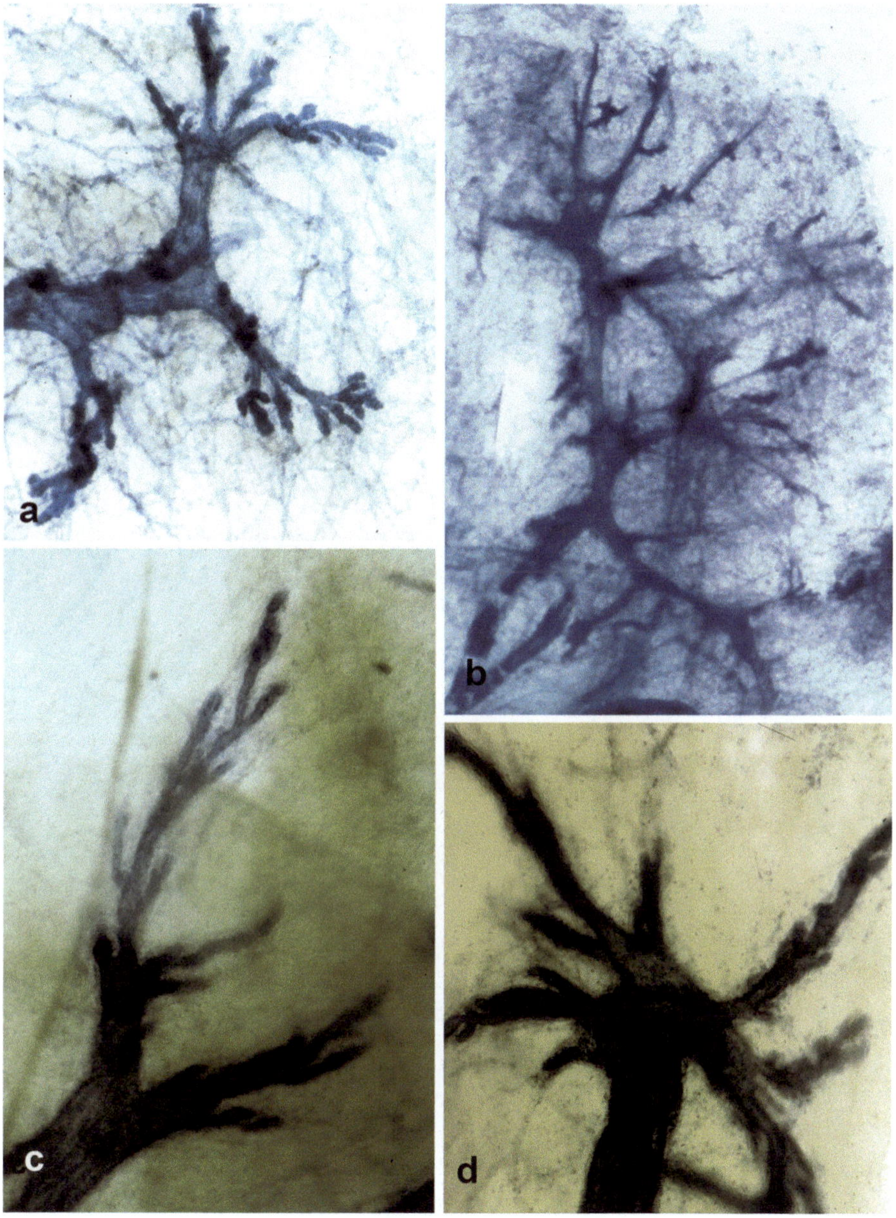

Fig. 1.1 Whole mount preparation stained with toluidine blue. (**a**–**d**) Terminal structures of the breast, ×25

and lobular carcinoma in situ. Lobules type 3 might originate hyperplastic or hypersecretory lobules, fibroadenomas, sclerosing adenosis, and apocrine cysts [10]. These observations suggest that the degree of differentiation or lobular development of the breast is influencing the type of tumors developed by this organ [10].

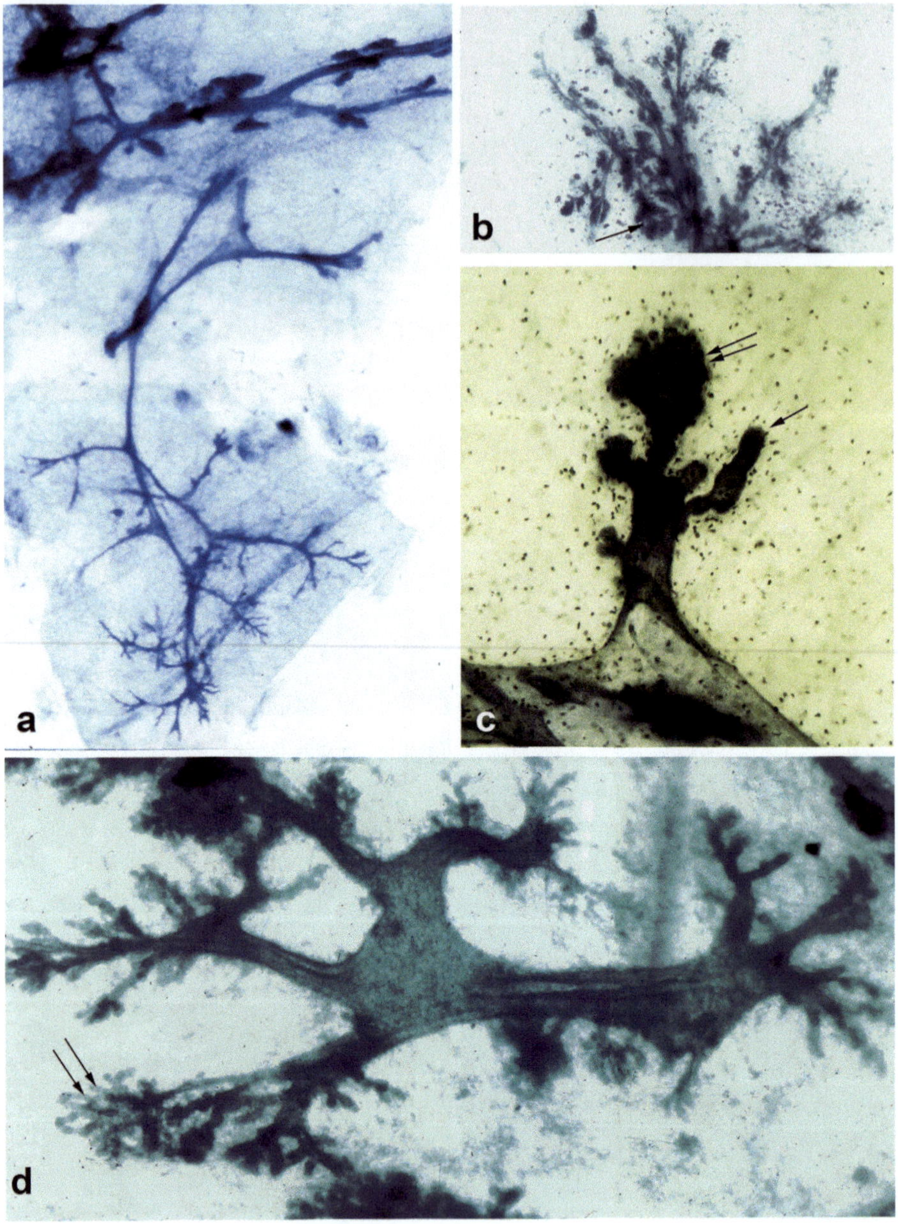

Fig. 1.2 Whole mount preparation Stained with toluidine blue. (**a**) Terminal structures of the breast, ×10. (**b**) Terminal structures and lobules type 1 (*arrow*), ×10. (**c, d**) Terminal structures (*arrow*) and lobules type 1 (*double arrows*), ×25

Table 1.1 Characteristics of the lobular structures of the human breast

Structure	Lobular area[a] (μm²)	No. of ductules/lobule[b]	No. of cells/cross section[c]
Lob 1	48±44	11.2±6.3	32.4±14.1
Lob 2	60±26	47.0±11.7	13.1±4.8
Lob 3	129±49	81.0±16.6	11.0±2.0

[a]Student's *T* tests were done for all possible comparisons. Lobular areas showed significant differences between Lob 1 vs. Lob 3 and Lob 2 vs. Lob 3 ($p<0.005$)
[b]The number of ductules per lobule was different ($p<0.01$) in all the comparisons
[c]The number of cells per cross section was significantly different in ductules of Lob 1 vs. 2 and 3 ($p<0.01$)

Table 1.2 Profile of the compartments of human breast

Lobular structure	Surface area[a]	No. of alveolar buds/structure
Type 1	$0.232\times10^{-2}\pm0.090\times10^{-2}$	11.2±6.34
Type 2	$0.167\times10^{-2}\pm0.035\times10^{-2}$	47.0±17.0
Type 3	$0.125\times10^{-2}\pm0.029\times10^{-2}$	81.0±16.6
Type 4	$0.120\times10^{-2}\pm0.050\times10^{-2}$	180.0±20.0

[a]Surface area expressed in mm. The area was measured in a Zeiss Video plan. Statistical difference was determined by the *t* test. Type 1 vs. Type 2, 3, and 4 lobules, $p<0.001$. Type 2 vs. 1, 3, and 4 lobules, $p<0.001$

The objective of this chapter is to describe the development of the human breast and determine how physiologic factors, such as aging and pregnancy, influence the distribution of lobular structures.

1.1.3 Use of the Whole Mounts for Studying the Role of Aging and Reproductive History in the Human Breast

1.1.3.1 Experimental Procedures

Fifty-one human breast samples obtained from bilateral or unilateral reduction mammoplasty (RM) were analyzed to determine the quantity and type of parenchymal structures present in them. These samples corresponded to 40 parous women ranging in age from 18 to 57 years and 11 nulliparous women ranging in age from 14 to 54 years. An average of 100 g of tissue was processed from every breast sample. The tissue was fixed in *10 %* neutral buffered formalin for a minimum of 24 h and processed for whole mount preparation by defattening in acetone and hydration in decreasing concentrations of ethanol. After rehydration the tissue was stained in toluidine blue for 2 h and differentially destained with pure methanol and 70 % ethanol for 30 min each, washed in distilled water, and fixed in 4 % ammonium molybdate for 30 min. After staining, the tissue was dehydrated and cut into 1 mm thick

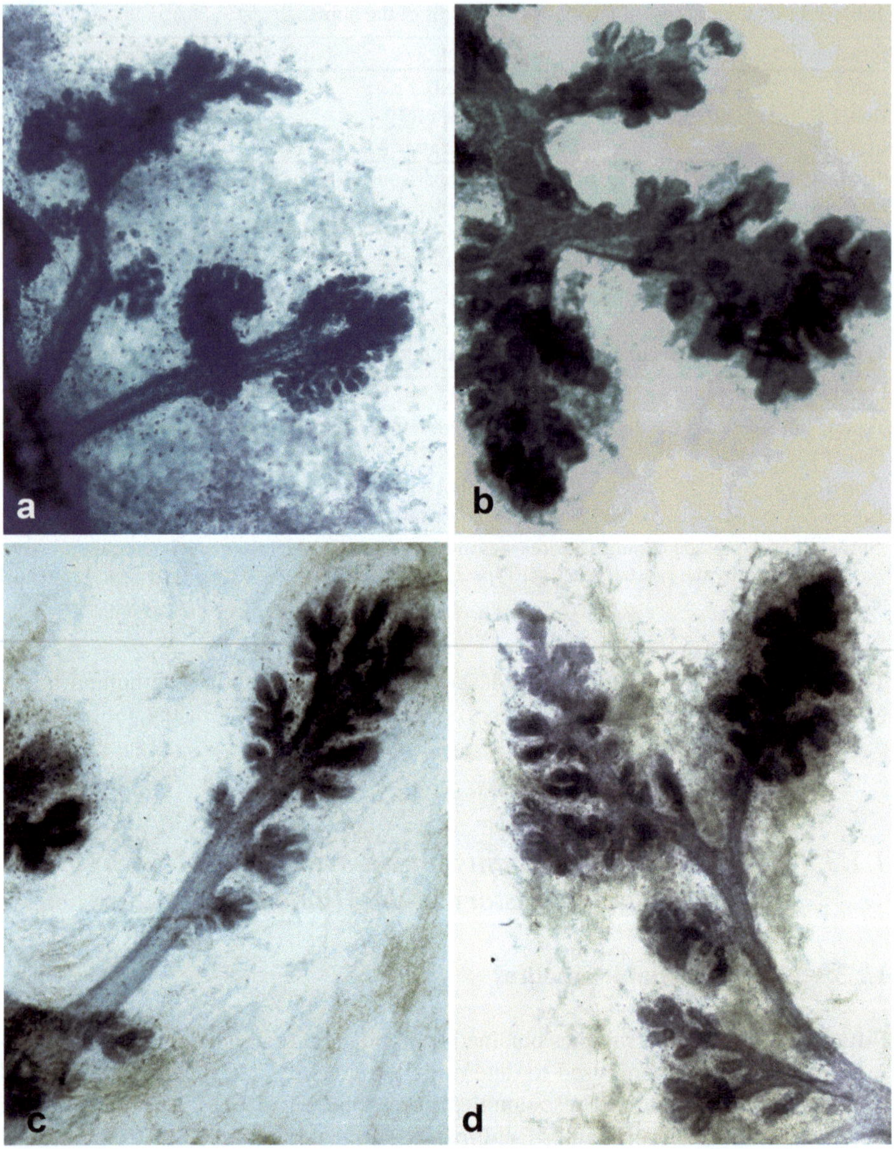

Fig. 1.3 Whole mount preparation stained with toluidine blue. (**a–d**) Lobules type 1 in the breast, ×25

slices, which were mounted on glass slides and cover slipped [8] (Fig. 1.9). An average of 12 slides per sample, or a total of 650 slides, were examined, and 31,222 structures were classified and counted. AII tissues were examined under a stereomicroscope for classification and quantization. The type and characteristics of lobular structures

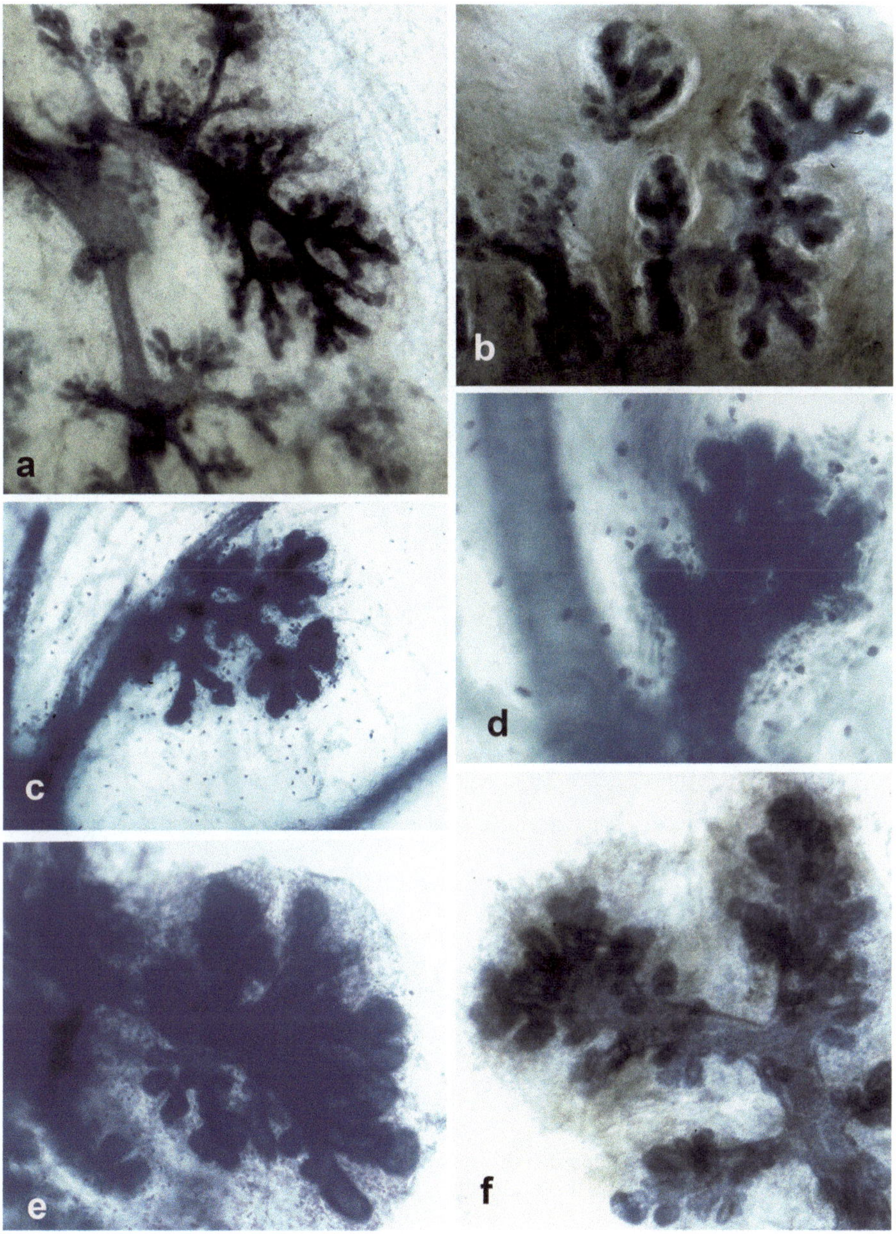

Fig. 1.4 Whole mount preparation stained with toluidine blue. (**a–f**) Lobules type 1 in the breast, ×25

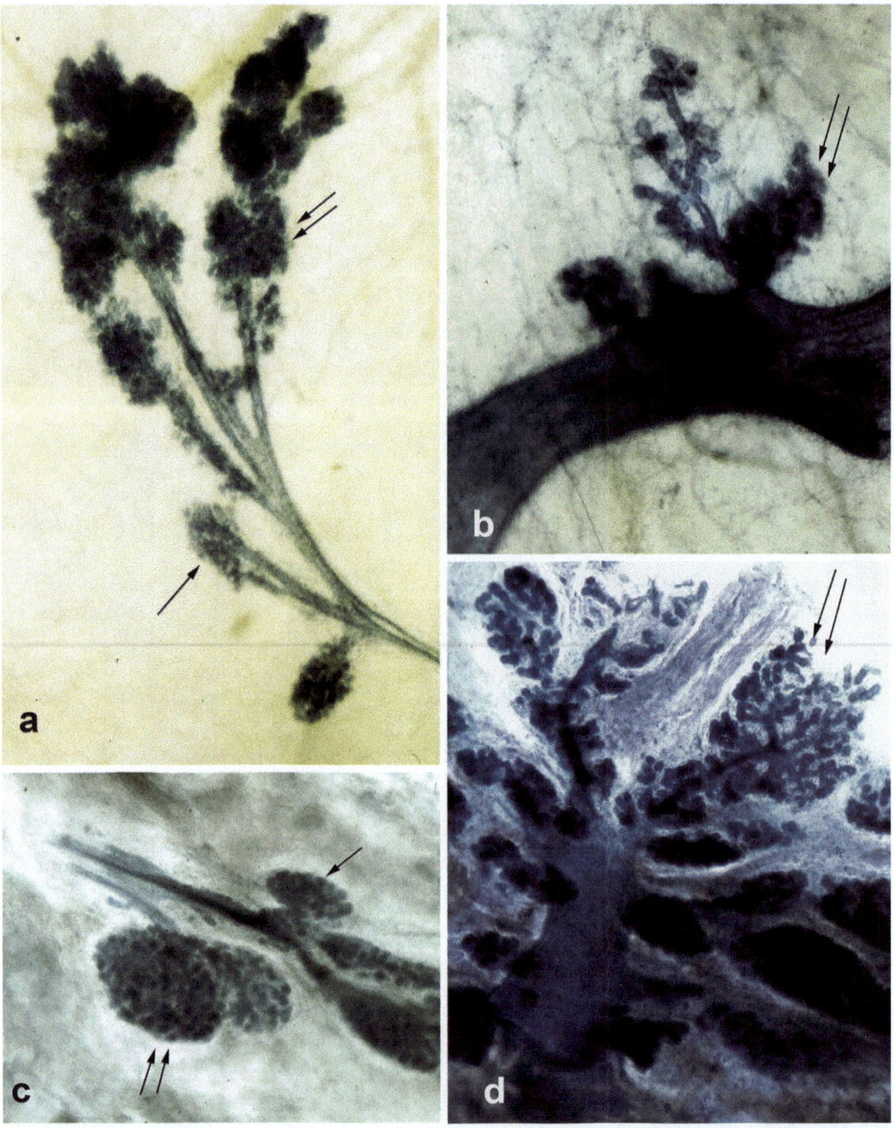

Fig. 1.5 Whole mount preparation stained with toluidine blue. (**a–d**) Lobules type 1 (*arrow*) and lobules type 2 (*double arrows*) in the breast, ×25

identified are summarized in Table 1.2. Terminal structures (TS) were defined as all straight blind-ending or club-shaped terminal ducts with no connection with lobular structures (Figs. 1.1 and 1.2). For every specimen the total number of TS and lobule types 1, 2, and 3 were counted, and the relative percentage of every structure was obtained. Analysis of variance was used to determine the differences related to age

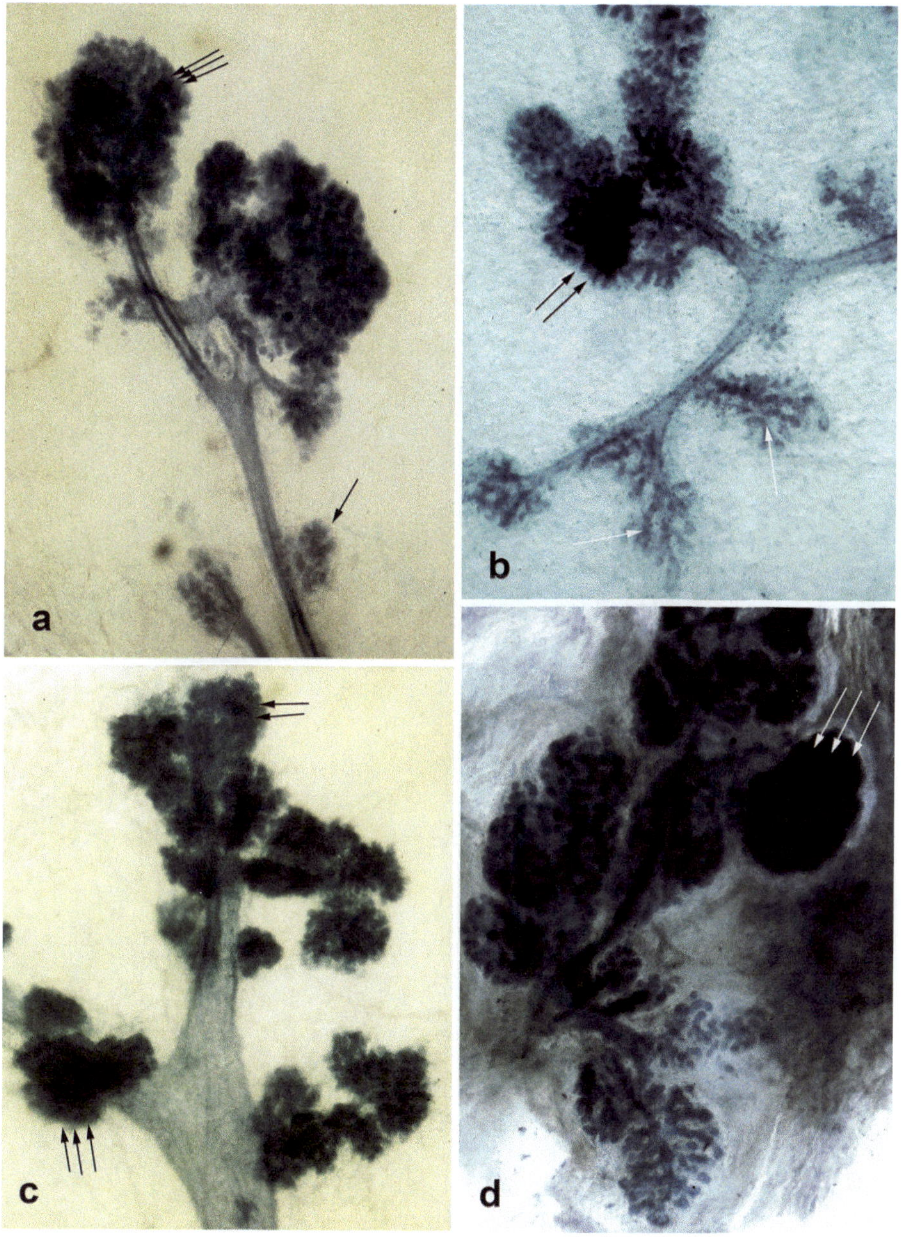

Fig. 1.6 Whole mount preparation stained with toluidine blue. (**a–d**) Lobules type 1 (*arrow*), lobules type 2 (*double arrows*), and lobules type 3 (*triple arrows*) in the breast, ×25

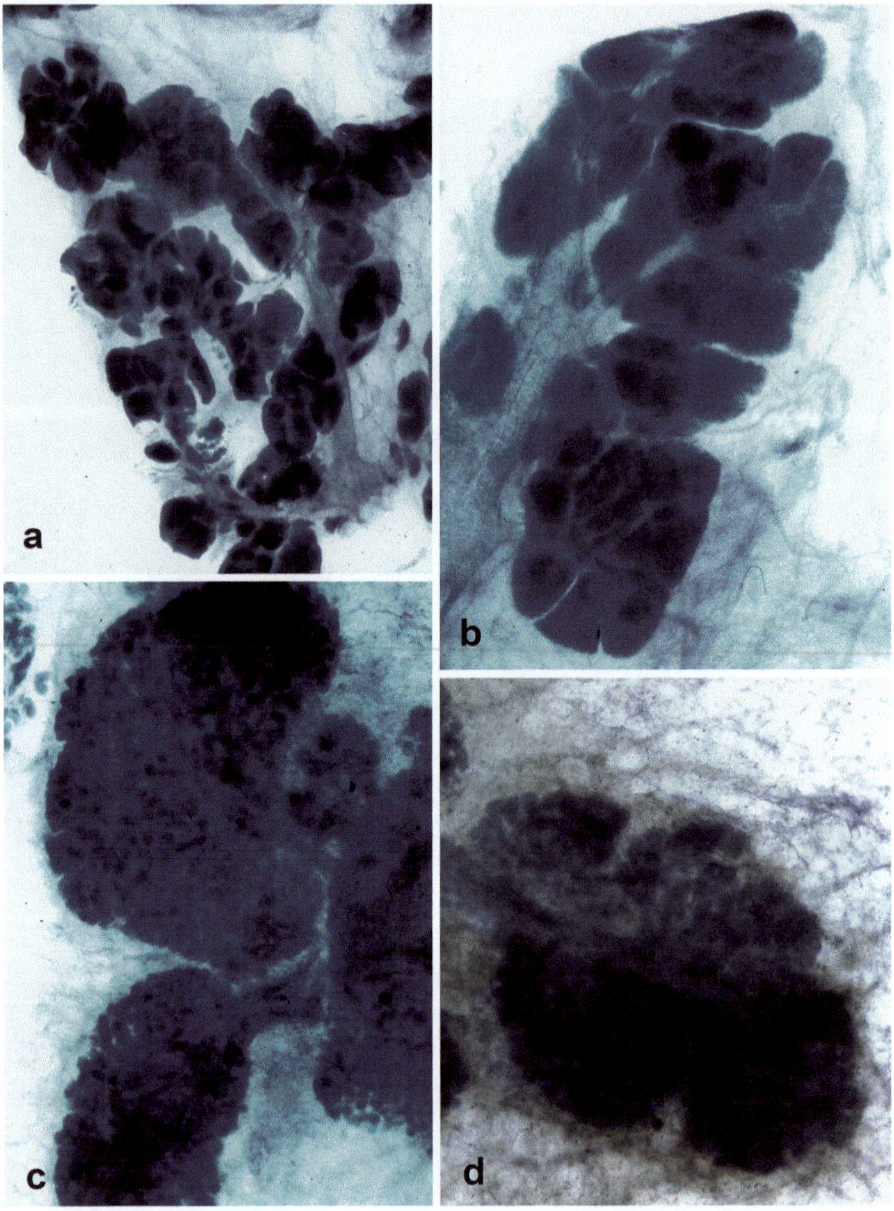

Fig. 1.7 Whole mount preparation stained with toluidine blue. (**a**–**d**) Lobules type 3 in the breast

and parity histology in the relative proportion of each one of the different structures of the breast tissue. The analysis of the relationship between age and reproductive history with the percentage of each individual lobular structure present in the breast was also tested using single regression analysis [25, 26].

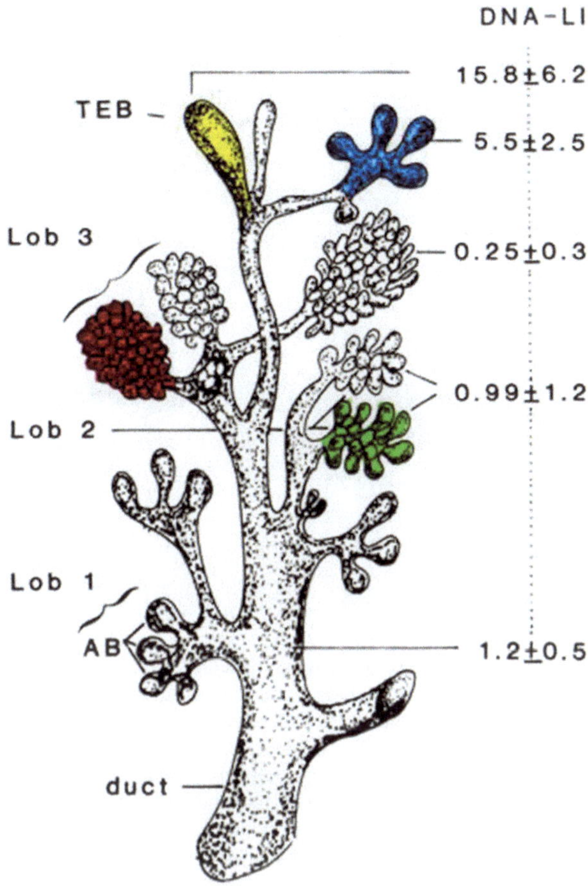

Fig. 1.8 Schematic representation of the different structures of the human breast and the proliferative activity, determined as DNA labeling index (Reprinted with permission of: Russo, J. and Russo, I.H. Development of Human Mammary Gland In: The Mammary Gland Development, Regulation, and Function. (M.C. Neville and C.W. Daniel, eds.) Plenum Pub. Corp., 1987 pp 67–93)

1.1.3.2 The Terminal Structures (TS)

The percentage of TS and of lobular structures identified using the parameters described in Table 1.1 was analyzed in all tissue samples and plotted against age from 14 to 58 years, separated at 4-year intervals, resulting in nine different age groups. Results were plotted for the total patient population without consideration of the parity history of the donor (Fig. 1.10) or separately in nulliparous (Fig. 1.11) and parous women (Fig. 1.12). In the total population of breast tissues studied, TS represented *10 %* of all the structures in tissues from women between the ages of 14 and 18 years, decreasing to less than 1 % between the ages of 19 and 40, and having a significant increase to the values found in younger women after 44 years of age ($p<0.05$) (Fig. 1.10) [26].

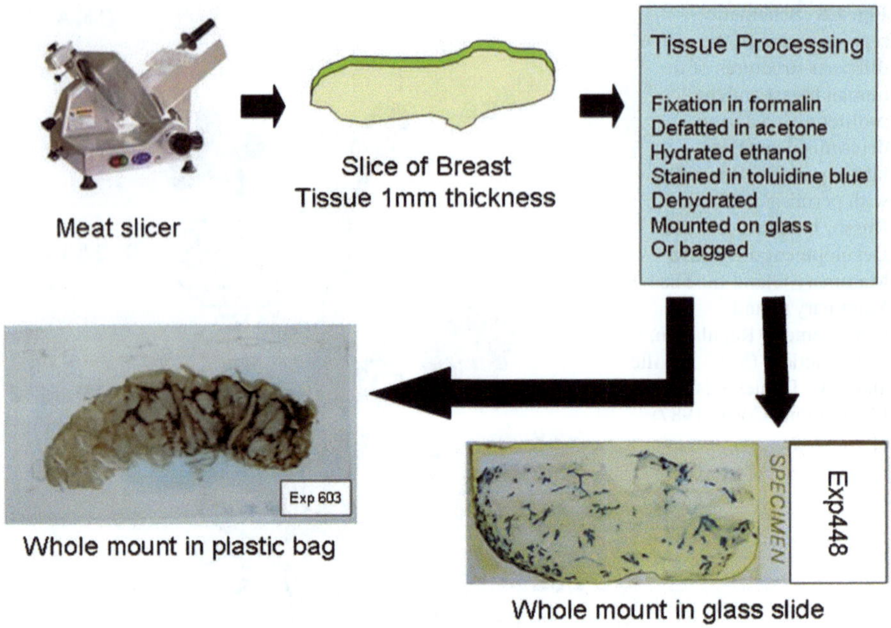

Fig. 1.9 Each breast was serially cut with a meat slicer into 1 mm thick slices. Each slice was sequentially numbered, labeled, and individually placed in a cloth bag for processing for whole mount by first defattening the tissues by immersion in acetone with continuous agitation for 2 or more days, and then hydrated in decreasing concentrations of ethanol, stored overnight in distilled water, and stained in 0.025 % toluidine blue solution at pH 7.6 for 2.5 h. The tissues were washed in distilled water, destained in methanol and 70 % ethanol, washed, fixed in ammonium molybdate, dehydrated, and cleared in xylene. Each slice of breast tissue was stored in glycerin within a polyethylene, 4-mL thick KapaklScotchpak pouch, and heat sealed with a Scotchpak sealer (Kapak, St. Louis Park, MN)

1.1.3.3 The Lobular Structures

The breasts of women between the ages of 14 and 18 years contained similar proportions of Lob 1, Lob 2, and Lob 3. Only after age 23 were there significant differences represented by a reduction in Lob 1, and an increase in Lob 3 ($p<0.01$), while no significant change in Lob 2 occurred (Fig. 1.10) [26]. These relative proportions remained unchanged until the fourth decade of life, after which Lob 1 increased to become the predominant structure, constituting almost 70 % of the total thereafter ($p<0.001$) (Fig. 1.10). Lobules type 2, which maintained a steady percentage (20 %) between the ages of 14 and 40 years, decreased to approximately 5 % after the fifth decade of life, but the differences were not significant. Lobules type 3, which started increasing significantly after 23 years of age, reached approximately 70 % of the total by the third decade of life, remaining at the same level until the mid-forties, decreasing significantly to about 15 % during the fifth and sixth decades of life ($p<0.001$) (Fig. 1.10). Regression analysis for age and percentage of structures

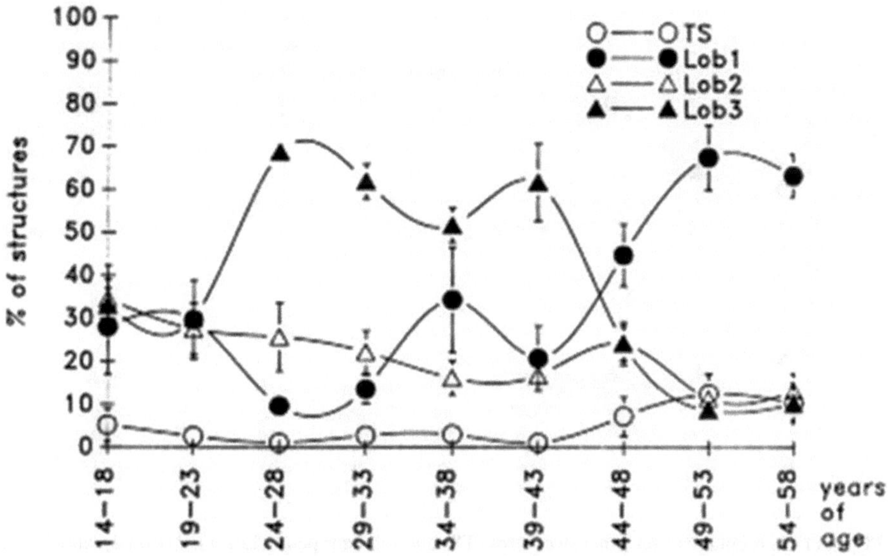

Fig. 1.10 Percentage of terminal structures (TS) and lobule types 1 (Lob 1), 2 (Lob 2), and 3 (Lob 3) in the breasts of both nulliparous and parous women between 14 and 58 years of age. Values represent the mean ± SE of 3 samples (14–18); 9 samples (19–23); 7 samples (24–28); 6 samples (29–33); 7 samples (34–38); 5 samples (39–43); 7 samples (44–48); 2 samples (49–53); and 4 samples (54–58) (Reprinted with permission of: Russo J, Rivera R, Russo IH (1992) Influence of age and parity on the development of the human breast. Breast Cancer Res Treat 23:211–218)

present in the mammary gland revealed that there was a positive correlation between age and percentage of TS and lobules type 1 (0.82 and 0.56, respectively) [26]. There was a negative correlation between age and percentage of lobules type 3 (−0.60). There was no correlation between age and the percentage of lobules type 2 present [26].

When the percentage of parenchymal structures was plotted separately for nulliparous and parous women, significant differences were found (Figs. 1.11 and 1.12). In nulliparous women the most frequent structures found at all ages were the lobules type 1, which constituted almost *50 %* of the structures present in the group of 14–18 years of age. Their number steadily, although slightly, increased between ages 20 and 30, although this increase was not statistically significant. Between the third and sixth decades of life their number remained constant (Fig. 1.11). The second most frequent structure was the lobule type 2, which constituted 20–30 % of the total between ages 14 and 23, decreasing progressively thereafter (Fig. 1.11). TS represented approximately 10 % of the total number of structures, and this percentage remained almost constant until the fourth decade of life, when an increase to 20 % was observed ($p < 0.05$) (Fig. 1.12).

Lobules type 3 were rarely seen in the mammary gland of nulliparous women, since they constituted less than 10 % of all the structures, a percentage that remained steadily low at all ages studied (Fig. 1.12). In the breast of parous women, on the

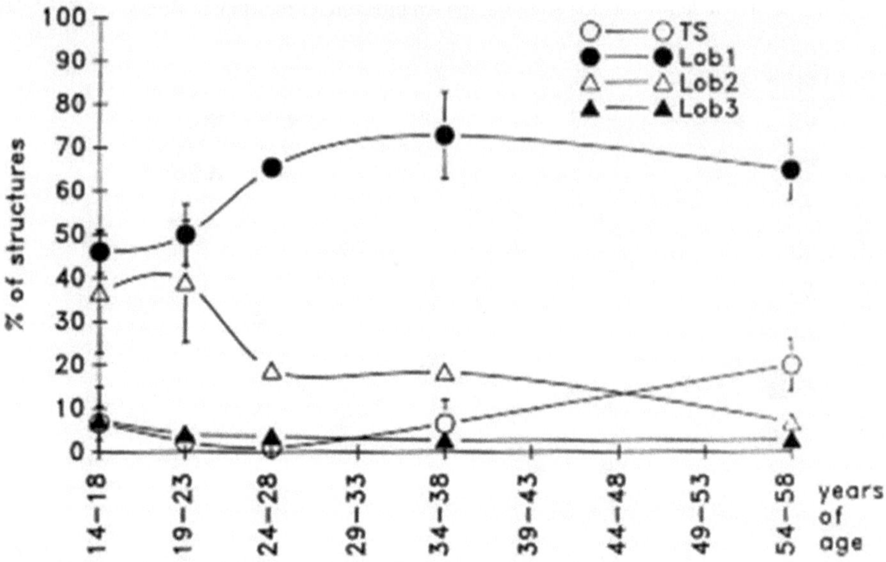

Fig. 1.11 Percentage of terminal structures (TS) and lobule types 1 (Lob 1), 2 (Lob 2), and 3 (Lob 3) in the breasts of nulliparous women between 14 and 58 years of age. Values represent the mean ± SE of 3 samples (14–18); 4 samples (19–23); 1 sample (24–28); 2 samples (34–38); 1 sample (54–58); no samples were available for the groups 29–33, 39–43, 44–48, or 49–53 (Reprinted with permission of: Russo J, Rivera R, Russo IH (1992) Influence of age and parity on the development of the human breast. Breast Cancer Res Treat 23:211–218)

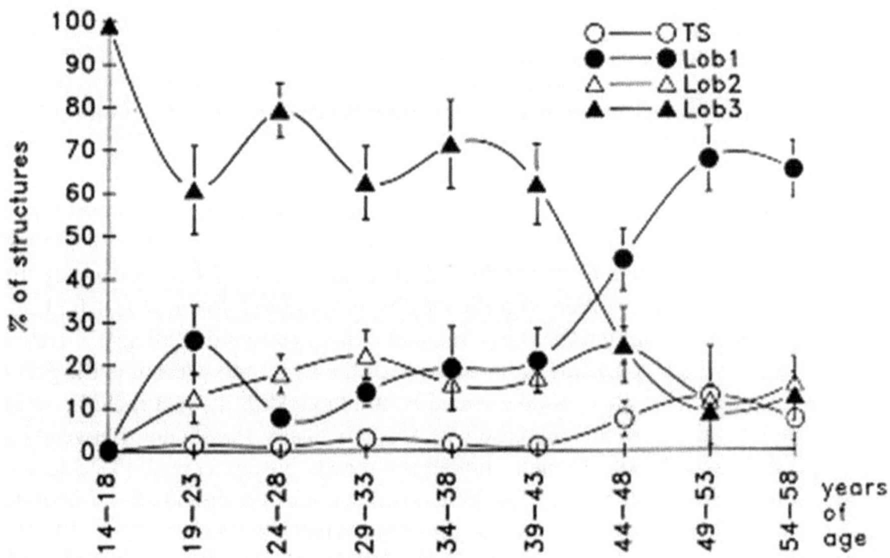

Fig. 1.12 Percentage of terminal structures (TS) and lobule types 1 (Lob 1), 2 (Lob 2), and 3 (Lob 3) in the breasts of nulliparous women between 14 and 58 years of age. Values represent the mean ± SE of 1 sample (14–18); 5 samples (19–23); 6 samples (24–28); 6 samples (29–33); 5 samples (34–38); 5 samples (39–43); 7 samples (44–48); 2 samples(49–53); and 3 samples (54–58) (Reprinted with permission of: Russo J, Rivera R, Russo IH (1992) Influence of age and parity on the development of the human breast. Breast Cancer Res Treat 23:211–218)

other hand, the most frequent structure found was the lobule type 3 (Fig. 1.12). The 14- to 18-year-old group was represented by a single sample of an 18-year-old parous woman in which almost 100 % of the structures were Lob 3. This percentage decreased to 70–80 % in the group of women between the second and the fourth decades of life [26].

After the fourth decade of life the number of lobules type 3 decreased further to 20 % ($p < 0.01$). It continued decreasing during the sixth decade of life, in which Lob 3 represented approximately 10 % of the total. Lob 1 and Lob 2 exhibited a similar curve, representing approximately 25 % of the total between the second and fourth decades of life. By the middle of the fourth decade, Lob 1 increased sharply ($p < 0.05$), whereas Lob 2 decreased, but not significantly (Fig. 1.12). It was found that there was a positive correlation between the percentage of TS and lobules type 1 with aging (0.74 and 0.87, respectively), and a negative correlation between the percentage of lobules type 3 and aging (–0.90). These observations suggested that with aging there was an involution of the lobules type 3 to type 1, whose percentage in parous women became similar to that of nulliparous women only after the fourth decade of life (Figs. 1.1 and 1.12) [26].

1.1.3.4 The Comparative Study of the Nulliparous and Parous Breast

The comparison between the architecture of the breast of nulliparous and parous women revealed that the breasts of nulliparous women contain almost exclusively lobules type 1, whose percentage remains nearly constant throughout the life-span of the individual, whereas the same structures in the parous women are in a very low percentage during the younger years, although they increase significantly, reaching the same level as that of the nulliparous women only after the fourth decade of life. These observations suggest that with aging the lobules type 3 regress to lobules type 1, whereas in nulliparous women lobules type 1 never reached the degree of differentiation found in women that have gone through pregnancy. The possible evolution of the lobular structures in the breasts of nulliparous and parous women is depicted in Fig. 1.13.

Both age and parity strongly influence the architecture of the mammary gland parenchyma. The breasts of nulliparous women contain more undifferentiated structures such as terminal ducts and lobules type 1, although occasional lobule types 2 and 3 are seen, an observation also described by other authors [27]. In parous women, on the other hand, the predominant structure is the most differentiated lobule type 3. In contrast to the predominant structure in nulliparous women, the lobule type 1, which remains constant throughout the life-span, the predominant lobule type 3 in parous women peaks during the early reproductive years and decreases after the fourth decade of life [26].

In the breasts of nulliparous women, lobules type 2 are present in moderate numbers during their early years and decrease sharply after age 23, whereas the number of lobules type 1 remains significantly higher. This observation suggests that a certain percentage of lobules type 1 might have progressed to lobules type 2, but the number of lobules progressing to type 3 remains significantly lower than in the parous woman. In the case of parous women it is interesting to note that a history of

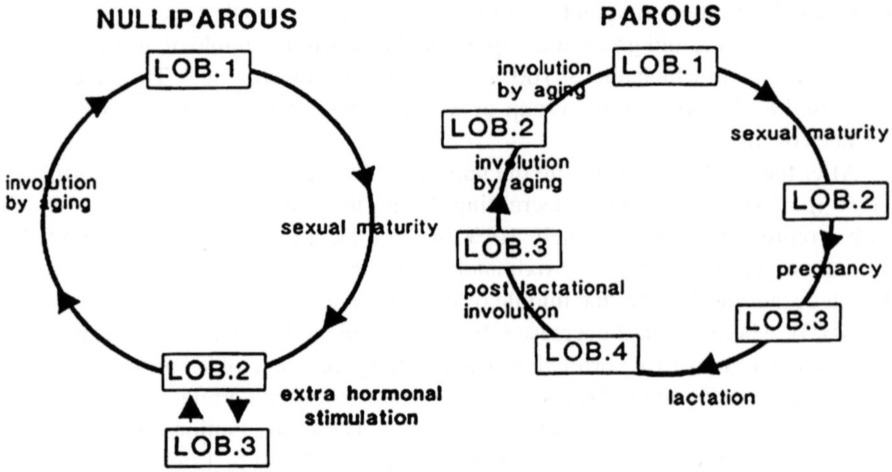

Fig. 1.13 Schematic representation of breast development based upon relative percentage of lobules present. Nulliparous women's breasts contain primarily lobules type 1 (Lob 1) with some progression to type 2 (Lob 2), and only minimal formation of lobules type 3 (Lob 3). Parous women undergo a complete cycle of development through the formation of lobules type 4 (Lob 4) which later regress (Reprinted with permission of: Russo J, Rivera R, Russo IH (1992) Influence of age and parity on the development of the human breast. Breast Cancer Res Treat 23:211–218)

parity between the ages of 14 and 20 correlates with a significant increase in the number of lobules type 3, which remain present as the predominant structure until the age of 40, the time at which a decrease in the number of lobules type 3 occurs, probably due to their involution predominantly to lobules type 2. It has been shown by several authors that postmenopausal involution is accompanied by diminution and atrophy of the parenchymal components [8, 28–30]. Among the changes described in the lobular structures is the increase in intralobuar fibrosis and hyalinization, which are interpreted as a response to the lack of hormonal stimulation [31]. We have observed that lobules type 1 are present in the breasts of both nulliparous and parous postmenopausal women, although the lobules type 1 found in the breasts of parous women present a higher frequency of hyalinization in the intralobular stroma than the lobules of nulliparous women [26]. During the postmenopausal years, both parous and nulliparous women have breasts with a preponderance of lobules type 1.

Although ductal breast cancer originates in lobules type 1, or TDLUs [10], the epidemiologic observation that nulliparous women exhibit a higher incidence of breast cancer than parous women [10, 11] indicates that lobules type 1 in these two groups of women might be biologically different, or exhibit different susceptibility to carcinogenesis [22, 32, 33]. The presence of lobules type 1 in the breasts of parous women has also been interpreted as a failure of the mammary parenchyma to respond to the influence of pregnancy and lactation [26, 34]. If this is the case, then parous women could contain in their breasts unstimulated as well as regressed lobules type 1. Therefore, the question that remains to be answered is whether the lobules type 1, which are present in the breasts of parous women, are as sensitive to carcinogenesis as the lobules found in the breasts of nulliparous women. We speculate that the lobules type 1 in the breasts of parous women are terminally differentiated structures that are the consequence of a regressive process, as depicted in Fig. 1.8. In addition, the proliferative activity of the lobules type 1 is lower in the breasts of parous women than in the breasts of nulliparous women [35–37].

The use of whole mounts (Fig. 1.9) [26] in the architectural analysis of the human breast has helped us understand that breast development requires a horizontal study in which all the different phases of growth are taken into consideration. For example, the analysis of breast structures at a single given point, i.e., ages 49–53 years, would lead us to conclude that the breasts of nulliparous and parous women appear identical. However, the phenomena occurring in prior years have imprinted permanent changes in the breast biology that affect the potential of the breast for neoplasia, but are no longer morphologically observable. This horizontal study allowed us to determine that parous women underwent lobular differentiation, whereas nulliparous women seldom reached the lobule type 3 stage (compare Figs. 1.11 and 1.12). This may further explain why parous women are more protected against carcinogenesis. In parallel studies, it has been shown that lobule types 1, 2, and 3 exhibit different cell kinetic characteristics: lobule types 1 and 2 grow faster in vitro and have a higher DNA labeling index (Fig. 1.8) and a shorter doubling time than lobules type 3 [33, 35]. They also exhibit different susceptibility to carcinogenesis. Lobule types 1 and 2 express higher survival efficiency in vitro when treated with chemical carcinogens like DMBA and MNU (Fig. 1.14), changes that are not manifested by lobule type 3 (Fig. 1.14) [32]. We have also shown that during the fourth and fifth decades of life there is a decrease in the number of both lobules type 2 and type 3 [26]. We have postulated that lobules type 2 are the site of origin of both lobular hyperplasia and carcinoma in situ [10]. Since it has been reported that the incidence of atypical lobular hyperplasia decreases significantly with advancing age [31], it is possible to postulate that the observed diminution in lobules type 2 is responsible for the decreased incidence of this type of lesion. In the experimental system we have found that pregnancy, as well as differentiation of the gland induced by the placental hormone human chorionic gonadotropin (hCG), follows a pattern of differentiation similar to the one reported here [38, 39]. The data discussed above are further supported by this experimental system, in which it has been observed that full-term pregnancy as well as treatment of virgin rats with hCG induces differentiation of the mammary gland which results in protection from chemically induced carcinogenesis. The stimulus of pregnancy or of exogenous hormones furthers the differentiation of the terminal end bud, stimulating lobule development [38, 39]. The fact that the protection conferred by these processes is maintained even after the termination of either pregnancy or hormonal treatment clearly indicates that the differentiation induced by these processes is a permanent modification of the biological characteristics of the mammary gland, even though the differentiated structures have regressed to seemingly more primitive conditions [26]. Collectively, the data presented here establish a baseline for understanding the evolution of glandular development, and how it is influenced by age and parity. This knowledge is of utmost importance for understanding the role of differentiation in the protection of the mammary gland against carcinogenesis [26, 36]. In addition,

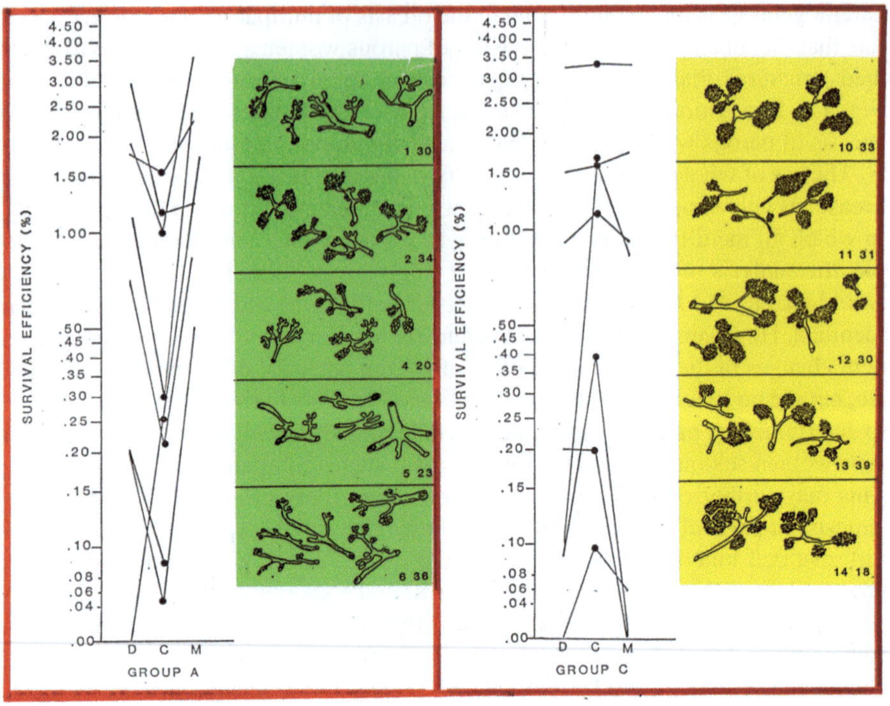

Fig. 1.14 Schematic representation of human breast samples classified in two groups: nulliparous (*left panel* or group A) and parous (*right panel* or group C). The first digit(s) indicates the sample number, and the second two digits are the age of the woman. The *left panel* or group A shows the Lob 1 and terminal structures of the breast tissue of nulliparous women that when placed in tissue culture showed an increased survival efficiency when treated with DMBA (D) and MNU (M). This effect was not observed in the samples obtained from parous women (*right panel* or group C) (Reprinted with permission from: Russo, J., and Russo Irma H. Role of differentiation on transformation of human breast epithelial cells. IN: Cellular and Molecular Biology of Mammary Cancer (edited by Medina D, Kidwell W., Heppner G., and Anderson E. Plenum Publishing Corporation 1997, pp399))

these data establish well-defined endpoints for studying the response of the mammary gland to hormonal or chemopreventive agents, which could be utilized in modulating the susceptibility of the breast epithelium to carcinogenesis.

1.2 The Architecture of the Different Quadrants of the Breast

1.2.1 Stating the Problem

Since breast cancer is the most common malignancy in the United States, and its incidence is progressively increasing [40, 41], an extensive search of risk factors has been undertaken due to the lack of identification of a given etiological agent(s) or of a plausible mechanism of initiation of the disease. This search has led to the identification of nulliparity as a risk factor [41]. The relative risk of nulliparous women of developing breast cancer ranges from 1.2 to 1.7 in comparison with parous women [42–54], especially in women whose malignancies are diagnosed at ages older than 40–45 years [46, 55, 56]. In order to be protective, pregnancy has to fulfill two main conditions: (a) it has to reach full term, since incomplete pregnancy does not offer protection, or even increases the risk of developing malignancies [56, 57]; and (b) the age of the woman at the time of the first full-term pregnancy. Maximal protection is observed when pregnancy occurs at ages younger than 20 years [42, 43, 45–52, 58–67]. It has not been clarified up to now what is the relationship of parity, age, and breast cancer risk.

Findings in an experimental animal model indicate that the induction of mammary cancer with chemical carcinogens is successful only when the carcinogen affects the undifferentiated terminal structures of the young virgin female rat gland. The same carcinogen and at the same dose fail to induce malignancies when the breast has completed its differentiation through a full-term pregnancy [16, 23, 37]. On the basis of these observations, the hypothesis that breast cancer arises in women whose breasts have failed to achieve an optimal degree of differentiation was developed. This concept places emphasis in the architecture of the breast and its level of differentiation, which might have been responsible because of its immaturity, for the initiation of the tumor.

The importance of the architecture of the normal breast and the influence of its development on the potential of the organ to develop malignancies have been indicated as early as 1934 by Dawson [68] and later emphasized by several other authors [35–39, 69–72]. One of the most comprehensive studies of the subgross and microscopic anatomy of the cancerous breast has been performed by Wellings et al. [24]. Through the study of whole mount preparations and histological sections of almost 200 breasts these authors identified the TDLU as the site of origin of mammary carcinomas, as well as of nonmalignant lesions [24]. In a comparative study with the experimental model described above, it was concluded that the terminal end bud, the site of origin of rat mammary carcinomas [16, 23], was equivalent to the Lob 1 or TDLU of the human female breast [10]. Since terminal end buds are present only in the undifferentiated virginal gland, but disappear during the process of differentiation occurring at pregnancy, which results in profuse lobular formation [16, 23], it is expected that the cancer-bearing breast will contain more Lob 1 or TDLU than the cancer-free breast, which might have been protected from neoplastic transformation by virtue of its more differentiated condition. The development of the human breast during the postpubertal years has been measured by quantization of the various types of lobules which represent different stages of development and differentiation [26] (see previous section).

1.2.2 Tridimensional Reconstruction of the Human Breast

We have studied whole breasts serially sectioned and prepared for whole mount to reconstruct them in a tridimensional fashion (Figs. 1.9, 1.15, and 1.16). The total number of ducts (TS) and lobules type 1, type 2, and type 3 (Figs. 1.1, 1.2, 1.3, 1.4,

1.5, 1.6, 1.7, and 1.8) were quantitated in each quadrant of the breast of parous women with cancer. In this study, the number of preneoplastic or neoplastic lesions is not reported, since this is not the objective of this work, and this topic has been excellently covered in other publications [10, 24]. The number of normal structures composing the architecture of the cancer-bearing breast in parous women is compared with the values obtained in nulliparous cancer patients, and these are compared with those present in the noncancerous breasts of parous and nulliparous women. One more factor taken into consideration in this study is the topographic location of the structures. Since almost 44 % of breast malignancies arise in the upper outer quadrant (UOQ; Fig. 1.17) [73–76, 81–84], all the breasts were analyzed by dividing whole organs into four quadrants (Fig. 1.15), thus being able to quantitate the total amount of stroma and parenchyma in each one of these regions of the breast. The patient population consisted of 16 women, of whom 6 were nulliparous and ranged in age from 20 to 61 years, and 10 were parous, ranging in age from 20 to 63 years (Table 1.3) [77]. Each one of these two major groups was subdivided according to the presence or absence of breast cancer (Table 1.3). The four nulliparous and the seven parous women whose breasts were free of mammary pathology were accident victims, whose autopsies were performed by the medical examiner. These tissues were generously provided by Dr. S. Bartow, who obtained them under NCI-RFP No. 1-CB-8423 1/NO 1-CN-23928. The breasts were removed at autopsy by s.c. mastectomy. The five women with breast cancer had had a diagnosis of infiltrating ductal carcinoma prior to modified radical mastectomy (MRM). The location of the tumors was determined at the time of biopsy (Table 1.3).

Serial sectioning of whole breasts of the women listed in Table 1.3 yielded a number of slices that ranged from 6 to 26 for each quadrant, depending upon the size of the breast (Figs. 1.16, 1.18, 1.19, and 1.20). The total area of each one of the slices obtained was determined and added to obtain the total area of the breast examined (Table 1.4). The total number of lobules type 1, 2, or 3 (Figs. 1.1, 1.2, 1.3, 1.4, 1.5, 1.6, 1.7, and 1.8) [77] present in each slice was counted and the total number of each lobule type was added to obtain the number for each specific quadrant and for the gland as a whole; the number of structures per cm^2 of tissue examined was obtained as a ratio between number of structures and size of the area examined. The analysis of results, taking into consideration the presence or absence of breast cancer in addition to parity is shown in Table 1.5.

1.2.2.1 The Architectural Structure of the Nulliparous Breast

The nulliparous women free of breast cancer (group I) (Fig. 1.21) and group II nulliparous women with breast cancer showed that breast architecture did not differ significantly, both having Lob 1 as the predominant structure with a lower percentage of Lob 2 and even lower Lob 3 (Table 1.5). In both groups the difference in relative percentage between Lob 1 and Lob 3 was highly significant, but there were no significant differences between the noncancerous and the cancer groups (Table 1.5) [77].

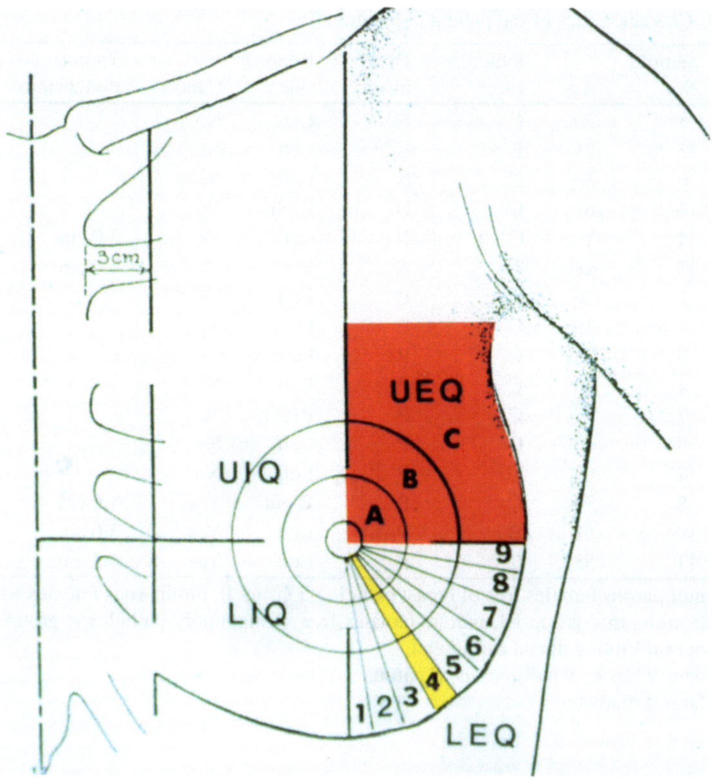

Fig. 1.15 Upon removal, each breast sample was fixed in 10 % neutral buffered formalin and then divided into four quadrants: UEQ (upper external quadrant), LEQ (lower external quadrant), LIQ (lower internal quadrant), and UIQ (upper internal quadrant); taking the nipple and the axillary tail of the breast as reference points. Each breast was serially cut with a meat slicer into 1 mm thick slices (see Fig. 1.9)

1.2.2.2 The Architectural Structure of the Parous Breast

The breast tissue of the seven parous women free of cancer (group III) contained the lowest percentage of Lob 1, a slightly higher percentage of Lob 2, with Lob 3 as the predominant structure (40.5 %) vs. Lob 1, 25.8 % ($p < 0.03$) (Table 1.5). Parous women with breast cancer (group IV), on the other hand, had a higher percentage of Lob 1 (52.6 %), whereas Lob 2 and Lob 3 represented 26.1 % and 21.2 %, respectively (Table 1.5). The fact that Lob 1 was the predominant structure in the breast of parous women with cancer, similar to what was observed in nulliparous women, suggests that pregnancy in these women did not elicit the degree of development reached by the parous breast of women free of cancer [77].

Table 1.3 Characteristics of the patient population[a]

Patient group	Sample No.	Age	Patient's race[b]	Parity history	Breast side	Cancer	Tumor dimensions[c]	Tumor location
I	5	20	C	N	Left	No		
	13	21	B	N	Left	No		
	6	30	C	N	Left	No		
	16	61	C	N	Right	No		
II	14	39	C	N	Left	Yes	3.0 cm	UOQ
	15	48	C	N	Right	Yes	4.0 cm	UOQ
III	1	20	C	P	Left	No		
	2	25	C	P	Left	No		
	3	28	C	P	Left	No		
	4	30	C	P	Left	No		
	7	37	C	P	Right	No		
	9	46	C	P	Left	No		
	12	63	C	P	Left	No		
IV	8	39	C	P	Right	Yes	4.0 cm	UOQ
	10	51	B	P	Left	Yes	3.0 cm	LOQ
	11	58	C	P	Left	Yes	5.5 cm	LOQ

[a]Group I, nulliparous females free of breast pathology; group II, nulliparous females with infiltrating ductal carcinoma; group III, parous females free of mammary pathology; group IV, parous females with infiltrating ductal carcinoma
[b]*C* Caucasian, *B* black, *N* nulliparous, *P* parous
[c]Tumor's largest diameter

1.2.3 Importance of the Quadrant in the Architecture of the Breast Tissue

When the total number of structures and their location within the four quadrants of the breast were taken into consideration, in nulliparous women of the noncancerous group (group I), it was found that the absolute number of the three types of lobules was higher in the UOQ and LOQ; however, Lob 1, which were predominant, were evenly distributed, with no significant differences among the four quadrants. The same applied to lobule types 2 and 3 (Table 1.5). Nulliparous women with breast cancer (group III) contained considerably lower total number of structures than the noncancerous group. Lobules type 1 were still the most numerous and the predominant structure in all the quadrants (Table 1.5) [77].

Parous women free of cancer (group III) contained a greater number of all the structures in the UOQ; however, their relative percentage was not different from the percentages in the other quadrants of the breast. The predominant structure was Lob 3 (Table 1.5). In the breast cancer group (group IV), the breast contained a lower overall number of structures than the noncancerous group with a higher percentage of Lob 1 than group III ($p<0.05$), but the differences in distribution among quadrants were not significant (Table 1.5). The relative percentage of Lob 1 and Lob 3 was in an inverse relationship with the values found in group III patients (Table 1.5). The breasts of women without cancer, either parous or nulliparous, contained a

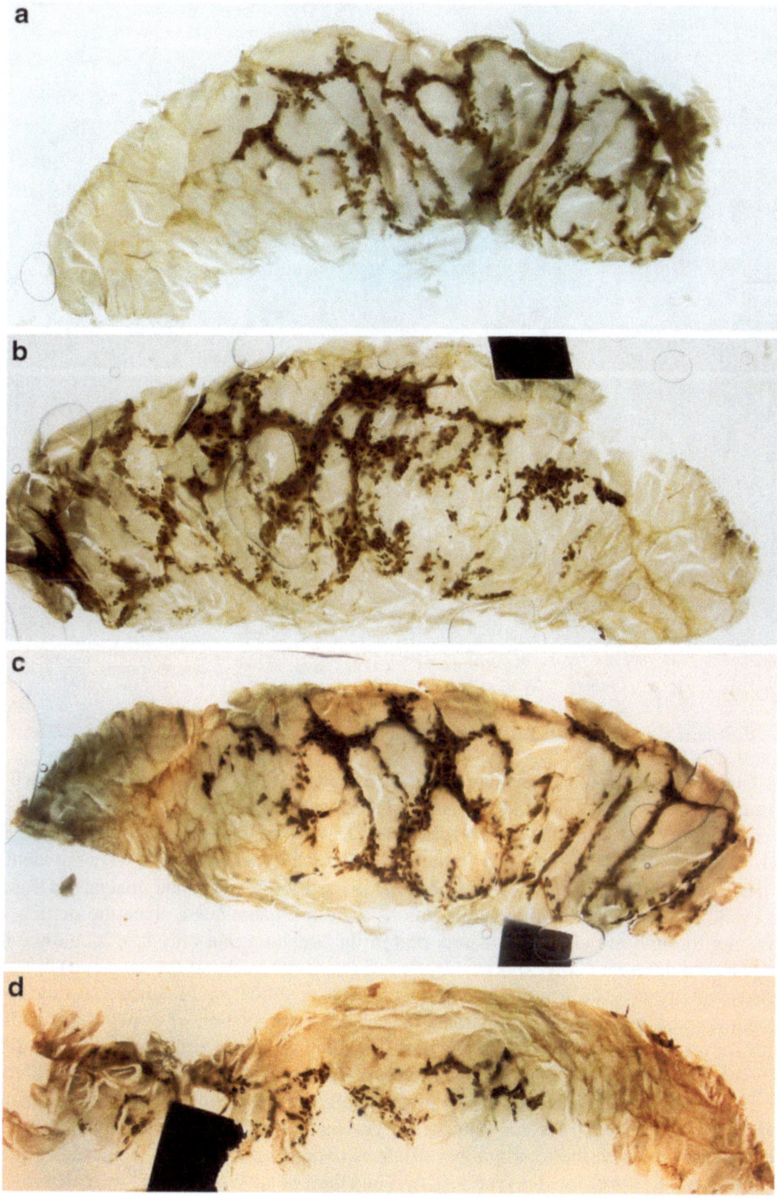

Fig. 1.16 (**a–d**) Slices of human breast prepared as described in Fig. 1.9

greater total number of structures than the breasts of women with cancer. Nulliparous women of group I contained a total of 21,205 structures in the four samples in a total area of 17,095 cm^2, averaging 1.24 structures/cm^2, whereas nulliparous women from the breast cancer group had 1,984 structures in three samples in a total area of 2,880 cm^2, averaging 0.68 structures/cm^2, which represented one-half of the values

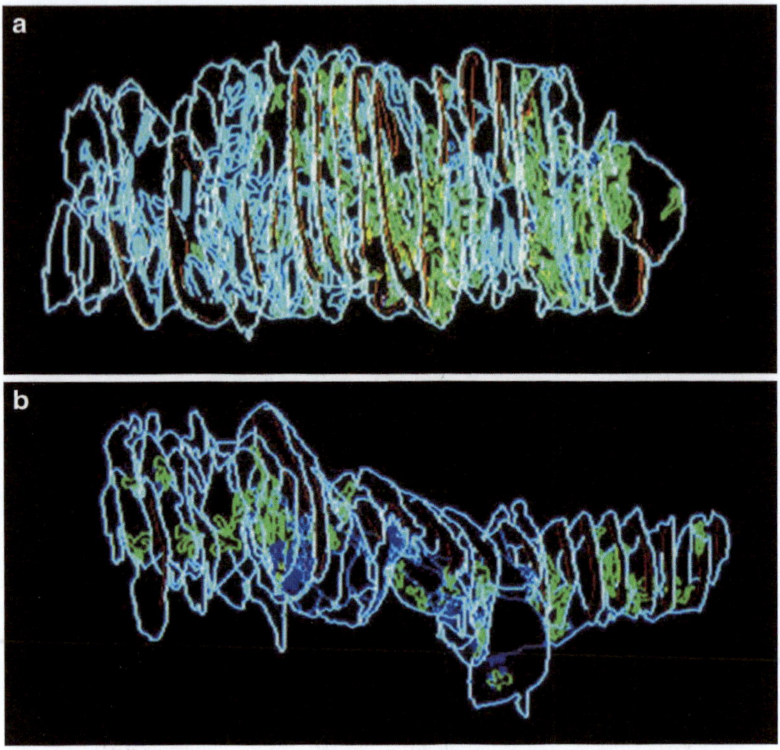

Fig. 1.17 (**a, b**) Tridimensional reconstruction of a human breast. The mammary parenchyma of each tissue slice was viewed under a Reichert Stereo Zoom microscope. The following normal structures were identified and quantitated: Lob 1, Lob 2, or Lob 3 (see Figs. 1.1, 1.2, 1.3, 1.4, 1.5, 1.6, 1.7, and 1.8). Lobules type 4, which are only found during pregnancy and lactation, were not identified in any of these samples; therefore, they were not considered in the statistical analysis of data. Solid tumors, cystic structures, or papillomatous lesions were not quantitated, since the purpose of this work was not to duplicate observations reported in the literature, but only to evaluate the level of development and differentiation expressed by the normal components of the cancerous breast for its comparison with the noncancerous one. Each lobule type identified was assigned a specific color and drawn with colored markers on transparency paper. After completion of drawing the outline and structures from each slice of breast, each lobular structure was traced with a stylus onto a digitizing tablet (Seiko Instrument, Inc., Japan), and represented by a polygon which could be displayed as a solid or transparent figure. Data were entered into a Dell computer using a three-dimensional reconstruction program (P C 13D) for computer-aided graphics (Uandel Scientific, Coret Madera, CA). For every breast specimen the total number of slices obtained per quadrant, the area of each slice (in cm^2), and the number of lobules type 1, 2, or 3 and their relative percentage were determined

found in women without cancer. The seven breast samples of parous women without cancer contained a total of 82,188 structures in an area of 12,823 cm^2, averaging 6.41 structures/cm^2. In group IV, the cancer-bearing parous women had a total of 22,626 structures in three breast samples with an area of 11,603 cm^2, averaging 1.95 structures/cm^2, a threefold reduction in number of structures in comparison with the parous noncancer group (Table 1.4) [77].

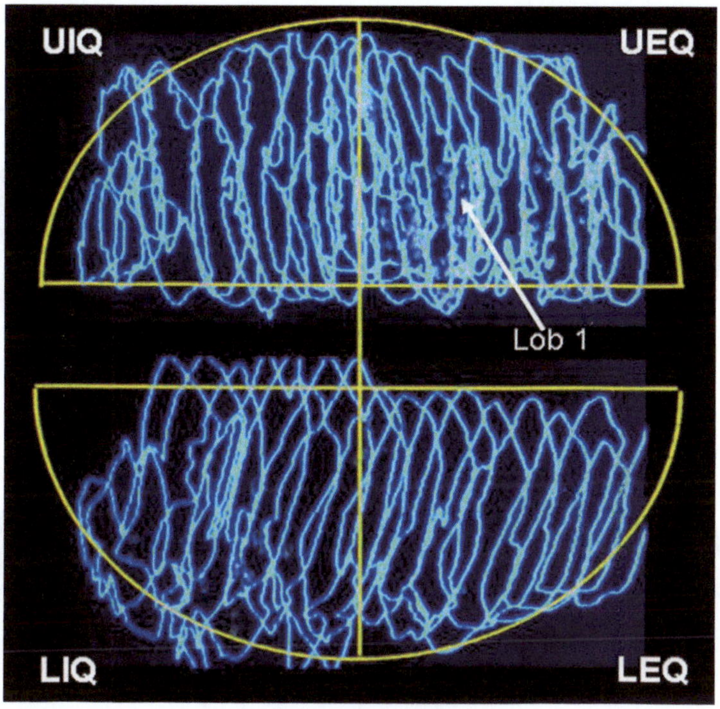

Fig. 1.18 Distribution of Lob 1 in the four quadrants of the breast

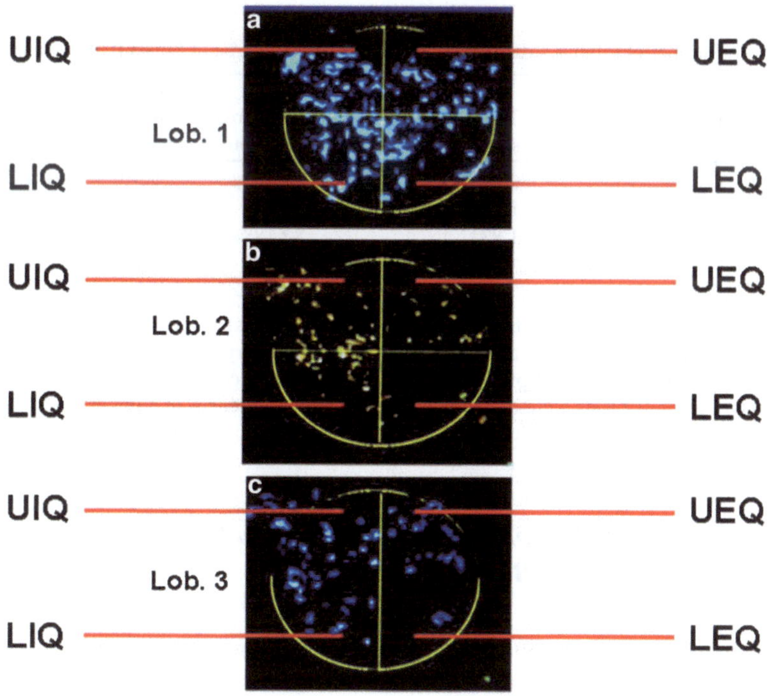

Fig. 1.19 Reconstruction of one of the breast samples showing the distribution of Lob 1 (**a**), 2 (**b**), and 3 (**c**) in each quadrant of the breast

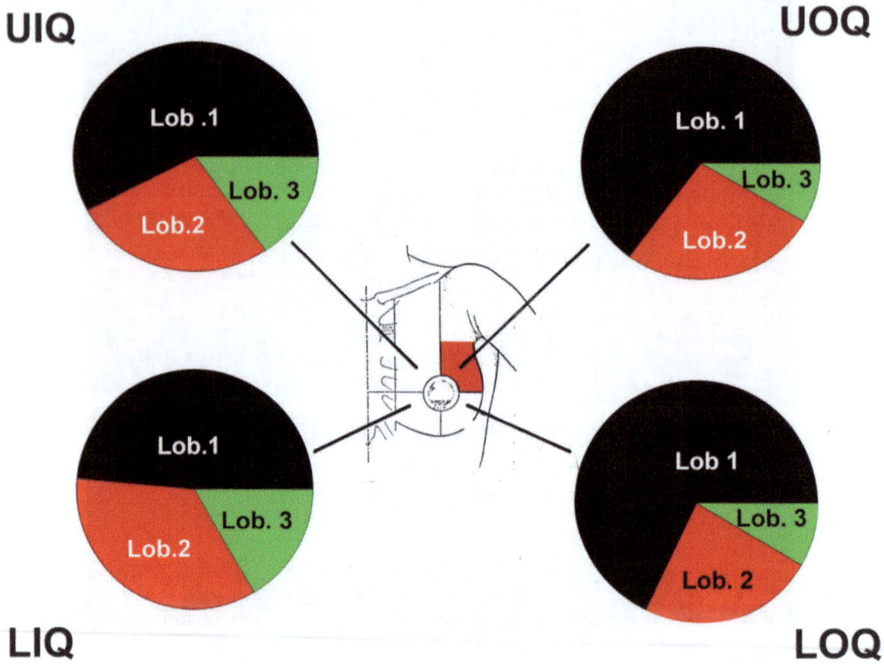

Fig. 1.20 Distribution of Lob 1, 2, and 3 in the breast tissue of parous women with cancer

Table 1.4 Number of lobular structures in relation to the total area of breast tissue examined

Patient group[a]	Total area counted/group[b] (cm^2)	Total area counted/sample[a] (cm^2)	No. of structures per group[c]	No. of structures per sample	No. of structures per cm^2
I	17,095	4,273	21,205	5,301	1.24 ± 0.20
II	2,880	1,440	1,984	661	0.68 ± 0.09
III	12,823	1,831	82,188	11,741	6.41 ± 1.20
IV	11,603	3,867	22,626	7,442	1.95 ± 0.15

[a]Patient groups as per Table 15
[b]Total area obtained by adding the total surface area of each slice of tissue examined under a stereomicroscope and added for each quadrant for each mammary gland and for the total group
[c]Summation of the total number of Lob 1, Lob 2, and Lob 3 added for each gland and each group of patients. Student's *t* tests were done for establishing the differences in number of structures per cm^2; significant differences between groups I and II ($p < 0.001$) and between groups III and IV ($p < 0.01$) were observed

1.2.4 Biological Importance of the Architecture of the Human Breast

The comparison of the normal architectural components of cancer-bearing breasts of parous and nulliparous women with those of noncancerous breasts leads us to conclude that the cancer-bearing breasts of parous women exhibit differences in

Table 1.5 Lobule type distribution by quadrants of the breast[a]

Patient group	Lobules Type	Lobules Total No.	Lobules %±SD	UOQ[b] lobules No.	UOQ[b] lobules %±SD	LOQ[c] lobules No.	LOQ[c] lobules %±SD	LIQ[d] lobules No.	LIQ[d] lobules %±SD	UIQ[e] lobules No.	UIQ[e] lobules %±SD
I	1	10,474	59.57[f]±18.32	3,935	64.62±17.41	3,325	67.72±24.8	1,077	48.42±13.98	2,137	57.52±16.7
	2	7,331	28.25[g]±10.38	2,998	27.00±11.85	2,534	23.72±15.28	821	35.05±7.17	978	27.22±4.62
	3	3,400	12.18[h]±10.15	1,096	8.40±7.34	1,252	8.52±9.61	467	16.55±11.43	585	15.27±12.79
II	1	1,180	57.30[i]±9.45	266	50.30±12.16	174	62.85±0.92	385	56.90±13.85	355	59.05±10.81
	2	445	26.55[j]±7.67	126	24.65±1.48	63	29.20±10.18	144	22.50±1.83	112	29.85±15.00
	3	359	16.15[k]±10.60	123	25.05±10.67	42	7.85±11.16	120	20.60±12.02	74	11.10±4.24
III	1	11,065	25.82[l]±23.73	3,640	15.77±18.95	2,942	26.60±25.89	1,413	33.34±26.82	3,070	27.57±34.44
	2	20,942	33.67[m]±17.31	7,716	42.98±19.51	6,065	38.34±24.56	2,890	27.27±9.38	4,271	26.08±5.92
	3	50,181	40.54[n]±26.68	16,221	47.55±22.16	12,411	35.04±27.73	9,996	39.24±29.92	11,553	46.13±25.33
IV	1	6,005	52.62[o]±31.15	1,063	55.86±39.55	1,448	48.36±44.74	1,784	56.37±30.02	1,710	49.90±28.31
	2	7,525	26.12[p]±16.18	1,497	26.63±23.93	2,018	26.80±24.30	2,203	23.00±6.89	1,807	28.06±15.45
	3	9,096	21.23[q]±17.81	1,472	17.45±20.15	2,494	24.83±92.17	3,311	20.63±23.78	1,819	22.03±15.47

Comparison of the number and relative percentage of lobules present in the mammary gland of patients in groups I, II, III, and IV (as per Table 1.4)

[a]Lobules type 1, 2, or 3 (*b*) UOQ, (*c*) LOQ, (*d*) LIQ, (*e*) UIQ of the breast. Student's *t* tests were done for all possible comparisons. There was a significant difference in the percentage of Lob 1, 2, and 3 present in the breast of nulliparous women (group I): *f* vs. *g* (p1.66E^{-6}); *g* vs. *h* (p1.18E^{-4}); and *f* vs. *h* (p4.67E^{-10})

In group II significant differences were: *I* vs. *j* (p4.99E^{-6}); *j* vs. *k* ($p<0.04$); and *I* vs. *k* (p1.03E^{-6})

In parous women without cancer (group III) lobules type 3 were in a significantly higher concentration than lobules type 1; *l* vs. *n* ($p<0.03$)

In parous women with cancer (group IV) lobules type 1 (*o*) were in a significantly higher concentration than lobules type 2 (*p*) and 3 (*q*) ($p<0.03$ and p6.15E^{-3}, respectively)

Comparison of groups III and IV revealed significant differences in *l* vs. *o* ($p<0.05$) and *n* vs. *q* ($p<0.02$)

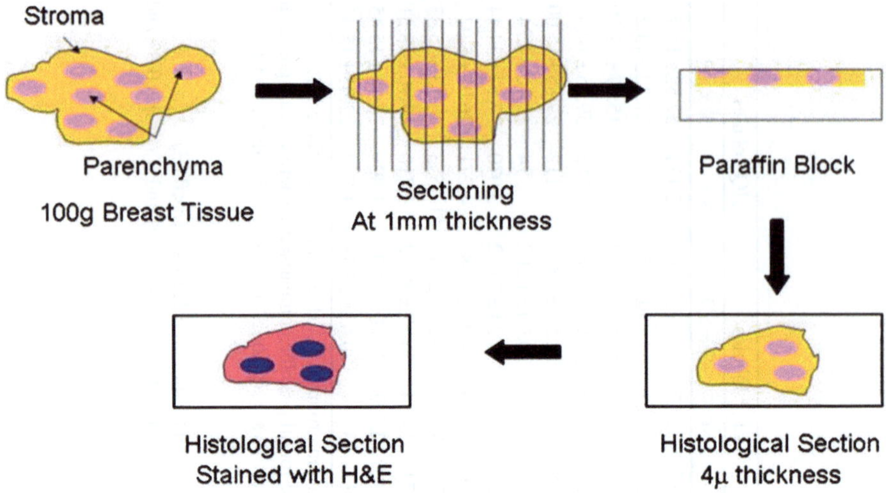

Fig. 1.21 Schematic of the methodological procedures followed for processing the samples obtained from 100 g of breast tissue from reduction mammoplasty (RM) specimens, familial breast cancer (FAM), and modified radical mastectomy (MRM)

developmental pattern in comparison with the noncancerous breasts of parous women, which contain a greater percentage of lobules type 3 with a concomitant reduction of lobules type 1 [77]. The breast of nulliparous women, with or without cancer, contains as the predominant structure the lobule type 1. In these two groups, the presence or absence of cancer, on the other hand, does not modify the basic architectural pattern of the breast. The observation that parous women with breast cancer exhibit a different mammary gland architecture from that of the noncancerous parous women (recall that in the first group the percentage of lobules type 1 is greater, approaching the percentages found in the nulliparous women) acquires relevance in light of the fact that mammary carcinomas in women initiate in the TDLU or lobules type 1, as supported by our own observations as well as by retrospective studies of surgical specimens removed for malignancy and in autopsy material [78, 79]. The greater proportion of lobules type 1 found in the breast of nulliparous women and of parous women with cancer suggests that those breasts are at a higher risk of developing malignancies if one considers each lobule type 1 as a putative target to a carcinogenic insult [16, 32].

Epidemiological observations indicate that an early full-term pregnancy reduces by fourfold the risk of developing breast cancer [80–82]. Parous women, nevertheless, develop breast cancer. A novel observation in our population of parous women with breast cancer is that the breast parenchyma not involved in the malignant process exhibits an architectural pattern similar to that of the nulliparous females. This fact might indicate that the presence of a greater proportion of lobules type 1 might have played a role in the susceptibility of those women to develop malignancies. It is possible to postulate that the mammary gland in these women had failed to respond to the hormones of pregnancy with full differentiation, thus remaining in an undifferentiated and therefore more susceptible stage for malignant transformation.

Although the study of the pathology and natural history of breast cancer has revealed that neoplasms develop in the breast with a certain pattern, with a greater proportion of both benign and malignant tumors arising in the upper external quadrant of the breast [73–76], there is no satisfactory explanation for this selective clustering of both benign and malignant lesions in this specific area of the breast. The present work was designed in part to understand this phenomenon by studying gland development through the analysis of breast architecture according to the four different quadrants of the mammary gland. We have found that in noncancerous breasts, lobules type 1 in nulliparous and lobule types 1, 2, and 3 in parous women are present in similar proportions in all the quadrants of the breast. An additional finding while comparing the architecture of the normal and of the diseased breast is that the total number of structures present in the breasts of both nulliparous and parous women with cancer is lower than the number in the corresponding control noncancerous breasts, despite the fact that similar or even a greater amount of tissue was examined, as determined by measuring the total area of the slices counted [77]. These observations indicate that a greater parenchymal mass is not required for developing breast cancer, but rather a certain specific ratio of lobules type 1 to type 3, or a certain specific ratio between the parenchymal and stromal components of the breasts. However, the possibility exists that these topographic and structural differences are associated with, or result from, variations in factors such as cell kinetics or hormone receptor content, which might ultimately result in an enhanced susceptibility to tumor initiation and progression, as has been shown in the experimental system [16, 23, 26, 37, 83]. Results reported here confirm previous observations that the presence of Lob 3 is associated with protection from breast cancer [7, 16, 26, 32, 84]. Therefore, it is possible to postulate that the architectural pattern of the breast, evaluated by the relative percentage of Lob 1, Lob 2, or Lob 3, and their density per surface area, could represent meaningful intermediate endpoints that could be a measure of the response of the gland to given chemopreventive agents.

1.2.5 Parenchyma:Stroma Relationship in the Nulliparous and Parous Breast

Breast development occurs through a process of ductal elongation, branching, and sprouting of ductules or alveoli, a process that requires extensive cell proliferation and penetration of the ductal epithelium into the stroma [85]. Both the intralobular and the interlobular stroma are affected simultaneously during development, pregnancy, lactation, and involution [86–88]. These processes occur, in turn, in a synchronous manner in response to specific hormonal and growth factor stimuli [87]. Two major mechanisms are considered to be involved in the interaction of the stroma and epithelial cells: the production of soluble growth factors and a modification of the composition of the extracellular matrix. This interaction seems to be bidirectional such that epithelial cells are also capable of influencing stromal cell behavior and governing gene expression [86]. The study of the stroma–parenchyma ratio in 14 breasts of pubertal, postpubertal, parous, and pregnant women showed that the relationship between parenchyma and stroma is a dynamic process. At puberty, almost 90 % of the mammary gland is made up of intralobular stroma, which represents 17 % of the total, and consists of the connective tissue that surrounds each individual alveolar bud, and the interlobular stroma, composed of fat and connective tissue, which separates one lobule from another. The parenchyma of these glands, representing 10 % of the mammary area, is made up almost exclusively of Lob 1 and ductal structures. In the glands of postpubertal and young nulliparous women, the parenchyma increases from 10 to 30 % of the total area of the gland (0–10 % is composed of Lob 1, 10–18 % of Lob 2, and 1–3 % of Lob 3). The intralobular stroma of these breasts represents approximately 28 % of the total. Parity induces significant differences in mammary gland development. The breast of parous women is mostly composed of Lob 3 and accounts for 24 % of the total parenchyma and a markedly reduced proportion of Lob 1 and Lob 2. Pregnancy induces dramatic changes in the parenchyma–stroma ratio. During the process of post-lactational involution, Lob 4 regresses to Lob 3. These lobules represent almost 60 % of the structures of the parous breast until the fourth decade of life, when Lob 2 and Lob 1 become preponderant structures and the stroma regains its prevalence. It is expected that the marked variations in epithelial–stromal ratio occurring during the various stages of breast development will influence the bidirectional connections between the components of cellular microenvironment (growth factors, hormones, and extracellular matrix) and the nucleus, leading to specific modifications in gene expression that may account for the differing susceptibility to, or risk of, developing breast cancer.

1.3 The Role of High Risk Breast Cancer in the Human Breast Architecture

We postulate that the induction of differentiation of the breast by the reproductive event is responsible for the inhibition of carcinogenic initiation, as it has been observed in the rodent experimental model [17–19, 36, 88–90] and in humans [91–93]. However, a certain percentage of parous women develop breast cancer [91]. We have observed that the architectural pattern of breast tissues from parous breast cancer-affected patients appears similar to that of nulliparous women, having Lob 1 as the prevalent structure. These characteristics indicate that the breast in these women has responded to the hormones of pregnancy differently than the breast of parous women who did not develop cancer [88–92]. The possibility that these breast tissues exhibit a defective response to the differentiating influence of the hormones of pregnancy warrants further investigation.

The pattern of breast development and differentiation is also influenced by inheritance [88]. Genetic influences, which are responsible for at least 5 % of the breast cancer cases, also seem to affect the architectural pattern of the breast. The studies of prophylactic mastectomy specimens obtained from women with a familial history of breast and breast/ovarian cancer or from women proven to be carriers of the *BRCA-1* gene by linkage analysis have revealed that morphologically and architecturally the breasts of nulliparous women are similar to those of parous women. Breast tissues from both groups of women are predominantly composed of Lob 1, with only a few specimens containing Lob 2 and Lob 3, in frank contrast to the predominance of Lob 3 found in parous women without a familial history of breast cancer [26, 88]. These observations suggest that genetic predisposition to breast cancer affects genes that control the branching pattern of the breast during lobular development. Therefore, although there is no explanation as of yet for the higher breast cancer risk exhibited by nulliparous, late parous, and genetically predisposed women, the fact that experimentally induced rat mammary carcinomas develop only when the carcinogen interacts with the undifferentiated and highly proliferating mammary epithelium of young nulliparous rats [17, 18, 93, 95] suggests that the breast of these "high breast cancer risk" women might exhibit some of the undifferentiated cell characteristics that have been shown to be essential for the initiation of cancer in rodents. The higher proliferative activity of the nulliparous women with Lob 1, in association with the higher breast cancer incidence in this group of women, suggests that these lobules are biologically different from those of early parous women [32, 36, 93–95].

1.3.1 Breast Architecture in the Women at High Risk for Breast Cancer

Lob 1, the most undifferentiated structure found in the breast of young nulliparous women, is the site of origin of ductal carcinomas [26]. The finding that the most undifferentiated structures originated the most aggressive neoplasms lends support to our hypothesis that the presence of Lob 1 explains the higher breast cancer risk of nulliparous women, as they represent the population with the highest concentration of undifferentiated structures in the breast [85]. In order to further the demonstration of this postulate, we have compared the breast tissue of three groups of women who underwent surgical procedures: The first group consisted of women who underwent reduction mammoplasty (RM) for cosmetic reasons. The second group included women who underwent prophylactic subcutaneous mastectomy after genetic counseling for either carrying the *BRCA-1* gene or belonging to a pedigree with familial breast cancer (FAM), and the final group included women who underwent MRM for the diagnosis of invasive carcinoma (Table 1.6).

The breasts of women who underwent RM contain the three types of lobules that follow the same morphologic characteristics that were described previously [8]. The three lobular structures are in general surrounded by a loose stroma that demarcates them from the interlobular stroma that may have a different ratio of connective and fat tissue. All of the lobules were very well demarcated, and no fibrous tissue was observed. Quantization of the three lobular structures in the overall population of breast tissue

Table 1.6 Lobular architecture of the breast tissue from RM, prophylactic mastectomy for FAM, and MRM

Group	No. of cases	Age mean ± SD (%)	Lob 1 mean ± SD (%)	Lob 2 mean ± SD (%)	Lob 3 mean ± SD (%)
RM (all)	33	29.4±8.2	22.5±23.7	37.3±28.6	38.4±34.2
RM (nulliparous)	9	22.9±6.7	45.9±27.4	47.2±22.0	6.9±7.0
RM (parous)	24	31.9±2.3	16.9±8.3	35.5±3.1	47.9±33.4
FAM (all)	17	37.0±2.9	47.9±37.3	39.9±31.3	9.91±4.41
FAM (nulliparous)	8	37.6±3.2	51.3±34.4	39.9±26.2	8.83±8.39
FAM (parous)	9	36.5±2.6	44.0±42.0	40.0±38.1	16.10±9.9
MRM (all)	43	35.4±3.9	74.3±25.8	22.3±22.1	3.35±10.0
MRM (nulliparous)	7	36.0±3.6	80.0±19.0	16.8±15.0	1.74±4.6
MRM (parous)	36	35.2±4.3	70.4±26.4	25.4±22.7	3.80±12.5

The lobular structures of both parous and nulliparous women, considered as a group, showed that Lob 1 of women with reduction mammoplasty (RM) was significantly lower than those of the FAM and MRM group ($t=7.6, p<0.0003$ and $t=11.0, p<0.00000$, respectively). The percentage of Lob 1 between the FAM and MRM was also significantly different ($t=3.38, p<0.001$). Lob 2 was significantly different between the RM and FAM ($t=3.58, p<0.0005$) and between FAM and MRM ($t=2.20$, $p<0.02$). Lob 3 was higher in the RM group than the FAM and MRM group ($t=3.83$, $p<0.0002$ and $t=8$, $p<0.0000$, respectively). Lob 3 was also different between FAM and MRM ($t=2.17$, $p<0.02$). In the RM group, nulliparous Lob 1 was significantly different from Lob 3 ($t=3.43, p<0.008$) and between Lob 2 and Lob 3 ($t=5.48, p<0.0005$). In the parous group, Lob 1 was significantly different from Lob 2 and Lob 3 ($t=2.78$, $p<0.009$ and $t=3.82$, $p<10.0000$, respectively). In the FAM group, nulliparous Lob 1 was significantly different from Lob 3 ($t=2.52$, $p<0.03$) and between Lob 2 and Lob 3 ($t=3.41, p<0.007$). In the parous group, Lob 1 was significantly different than Lob 3 and Lob 2 to Lob 3 ($t=2.83, p<0.01$ and $t=2.81, p<0.01$, respectively). In the MRM group, Lob 1 was significantly different than Lob 2 and Lob 3 ($t=3.00, p<0.005$ and $t=3.80, p<0.0002$). Lob 2 was significantly different from Lob 3 ($t=2.7, p<0.007$). In the parous group, Lob 1 was significantly different from Lob 2 and Lob 3 ($p<0.005$ and $p<0.001$, respectively) and between Lob 2 and Lob 3 ($p<0.005$). For groups RM and FAM, an average of 100 g of tissue was processed from every specimen. The tissues were fixed in 10 % neutral buffered formalin for a minimum of 24 h and processed for whole mount preparation and histological sections for microscopic evaluation. For whole mount preparation, the tissues were defatted in acetone and hydrated in a decreasing concentration of ethanol. After rehydration, the tissues were stained in toluidine blue for 2 h and differentiated and destained with pure methanol and 70 % ethanol for 30 min each, washed in distilled water, and fixed in 4 % ammonium molybdate for 30 min. After staining, the tissue was dehydrated and cut into 1 mm thick slices, which were mounted on glass slides and cover slipped. Histological sections were prepared from paraffin-embedded tissue sectioned at 5 μm each and stained with hematoxylin and eosin. An average of 12 slides per sample was examined, and 2,826 structures were classified and counted. The entire whole mounts and histological sections were examined under a stereomicroscope (Olympus SZH) or a bright field Olympus microscope (BH-2), respectively, for classification and quantification. For the MRM group, all tissues were fixed in formalin, embedded in paraffin, sectioned at 5 μm thickness, and processed for light microscopy. An average of 12 histological sections per sample was examined. These histological sections were utilized for characterizing the type of lobular structures and the number and type of pathological lesions present by applying criteria previously described

studied indicated that lobules type 1 represented 22.5 % of the structures, whereas Lob 2 was 37.3 % and Lob 3 was 38.4 % of the total number of structures. The differences are statistically significant (Table 1.6). The separation of the breast samples, based on the pregnancy history of the host, such as nulliparity and parity, showed a different pattern

Table 1.7 Profile of the lobular structures in the breast tissues obtained from RM, prophylactic mastectomy for FAM, and MRM for invasive cancer

Group	No. of cases	Well-defined lobules (%)	Poorly defined lobules (%)	None	Fibrosis mild to moderate	Marked
RM	33	33 (100)	0	33 (100)	0	0
FAM	17	8 (47)	9 (53)	0	3 (17.6)	14 (82.4)
MRM	43	40 (93)	3 (7)	1 (2.3)	39 (90.3)	3 (7.4)

of lobular development. The breasts of nulliparous women contained a significantly higher number of Lob 1 and Lob 2, with 45.9 % and 47.2 %, respectively, and a highly significant lower number of Lob 3 (6.9 %) (Table 1.6). In the breasts of parous women, the pattern was inverse, with Lob 2 and Lob 3 being the most abundant at 35.5 % and 47.9 %, respectively, whereas Lob 1 comprised only 16.9 % of the total.

The breasts of women with familial breast cancer were obtained from prophylactic mastectomies [94]. Seventeen samples that have been verified to be either *BRCA-1* positive or a carrier of genetic abnormalities are the ones entered in this analysis (Tables 1.7 and 1.8). The average age of these women was 37.0±2.9 (Table 1.6). The histological appearance of the lobular structures was different from that observed in the breast tissue of women who underwent RM. Eight out of 17 breast samples presented well-demarcated lobular structures, but all of them had moderate or marked fibrosis of the intralobular stroma (Table 1.7). Ductal hyperplasia (mild to severe) in the Lob 1 or Lob 2 was observed in seven cases and carcinoma in situ (solid, cribriform, and papillary) in one case, and invasive carcinomas ipsilaterally or contralaterally was observed in nine cases (Table 1.9) [94].

The distribution of Lob 1, Lob 2, and Lob 3 in the breast tissue derived from women with *BRCA-1*, or being carriers of genetic abnormalities by linked analysis was 47.9 %, 39.9 %, and 9.9 %, respectively (Table 1.6). This pattern was significantly different than that observed in the RM group I, containing a higher percentage of lobules type 1 ($p<0.0008$), whereas Lob 3 was significantly lower ($p<0.00004$; Table 1.6).

The separation of the breast samples, based on the pregnancy history of the host, such as nulliparity and parity, indicated that in both subgroups, the percentage of Lob 1 was significantly higher than Lob 3 (Table 1.6) and that the differences between nulliparous and parous observed in the control or RM group were not present in the breast tissue derived from women with familial breast cancer (Table 1.6). Lob 1 represents 51.3 % and 44.0 % in the nulliparous and parous women, respectively. This indicates a reversion of the pattern observed in the parous in which Lob 1 is less frequent. In the familial cases, the comparison of the nulliparous from the RM group with those of the FAM group is not statistically different. Instead, the parous breast tissue of the RM group was significantly different than those of the FAM group (Table 1.8) [94].

In order to determine whether those *BRCA-1*-positive (+) breast tissues were different from those designated to be carriers, but in which *BRCA* status was not

Table 1.8 Profile of the breast tissue obtained from prophylactic mastectomy in *BRCA-1* positive or carrier of familial breast cancer

Sample No.	Pedigree No.	*BRCA-1*	Age	Parity	Breast with cancer	Breast studies
217	1973-11	Positive	33	Yes	Right	Right and left
231	3312-18	Carriers	33	No	Left	Left
234	1234-68	Positive	33	Yes	–	Right and left
222	3703-1	Carriers	34	No	–	Right and left
221	3300-1	Carriers	35	Yes	–	Right and left
230	3619-5	Carriers	36	No	–	Right
205	2850-19	Carriers	36	No	Left	Right
203	3481-1	Carriers	37	Yes	Right	Right and left
202	3481-10	Carriers	37	Yes	Right	Left
232	2887-3	Unlinked	38	No	–	Right and left
229	1816-62	Positive	38	No	Left	Left
206	1815-1	Carriers	38	Yes	Right	Left
228	2552-13	Carriers	39	Yes	Right	Left
210	1086-28	Carriers	40	Yes	Right	Right
218	1816-680	Positive	43	No	Left	Right
223	1816-686	Positive	37	No		Right and left
212	3386-1	Carriers	43	No		Right and left

Table 1.9 Type of lesions found in the breast tissue studied from RM, prophylactic mastectomy for FAM and MRM

			Ductal hyperplasia		Ductal cancer in situ	
Group	No. of cases	No. of lobules counted	#	(%)	#	(%)
RM	33	31,220	0	(0)	0	(0)
FAM	17	3,162	7	(41.2)	1	(5.9)
MRM	43	2,901	27	(63.0)	5	(11.4)

determined yet (Table 1.6), we separated these two groups (*BRCA*+ and carriers), and we found that the percentage of lobular structures was not significantly different.

The age of women from RM group was different from those of the FAM group; the average was 29.4 and 37.0 for the first group and second group, respectively. This difference is significant. In order to determine whether age may be contributory to the differences observed, we retabulated the data of the RM group for those women with matching age to those of the FAM group, and we found that the difference between both groups still persists, indicating that the familial factor could be in itself a deterrent in the pattern of architectural development of the breast [94].

The architectural pattern of the breast tissue obtained from MRM was from 43 breast samples. The average age for this group was 35.4 years, with no significant difference between the age for the nulliparous and parous women (Table 1.6). Quantization of the three lobular structures in the overall population of breast tissue studied indicated that Lob 1 represented 74.25 % of the structures, whereas Lob 2 and Lob 3 were 22.3 % and 3.4 %, respectively, of the total number of structures. The differences are statistically significant (Table 1.6).

The separation of the breast samples, based on the pregnancy history of the host, such as nulliparity and parity, showed no different pattern of lobular development. The breasts of nulliparous women contained a significantly higher number of Lob 1 and Lob 2, 80.0 % and 16.8 %, respectively, and a highly significant lower number of Lob 3 (1.7 %; Table 1.6). In the breasts of parous women, the pattern was similar, with Lob 1 and Lob 2 being the most abundant at 70.4 % and 25.4 %, respectively, whereas Lob 3 comprised only 3.8 % of the total. The differences between nulliparous and parous were not statistically significant (Table 1.6) [94].

The histological appearance of the lobular structures was not as different from that observed in the breast tissue of women who underwent RM, but it was when compared with the FAM group; 92.8 % of the lobules were well defined in the RM group when compared with only 47.0 % in the FAM group (3 out of 43 breast samples were markedly fibrotic (Table 1.7), which was significantly lower than the FAM group, in which most of the lobules presented marked intralobular fibrosis). Ductal hyperplasia in Lob 1 or Lob 2, on the other hand, was observed in 62.9 % of the cases and carcinoma in situ was observed in 11.4 % of the cases (Table 1.9). Invasive carcinomas ipsilaterally or contralaterally were observed in 88.6 % of the cases. The age of women from MRM group was not different from those of the FAM group (Table 1.6) [94].

Altogether, these results show that the breasts of parous women from the FAM and the MRM groups exhibited a different architectural pattern from those of parous women in the RM group, which can be considered the normal or control population [26, 94]. The observation that Lob 1 of the breast of both nulliparous and parous women of the FAM and MRM groups is the most frequent structure is in agreement with the knowledge that the cancer in the breast starts in Lob 1 [10, 24, 89]. The greater proportion of Lob 1 found in the breast of nulliparous and parous women of the FAM and MRM groups suggests that these breasts were at higher risk of developing malignancies due to the fact that each Lob 1 is the target of carcinogenic insult [10, 24].

1.3.2 Specific Considerations on the Relationship Between Lobular Development and Familial Breast Cancer-Related Genes

It has been postulated that *BRCA-1* and/or *BRCA-2* may serve to control cell proliferation and differentiation during developmental stages characterized by rapid growth [96]. This model predicts that individuals possessing germline mutations in *BRCA-1* and/or *BRCA-2* may be particularly susceptible to early events in mammary carcinogenesis during pregnancy [96]. Epidemiological observations suggest that women with a positive family history of breast cancer may experience a significantly greater increase in breast cancer risk associated with their first pregnancy relative to women without a family history of breast cancer [97]. How *BRCA-1* and/or *BRCA-2* control breast differentiation is unknown. In this work, we indicate that in both sporadic and familial breast cancer, the pattern of lobular development is very similar. In rodents as well as in the breast tissue of women who underwent plastic surgery for cosmetic reasons, parity is associated with lobular differentiation [17, 89]. In both cases, lobular differentiation makes the mammary tissue refractory to neoplastic transformation by chemical carcinogens [98]. Moreover, the relationship of the differentiation effect induced by pregnancy and the induced protection against breast cancer in women who have undergone this FFTP early in life [91] are indications that the same operational events are modified in both familial and sporadic cases of breast cancer.

In addition to the overall architectural differences described previously in this chapter, the breast tissues from women with hereditary breast cancer present histological differences in the intralobular stroma. The intralobular stroma, different from the more dense collagenized interlobular stroma, is a dynamic compartment of the breast composed of loosely arranged connective tissue containing cells such as fibroblasts, blood vessels, and inflammatory cells such as lymphocytes, mast cells, and macrophages [99, 100]. The intralobular stroma contrasts with the interlobular stroma, which has fewer cells separated by larger quantities of more compact collagen. The role of intralobular stroma during breast development from adolescence to premenopausal maturity, pregnancy and lactation, involution, and postmenopausal changes has been implicated; however, how the interaction with the epithelial cells takes place is unknown. Most of our understanding of the interaction between epithelial cells and stroma in the breast is from the experiments of Sakakura [101]. In this work, we indicate that the intralobular stroma of the Lob 1 of the breast of women with familial breast cancer has lost the loosely arranged connective tissue for a denser stroma that erases its demarcation from the intralobular stroma. The intralobular stroma of the breast tissue from the FAM group was more fibrotic and dense. These findings suggest that in the breast cancer families either the development of the breast parenchyma has failed to respond to the normal physiological stimuli that determine the formation of lobular structures, indicative of differentiation, or the involution pattern of lobular structures type 3 after pregnancy is more rapid in these women than in those in the control. It is noteworthy that early pregnancies influence breast cancer risk by altering the structure of the mammary parenchyma [10, 26]. It has been hypothesized that late pregnancies could likewise influence breast cancer risk via alterations in the mammary parenchyma by delaying or interrupting the normal process of involution of glandular tissue of the breast [102]. In the mouse, it has been observed that *BRCA-1* is induced during puberty, pregnancy, and following treatment of ovariectomized animals with 17-β-estradiol and progesterone. Therefore, it is not surprising that in the human breast, alteration of this gene may explain the altered morphological pattern observed. Our finding that the intralobular stroma is more fibrotic in the FAM group than in the MRM and RM groups may explain in part the increased mammographic density in women with familial breast cancer. Although the intralobular stroma is only a small component of all of the factors that determine the mammographic pattern, the mammographic breast density reflects proliferation of breast stroma through collagen formation and fibrosis. The factors that determine breast densities depend on the interplay of hormones, such as estrogen and growth factors, such as epidermal growth factor, transforming growth factor, and insulin growth factors I and II. How all of these factors and the genes related to familial breast cancer interrelate in the biology of the intralobular stroma is not known.

The development of mammary ductal structures involves a complex interplay between epithelium and mesenchyme, and the branching of the mammary ducts is dependent on circulating hormones for stimulation and synchronization with reproductive events but is also influenced by local factors to provide signals that influence glandular growth, differentiation, and morphogenesis. The matrix-degrading metalloproteinases stromelysin-1, stromelysin-3, and gelatinase A are expressed during ductal branching morphogenesis of the murine mammary gland. Whereas the role of metalloproteinases in the branching pattern of the mammary gland and its relationship with *BRCA-1* requires further investigation, on the basis of these data it is possible to postulate that the breast tissue from women with hereditary breast cancer suffers from an alteration of the interaction between the epithelium and the stroma, resulting in a modified interaction between the stroma and parenchyma as described here.

Our work also confirmed the previous observation reported in the literature [103] that breast tissue of women from sporadic breast cancer contained higher numbers of ductal carcinoma in situ and ductal hyperplasias than those from the familial breast cancer group. Overall, our data indicate that *BRCA-1* or related genes associated with familial breast cancer play a role in the lobular pattern of the breast mainly by altering the pattern of involution after pregnancy, with a consistent increase in Lob 1 as compared to the control population. We concluded that the developmental pattern of the breast of parous women from the familial breast cancer group was similar to that of nulliparous women from the same group and was less developed than the breasts of parous women without history of familial breast cancer. The breasts of women belonging to the familial breast cancer group also presented differences in the branching pattern of the ductal epithelium, observations that suggested that the genes that control lobular development might have been affected in those women belonging to families with a history of breast and breast/ovarian cancer [104]. Supporting evidence of this fact is the poor milk production reported in carriers of the *BRCA-1* mutation compared with female relatives without mutation [104] and the poor differentiation of the mammary gland of mice with *BRCA-1* mutations [105].

1.4 Comparative Study of Mammary Gland Development by Magnetic Resonance Imaging and Subgross Anatomy

The study of human breast development and differentiation is critical to understanding the pathogenesis of breast cancer. This study was sought to determine whether mammary parenchymal structures characterized in whole mount preparations of fixed tissues could be identified by magnetic resonance imaging (MRI). For this purpose, human breast samples were studied by MRI, and results were compared with subgross anatomic findings, namely the characterization of mammary structures representative of stages of differentiation such as lobules type 1, 2, or 3. Using a 512×512 matrix and a variety of MRI contrast techniques (spin echo, fast gradient echo, and inversion recovery) permitted visualization of gland structures as small as 0.7 mm (Fig. 1.22). This value corresponds to the size of the lobule type 3, which is one of the most differentiated structures of the breast. These findings indicate that it may be possible to study human mammary gland development and to detect fine structural changes indicative of differentiation [106].

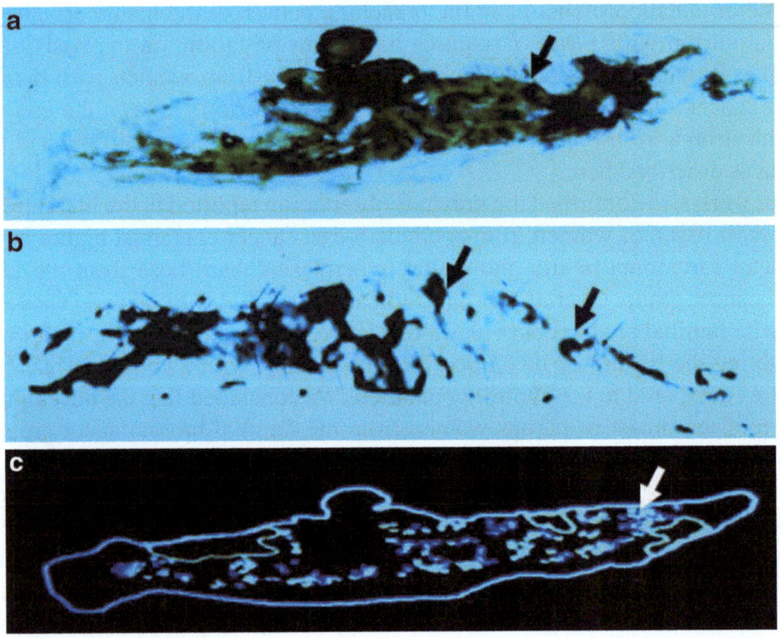

Fig. 1.22 (**a**) Slice of breast tissue from a parous woman showing predominance of lobules type 3. (**b**) Magnetic imaging resonance obtained from sample shown in (**a**). The *arrows* identify the Lob 3 found in the whole mount preparation. (**c**) Computer reconstruction of the slice of breast tissue shown in (**a**) and depicting the Lob 3

1.5 Unifying Concepts

Breast cancer originates in undifferentiated terminal structures of the mammary gland. The terminal ducts of Lob 1 of the human female breast, which are the sites of origin of ductal carcinomas, are at their peak of cell replication during early adulthood, a period during which the breast is more susceptible to carcinogenesis. The susceptibility of Lob 1 to undergo neoplastic transformation has been confirmed by in vitro studies, which have shown that this structure has the highest proliferative activity and rate of carcinogen binding to the DNA. More importantly, when treated with carcinogens in vitro, its epithelial cells express phenotypes that are indicative of cell transformation. These studies indicate that in the human breast, the target cell of carcinogens is found in a specific compartment in which the characteristics are the determinant factors in the initiation event. These target cells will become the stem cells of the neoplastic event, depending on (a) topographic location within the mammary gland tree, (b) age at exposure to a known or putative genotoxic agent, and (c) reproductive history of the host. The higher incidence of breast cancer observed in nulliparous women supports this concept because it parallels the higher cancer incidence elicited by carcinogens in rodents when exposure occurs at a young age.

In addition, it has been shown that an increase in parity is associated with a pronounced decrease in the risk of breast cancer, each additional live birth conferring a 10 % risk reduction. Thus, the protection afforded by early full-term pregnancy in women could be explained by the higher degree of differentiation of the mammary gland at the time in which an etiologic agent or agents act. Even though differentiation significantly reduces cell proliferation in the mammary gland, the mammary epithelium remains capable of responding with proliferation to given stimuli, such as a new pregnancy. Under these circumstances, however, the cells that are stimulated to proliferate are from structures that have already been primed by the first cycle of differentiation, thus creating a second type of stem cell that is able to metabolize the carcinogen and repair the DNA damage induced more efficiently than the cells of the virginal gland and is less susceptible to carcinogenesis, as it has been demonstrated in the rodent experimental system. However, a carcinogenic stimulus powerful enough may overburden the system, successfully initiating a neoplastic process. These conditions might explain the small fraction of women who develop breast cancer after an early FFTP, meaning completion of the first cycle of differentiation. The relevance of our work lies in the vis-à-vis comparison of in vivo and in vitro studies in the human breast that validates experimental data for extrapolation to the human situation. The finding that differentiation is a powerful inhibitor of cancer initiation provides a rationale in identifying the genes controlling this process. The knowledge gained will provide novel tools for developing rational strategies for breast cancer prevention.

Literature Cited

1. Greenlee RT, Murray T, Bolden S, Wingo PA (2000) Cancer statistics, 2000. CA Cancer J Clin 50:7–33
2. Henderson BE, Ross RK, Pike MC (1993) Hormonal chemoprevention of cancer in women. Science 259:633–638
3. MacMahon B, Cole P, Lin TM, Lowe CR, Mirra AP, Ravnihar B, Salber EJ, Valaoras VG, Yuasa S (1970) Age at first birth and breast cancer risk. Bull World Health Organ 43: 209–221
4. Vessey MP, McPherson K, Roberts MM, Neil A, Jones L (1985) Fertility in relation to the risk of breast cancer. Br J Cancer 52:625–628
5. Trapido EJ (1983) Age at first birth, parity, and breast cancer risk. Cancer 51:946–948
6. Salazar H, Tobon H (1974) Morphologic changes of the mammary gland during development, pregnancy and lactation. In: Josimovich JB (ed) Lactogenic Hormones, Fetal Nutrition and Lactation (Problems of human reproduction). Wiley, New York, pp 221–277
7. Vorherr H (1974) The breast. Academic, New York, pp 1–18
8. Russo J, Russo IH (1987) Development of the human mammary gland. In: Neville MC, Daniel C (eds) The mammary gland. Plenum, New York, pp 67–93
9. Dabelow A, Horstmann E (1957) Die Haut. Die Milchdrüse. (Handbuch der mikroskopischen Anatomie des Menschen Handbook of Mikroscopic Anatomy/Haut- und Sinnesorgane) (German Edition) [Hardcover]. Springer, Berlin, pp 277–485
10. Russo J, Gusterson BA, Rogers AE, Russo IH, Wellings SR, van Zwieten MJ (1990) Comparative study of human and rat mammary tumorigenesis. Lab Invest 62:244–278
11. Cole P, MacMahon B (1969) Oestrogen fractions during early reproductive life in the aetiology of breast cancer. Lancet 1:604–606
12. Valaoras VG, MacMahon B, Trichopoulos D, Polychronopoulou A (1969) Lactation and reproductive histories of breast cancer patients in greater Athens, 1965–67. Int J Cancer 4: 350–363
13. MacMahon B, Cole P, Brown J (1973) Etiology of human breast cancer: a review. J Natl Cancer Inst 50:21–42
14. Wynder EL, Bross IJ, Hirayama T (1960) A study of the epidemiology of cancer of the breast. Cancer 13:559–601
15. Bonser GM, Dossett JA, Jull JW (1961) Human and experimental breast cancer. Charles C. Thomas, Springfield, IL, pp 316–363
16. Russo J, Tay LK, Russo IH (1982) Differentiation of the mammary gland and susceptibility to carcinogenesis. Breast Cancer Res Treat 2:5–73
17. Russo IH, Russo J (1978) Developmental stage of the rat mammary gland as determinant of its susceptibility to 7,12-dimethylbenz[a]anthracene. J Natl Cancer Inst 61:1439–1449
18. Russo J, Russo IH (1978) DNA labeling index and structure of the rat mammary gland as determinants of its susceptibility to carcinogenesis. J Natl Cancer Inst 61:1451–1459
19. Russo J, Wilgus G, Russo IH (1979) Susceptibility of the mammary gland to carcinogenesis: I Differentiation of the mammary gland as determinant of tumor incidence and type of lesion. Am J Pathol 96:721–736
20. Russo J, Russo IH (1980) Influence of differentiation and cell kinetics on the susceptibility of the rat mammary gland to carcinogenesis. Cancer Res 40:2677–2687
21. Russo J, Russo IH (1980) Susceptibility of the mammary gland to carcinogenesis. II. Pregnancy interruption as a risk factor in tumor incidence. Am J Pathol 100:497–512
22. Russo J, Wilgus G, Tait L, Russo IH (1981) Influence of age and parity on the susceptibility of rat mammary gland epithelial cells in primary cultures to 7,12-dimethylbenz(a)anthracene. In Vitro 17:877–884
23. Russo J, Russo IH (1982) Is differentiation the answer to breast cancer prevention? IRCS Med Sci 10:935–941

24. Wellings SR, Jensen HM, Marcum RG (1975) An atlas of subgross pathology of the human breast with special reference to possible precancerous lesions. J Natl Cancer Inst 55:231–273
25. Neter J, Wasserman W (1974) Applied linear Statistical models. Richard D. Irwin, Homewood, IL, pp 722–754
26. Russo J, Rivera R, Russo IH (1992) Influence of age and parity on the development of the human breast. Breast Cancer Res Treat 23:211–218
27. Monaghan P, Perusinghe NP, Cowen P, Gusterson BA (1990) Peripubertal human breast development. Anat Rec 226:501–508
28. Prechtel K (1971) Mastopathie und altersabhangige brustdrusenveranderungen. Fortschr Med 89:1312–1315
29. Hutson SW, Cowen PN, Bird CC (1985) Morphometric studies of age related changes in normal human breast and their significance for evolution of mammary cancer. J Clin Pathol 38:281–287
30. Meyer JS (1977) Cell proliferation in normal human breast ducts, fibroadenomas, and other ductal hyperplasias measured by nuclear labeling with tritiated thymidine. Effects of menstrual phase, age, and oral contraceptive hormones. Hum Pathol 8:67–81
31. Page DL, Anderson TJ (1987) Diagnostic histopathology of the breast. Churchill Livingstone, New York, 22
32. Russo J, Reina D, Frederick J, Russo IH (1988) Expression of phenotypical changes by human breast epithelial cells treated with carcinogens in vitro. Cancer Res 48:2837–2857
33. Russo J, Mills MJ, Moussalli MJ, Russo IH (1989) Influence of human breast development on the growth properties of primary cultures. In Vitro Cell Dev Biol 25:643–649
34. Levin ML, Sheehe PR, Graham S, Glidewell O (1964) Lactation and menstrual function as related to cancer of the breast. Am J Public Health Nations Health 54:580–587
35. Russo J, Calaf G, Roi L, Russo IH (1987) Influence of age and gland topography on cell kinetics of normal human breast tissue. J Natl Cancer Inst 78:413–418
36. Russo J, Russo IH (1987) Biological and molecular bases of mammary carcinogenesis. Lab Invest 57:112–137
37. Russo IH, Calaf G, Russo J (1990) Hormones and proliferative activity in breast tissue. In: Stoll BA (ed) Approaches to breast cancer prevention. Kluwer, Dordrecht, pp 35–51
38. Russo IH, Koszalka M, Russo J (1990) Human chorionic gonadotropin and rat mammary cancer prevention. J Natl Cancer Inst 82:1286–1289
39. Russo IH, Koszalka M, Russo J (1991) Comparative study of the influence of pregnancy and hormonal treatment on mammary carcinogenesis. Br J Cancer 64:481–484
40. Kelsey JL, Horn-Ross PL (1993) Breast cancer: magnitude of the problem and descriptive epidemiology. Epidemiol Rev 15:7–16
41. Kelsey JL, Gammon MD, John EM (1993) Reproductive factors and breast cancer. Epidemiol Rev 15:36–47
42. Yuan JM, Yu MC, Ross RK, Gao YT, Henderson BE (1988) Risk factors for breast cancer in Chinese women in Shanghai. Cancer Res 48:1949–1953
43. Ewertz M, Duffy SW (1988) Risk of breast cancer in relation to reproductive factors in Denmark. Br J Cancer 58:99–104
44. Paffenbarger RS Jr, Kampert JB, Chang HG (1980) Characteristics that predict risk of breast cancer before and after the menopause. Am J Epidemiol 112:258–268
45. Lubin JH, Burns PE, Blot WJ, Lees AW, May C, Morris LE, Fraumeni JF Jr (1982) Risk factors for breast cancer in women in northern Alberta, Canada, as related to age at diagnosis. J Natl Cancer Inst 68:211–217
46. Ewertz M, Duffy SW, Adami HO, Kvale G, Lund E, Meirik O, Mellemgaard A, Soini I, Tulinius H (1990) Age at first birth, parity and risk of breast cancer: a meta-analysis of 8 studies from the Nordic countries. Int J Cancer 46:597–603
47. Tulinius H, Sigvaldason H, Hrafnkelsson J, Olafsdottir G, Tryggvadottir L, Sigurthsson K (1990) Reproductive factors and breast cancer risk in Iceland. A second cohort study. Int J Cancer 46:972–975

48. Carter CL, Jones DY, Schatzkin A, Brinton LA (1989) A prospective study of reproductive, familial and socioeconomic risk factors for breast cancer using NHANES I data. Public Health Rep 104:45–50
49. Leon DA (1989) A prospective study of the independent effects of parity and age at first birth on breast cancer incidence in England and Wales. Int J Cancer 43:986–991
50. MacMahon B, Purde M, Cramer D, Hint E (1982) Association of breast cancer risk with age at first and subsequent births: a study in the population of the Estonian Republic. J Natl Cancer Inst 69:1035–1038
51. Salber EJ, Trichopulos D, MacMahon B (1969) Lactation and reproductive histories of breast cancer patients in Boston, 1965–66. J Natl Cancer Inst 43:1013–1024
52. Vatten LJ, Kvinnsland S (1992) Pregnancy-related factors and risk of breast cancer in a prospective study of 29,981 Norwegian women. Eur J Cancer 28A:1148–1153
53. Pathak DR, Speizer FE, Willett WC, Rosner B, Lipnick RJ (1986) Parity and breast cancer risk: possible effect on age at diagnosis. Int J Cancer 37:21–25
54. La Vecchia C, Negri E, Boyle P (1989) Reproductive factors and breast cancer: an overview. Soz Praventivmed 34:101–107
55. Parazzini F, La Vecchia C, Negri E (1991) Spontaneous and induced abortions and risk of breast cancer. Int J Cancer 48:816–820
56. Remennick LI (1990) Induced abortion as cancer risk factor: a review of epidemiological evidence. J Epidemiol Community Health 44:259–264
57. Wang QS, Ross RK, Yu MC, Ning JP, Henderson BE, Kimm HT (1992) A case-control study of breast cancer in Tianjin, China. Cancer Epidemiol Biomarkers Prev 1:435–439
58. Negri E, La Vecchia C, Bruzzi P, Dardanoni G, Decarli A, Palli D, Parazzini F, Rosselli del Turco M (1988) Risk factors for breast cancer: pooled results from three Italian case-control studies. Am J Epidemiol 128:1207–1215
59. Helmrich SP, Shapiro S, Rosenberg L, Kaufman DW, Slone D, Bain C, Miettinen OS, Stolley PD, Rosenshein NB, Knapp RC, Leavitt T Jr, Schottenfeld D, Engle RL Jr, Levy M (1983) Risk factors for breast cancer. Am J Epidemiol 117:35–45
60. Lund E (1991) Breast cancer mortality and the change in fertility risk factors at menopause: a prospective study of 800,000 married Norwegian women. Epidemiology 2:285–288
61. Negri E, La Vecchia C, Duffy SW, Bruzzi P, Parazzini F, Day NE (1990) Age at first and second births and breast cancer risk in biparous women. Int J Cancer 45:428–430
62. Tulinius H, Day NE, Johannesson G, Bjarnason O, Gonzales M (1978) Reproductive factors and risk for breast cancer in Iceland. Int J Cancer 21:724–730
63. Layde PM, Webster LA, Baughman AL, Wingo PA, Rubin GL, Ory HW (1989) The independent associations of parity, age at first full term pregnancy, and duration of breastfeeding with the risk of breast cancer. Cancer and Steroid Hormone Study Group. J Clin Epidemiol 42: 963–973
64. Hunt SC, Williams RR, Skolnick MH, Lyon JL, Smart CR (1980) Breast cancer and reproductive history from genealogical data. J Natl Cancer Inst 64:1047–1053
65. Vessey MP, McPherson K, Yeates D, Doll R (1982) Oral contraceptive use and abortion before first term pregnancy in relation to breast cancer risk. Br J Cancer 45:327–331
66. Bain C, Willett W, Rosner B, Speizer FE, Belanger C, Hennekens CH (1981) Early age at first birth and decreased risk of breast cancer. Am J Epidemiol 114:705–709
67. Brinton LA, Hoover R, Fraumeni JF Jr (1983) Reproductive factors in the aetiology of breast cancer. Br J Cancer 47:757–762
68. Dawson EK (1934) A histological study of the normal mamma in relation to tumour growth. Edinb Med J 41:653–682
69. Ferguson DJ, Anderson TJ (1981) Morphological evaluation of cell turnover in relation to the menstrual cycle in the "resting" human breast. Br J Cancer 44:177–181
70. Geschickter CF (1945) Normal development and functional changes in the mammary gland. In: Geschickter CF (ed) Diseases of the breast: diagnosis, pathology, treatment. Lippincott, Philadelphia, pp 3–41

71. Ingleby H, Gersohn-Cohen J (1960) The normal breast. In: Ingleby H, Gersohn-Cohen J (eds) Comparative anatomy, pathology and roentgenology of the breast. Univ Penn Press, Philadelphia, pp 3–119
72. Dawson EK (1935) A histological study of the normal mamma in relation to tumor growth: the mature gland in pregnancy and lactation. Edinb Med J 42:569–598
73. Haagensen CD (1986) Results with Halsted's radical mastectomy. In: Haagensen CD (ed) Diseases of the breast. W.B. Saunders, Philadelphia, pp 903–932
74. Smithers DW, Rigby-Jones P (1959) Clinical evidence of parasternal lymph node involvement in neoplastic disease. Acta Radiol Suppl 188:235–247
75. Truscott BM (1947) Carcinoma of the breast; an analysis of the symptoms, factors affecting prognosis, results of treatment and recurrences in 1211 cases treated at the Middlesex Hospital. Br J Cancer 1:129–145
76. Harnett WL (1948) A statistical report on 2529 cases of cancer of the breast. Br J Cancer 2: 212–239
77. Russo J, Romero AL, Russo IH (1994) Architectural pattern of the normal and cancerous breast under the influence of parity. Cancer Epidemiol Biomarkers Prev 3:219–224
78. Sandison AT (1962) An autopsy study of the adult human breast: with special reference to proliferative epithelial changes of importance in the pathology of the breast. Natl Cancer Inst Monogr 4:1–145
79. Gallagher HS (1969) Early phases in the development of breast cancer. Cancer 24:1170–1178
80. McGregor H, Land CE, Choi K, Tokuoka S, Liu PI, Wakabayashi T, Beebe GW (1977) Breast cancer incidence among atomic bomb survivors, Hiroshima and Nagasaki, 1950–69. J Natl Cancer Inst 59:799–811
81. Yuasa S, MacMahon B (1970) Lactation and reproductive histories of breast cancer patients in Tokyo, Japan. Bull World Health Organ 42:195–204
82. Trichopoulos D, Brown JB, Garas J, Papaioannou A, MacMahon B (1981) Elevated urine estrogen and pregnanediol levels in daughters of breast cancer patients. J Natl Cancer Inst 67:603–606
83. Russo J, Gimotty PA, Russo IH (1990) Topographic variations in breast development. Breast Cancer Res Treat 16:95
84. Russo J, Calaf G, Russo IH (1993) A critical approach to the malignant transformation of human breast epithelial cells with chemical carcinogens. Crit Rev Oncol 4:403–417
85. Russo J, Russo IH (1999) Cellular basis of breast cancer susceptibility. Oncol Res 11: 169–178
86. Xie J, Haslam SZ (1997) Extracellular matrix regulates ovarian hormone-dependent proliferation of mouse mammary epithelial cells. Endocrinology 138:2466–2473
87. Russo J, Russo IH (1998) Development of the human breast. In: Knobil E, Neill JD (eds) Encyclopedia of reproduction. Academic, New York, pp 71–80
88. Russo J, Russo IH (1997) Toward a unified concept of mammary carcinogenesis. Prog Clin Biol Res 396:1–16
89. Russo J, Russo IH (1994) Toward a physiological approach to breast cancer prevention. Cancer Epidemiol Biomarkers Prev 3:353–364
90. Russo J, Saby J, Isenberg WM, Russo IH (1977) Pathogenesis of mammary carcinomas induced in rats by 7,12-dimethylbenz(a)anthracene. J Natl Cancer Inst 59:436–445
91. Lambe M, Hsieh CC, Chan HW, Ekbom A, Trichopoulos D, Adami HO (1996) Parity, age at first and last birth, and risk of breast cancer: a population-based study in Sweden. Breast Cancer Res Treat 38:305–311
92. Russo J, Russo IH (1998) Differentiation and breast cancer development. In: Heppner G (ed) Advances in oncobiology. JAI Press Inc, Greenwich, CN, pp 1–10
93. Russo J (1983) Basis of cellular autonomy in the susceptibility to carcinogenesis. Toxicol Pathol 11:149–166
94. Russo J, Lynch H, Russo IH (2001) Mammary gland architecture as a determining factor in the susceptibility of the human breast to cancer. Breast J 7:278–291

95. Russo J, Ao X, Grill C, Russo IH (1999) Pattern of distribution of cells positive for estrogen receptor alpha and progesterone receptor in relation to proliferating cells in the mammary gland. Breast Cancer Res Treat 53:217–227
96. Rajan JV, Marquis ST, Gardner HP, Chodosh LA (1997) Developmental expression of Brca2 colocalizes with Brca1 and is associated with proliferation and differentiation in multiple tissues. Dev Biol 184:385–401
97. Colditz GA, Rosner BA, Speizer FE (1996) Risk factors for breast cancer according to family history of breast cancer. For the Nurses' Health Study Research Group. J Natl Cancer Inst 88:365–371
98. Hu YF, Russo IH, Zalipsky U, Lynch HT, Russo J (1997) Environmental chemical carcinogens induce transformation of breast epithelial cells from women with familial history of breast cancer. In Vitro Cell Dev Biol Anim 33:495–498
99. Eyden BP, Watson RJ, Harris M, Howell A (1986) Intralobular stromal fibroblasts in the resting human mammary gland: ultrastructural properties and intercellular relationships. J Submicrosc Cytol 18:397–408
100. Ozzello L (1970) Epithelial-stromal junction of normal and dysplastic mammary glands. Cancer 25:586–600
101. Sakakura T, Sakagami Y, Nishizuka Y (1979) Persistence of responsiveness of adult mouse mammary gland to induction by embryonic mesenchyme. Dev Biol 72:201–210
102. Henson DE, Tarone RE (1993) On the possible role of involution in the natural history of breast cancer. Cancer 71:2154–2156
103. Sun CC, Lenoir G, Lynch H, Narod SA (1996) In-situ breast cancer and BRCA1. Lancet 348:408
104. Jernström H, Johannsson O, Borg Å, Olsson H (1998) Do BRCA1 mutations affect the ability to breast-feed? Significantly shorter length of breast-feeding among BRCA1 mutation carriers compared with their unaffected relatives. Breast 7:320–324
105. Xu X, Wagner KU, Larson D, Weaver Z, Li C, Ried T, Hennighausen L, Wynshaw-Boris A, Deng CX (1999) Conditional mutation of Brca1 in mammary epithelial cells results in blunted ductal morphogenesis and tumour formation. Nat Genet 22:37–43
106. Arlas·Mendoza F, Romero AL, Rivera MR, Carvajal L, Willard T, Liang JD, Russo IH, Brown TR, Russo J (1992) Comparative study of mammary gland development by magnetic resonance Imaging and subgross anatomy. Proc Am Assoc Cancer Res 33:270a

Chapter 2
Histological Evaluation of the Normal Breast

Jose Russo and Irma H. Russo

Keywords Histological evaluation • Procurement of normal breast tissue • Reduction mammoplasty • Surgical tissue • Autopsy material • Needle core biopsies • The stroma of the breast • Cell types in the normal breast • Immunocytochemical markers

2.1 Introduction

The breast is a bilateral organ that in the female undergoes dramatic changes in size, shape, and function in association with infantile growth, puberty, pregnancy, lactation, and postmenopausal regression [1, 2]. The development of the human breast is a progressive process initiated during embryonic life. This process has been discussed extensively in previous publications [3] and will, therefore, not be addressed in this chapter. In this chapter, we describe the parameters that need to be considered in order to evaluate the normality of the human breast. In addition, we address the different morphological patterns observed which depend on the sampling procedures utilized.

2.2 Procurement of Normal Breast Tissue

2.2.1 Reduction Mammoplasty

Reduction Mammoplasty is a surgical procedure of breast reduction by which excess breast fat, glandular tissue, and skin are removed to achieve a breast size in proportion to body size. It is a cosmetic surgery, but also therapeutic for those women who suffer the discomfort associated with overly large breasts.

J. Russo, M.D. (✉) • I.H. Russo, M.D.
Irma H Russo MD Breast Cancer Research Laboratory, Fox Chase Cancer Center, Cottman Avenue 333, Philadelphia, PA 19111, USA
e-mail: jose.russo@fccc.edu; Irma.Russo@fccc.edu

J. Russo and I.H. Russo, *Techniques and Methodological Approaches in Breast Cancer Research*, DOI 10.1007/978-1-4939-0718-2_2,

Table 2.1 Profile of the population under study

Group	Glandular differentiation	Sample	Age[a]	Parity history			Glandular development[f]
A	Low	1	30	G0[b]	P0[c]	A0[d]	Lobule Type 1
		2	34	G0	P0	A0	Lobule Type 1, alveolar bud, ducts
		3	48	G0	P0	A0	Lobule Type 1, ducts
		4	20	G0	P0	A0	Lobule Type 1
		5	23	G0	P0	A0	Ducts only
		6	36	G0	P0	A0	Lobule Type 1
		7	53	G0	P0	A0	Lobule Type 1, ducts
		8	24	G0	P0	A0	Lobule Type 1, ducts
B	Medium	16[e]	60	G1	P1	A0	Lobule Type 1, ducts
		17	30	G1	P1	A0	Lobule Types 1, 2, ducts
		18	63	G4	P4	A0	Lobule Type 1, ducts
		19	31	G3	P1	A2	Lobule Type 1, ducts
		20	44	G5	P5	A0	Lobule Types 1, 2
		21	31	G3	P2	A1	Lobule Type 1
		22	36	G3	P2	A1	Lobule Type 1
C	High	9	52	G1	P1	A0	Lobule Types 2, 3
		10	33	G1	P1	A0	Lobule Type 3
		11	31	G3	P3	A0	Lobule Type 3
		12	30	G3	P3	A0	Lobule Type 3
		13	39	G2	P2	A0	Lobule Type 3
		14	18	G0	P0	A0	Lobule Type 3
		15	50	G2	P2	A0	Lobule Types 2, 3

[a]Age in years at time of surgery
[b]G, gravidity, number of pregnancies
[c]P, parity, number of deliveries
[d]A, pregnancies ending in abortion
[e]Patients who developed infiltrating ductal carcinoma in contralateral breast
[f]*Lob* lobule, *AB* alveolar bud

Table 2.2 Histopathological criteria of normality for acceptance of breast tissue in this study

Presence of ductal structures alone or in combination with ductule forming lobules surrounded by identifiable intralobular loose connective tissue
The interlobular connective tissue must contain fat and connective tissue but not fibrosis
The ductal and ductular structures must be covered by an identifiable cuboidal or low cuboidal epithelium resting on basement membrane and a discontinuous layer of myoepithelial cells
Breast tissues that contained hyperplastic epithelium in the ductal structures, lobular hyperplasia, cystic formation, apocrine metaplasia, or other metaplastic changes, or another morphological indication of derangement of the normal structure as described above were deleted from this study

We have used 22 reduction mammoplasties performed on patients whose clinical, obstetric, and gynecological histories were available (Table 2.1) [4]. All tissues were morphologically evaluated following the criteria outlined in Table 2.2 [4]. These criteria are what determine the normality of breast tissue by its ductal and lobular composition and state that no pathological conditions are present in the tissue under study (Table 2.2). The histological criteria were correlated with the age and pregnancy history and scored on three levels of glandular differentiation: low, medium, and high (Table 2.1). In our experience, working in conjunction with a plastic surgeon is vital for obtaining this tissue. Every reduction mammoplasty specimen obtained weighed between 300 and 500 g and the tissues for morphological examination were taken from ten different areas of the gland, fixed in 10 % neutral buffered formalin, embedded in paraffin, sectioned at 5 μm in thickness, and stained with hematoxylin and eosin (H&E) [4].

The breast tissue from premenopausal nulliparous women is mainly characterized by the presence of fat tissue, separating areas of interlobular stroma composed of fibro connective tissue. The glandular or parenchyma is formed mainly by Lobules Type 1 (Fig. 2.1a) as opposed to the breast tissue from premenopausal parous women in which the amount of fat is significantly less prominent and the interlobular stroma separates Lobules Type 3 (Sample 9 from Table 2.1 and Fig. 2.1b). In general, the glands of nulliparous women are mainly composed of Lobules Type 1 and ducts (Samples 3 and 8 in Table 2.1 and Fig. 2.1c, d), whereas the breast tissue of parous women is mainly formed by Lobules Type 3 (Table 2.1) (see also Chap. 1).

2.2.2 Surgical Tissue Adjacent to Benign and Malignant Lesions

We have used normal breast tissue samples from 15 women who underwent surgery for benign and suspicious breast lesions [5]. Excision biopsies were performed and provided 2–6 g of tissue. The normal breast tissue samples were dissected 2 cm distant to the lesion. The donors ranged in age from 21 to 55 years and were subdivided into 2 groups: young women, composed of 5 premenopausal women ranging in age from 21 to 32 years (26.6 + 5 years), and older women composed of 10 women, 4 premenopausal, 4 postmenopausal, and 2 of menopausal status unknown, ranging in age from 42 to 57 years (49.6 + 5.5 years). Table 2.3 depicts the age and reproductive profile of the population under study [5]. The morphological profile of each of these samples depends significantly on the reproductive history and the age at the time of sample collection (Fig. 2.2). The breast tissue from premenopausal parous women contains islands of fat tissue separating areas of interlobular stroma composed of fibro connective tissue. The glandular or parenchyma is formed by Lobules Type 1, 2, or 3 in premenopausal women with history of pregnancy (Fig. 2.2a, b, d) (Samples 1, 2, and 4 in Table 2.3). Whereas the normal architecture of the breast is clearly defined from the benign lesions as depicted in Fig. 2.2a or d, it is better to follow the rule of 2 cm from any kind of lesion when the tissue is planned to be used as a normal control either for tissue microarray or for laser capture micro-dissection (see Chaps. 3 and 4). The breast tissue from premenopausal nulliparous woman (Sample 5 in Table 2.3) is also composed of interlobular stroma formed by fibro connective tissue made up of collagen and elastic fibers and fibroblasts. In the specific

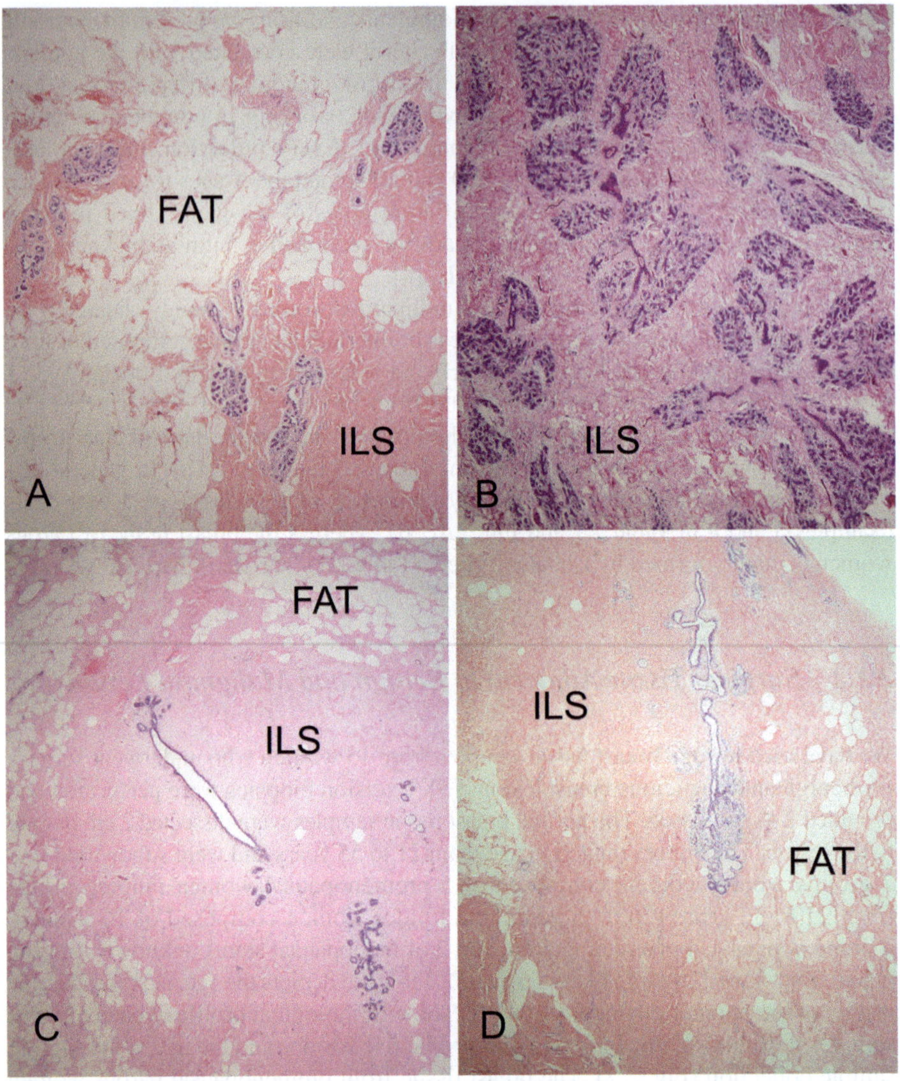

Fig. 2.1 Histological sections of reduction mammoplasty specimens. Stained with hematoxylin and eosin, ×2. (**a**) Breast tissue from a premenopausal nulliparous woman. Fat tissue (FAT) is separating areas of interlobular stroma (ILS) composed of fibro connective tissue. The glandular or parenchyma is formed by Lobules Type 1. (**b**) Breast tissue from a premenopausal parous woman. The interlobular stroma (ILS) composed of fibro connective tissue is separating Lobules Type 3. (**c**, **d**) Breast tissue from premenopausal nulliparous woman. Fat tissue (FAT) is separating areas of interlobular stroma (ILS) composed of fibro connective tissue. The glandular or parenchyma is formed by Lobules Type 1

case of Sample 5 the adjacent lesion was in an intraductal proliferation that was found 2 cm from the margin of resection. Fresh tissues obtained from excisional biopsies are excellent material for in vitro study either for developing primary culture [4] or for performing organ culture in which cell kinetics studies are performed [5].

Table 2.3 Age, menopausal status, and reproductive histories of the population under study[a]

Group	Patient No.	Age, year[b]	Menopausal status	Last menstrual period	Full-term pregnancies
Young women	1	21	Premenopausal	1 day	2
	2	31	Premenopausal	14 days	1
	3	27	Premenopausal	14 days	1
	4	32	Premenopausal	14 days	4
	5	22	Premenopausal	13 days	0
Older women	6	50	Premenopausal	10 days	3
	7	46	Premenopausal	12 days	1
	8	42	Premenopausal	1 day	1
	9	55	Postmenopausal	1 year	3
	10	42	Premenopausal	1 day	1
	11	47	NA	NA	NA
	12	53	Postmenopausal	5 years	0
	13	56	Postmenopausal	12 years	3
	14	57	Postmenopausal	15 years	2
	15	48	NA	NA	NA

[a]NA, information not available
[b]Mean age ± SD for young women was 26.6±5.0 years; it was 49.6±5.5 years for older women

2.2.3 *Autopsy Material from Accidental Death*

We have utilized breast tissues obtained from 16 accident victims whose autopsies were performed by the medical examiner [6]. These tissues were provided by Dr. S. Bartow who obtained them under NCI-RFP No 1-CB-84231/NO1-CN-23928. The breasts were removed at autopsy by subcutaneous mastectomy. Five of these samples contained invasive ductal carcinomas (Table 2.4). Upon removal from the body the tissue was fixed in 10 % neutral buffered formalin and embedded in paraffin for histological examination. Figure 2.3a–c depicts the histological appearance of Samples 3, 6, and 13 (Table 2.4), which were taken from premenopausal nulliparous women and contained a fat stroma, or combination, with connective tissue mainly formed by collagen fibers. Another sample, also belonging to a premenopausal nulliparous woman (Sample 14, Table 2.4), was characterized by an interlobular stroma composed of fibro connective tissue surrounding Lobules Type 1 and ductal structures (Fig. 2.3d). The breast tissue obtained from autopsy material is excellent for either performing morphological studies by whole mounts, as described in Chap. 1, or histological observations. However, in our hands it did not result as optimal for immunocytochemical studies, due to the long storage in formalin.

2.2.4 *Needle Core Biopsies*

Breast needle core biopsies must be obtained by a qualified physician and specimens obtained after an authorization for use and disclosure of protected health information for research approved by the Institutional Review Board (IRB) in compliance

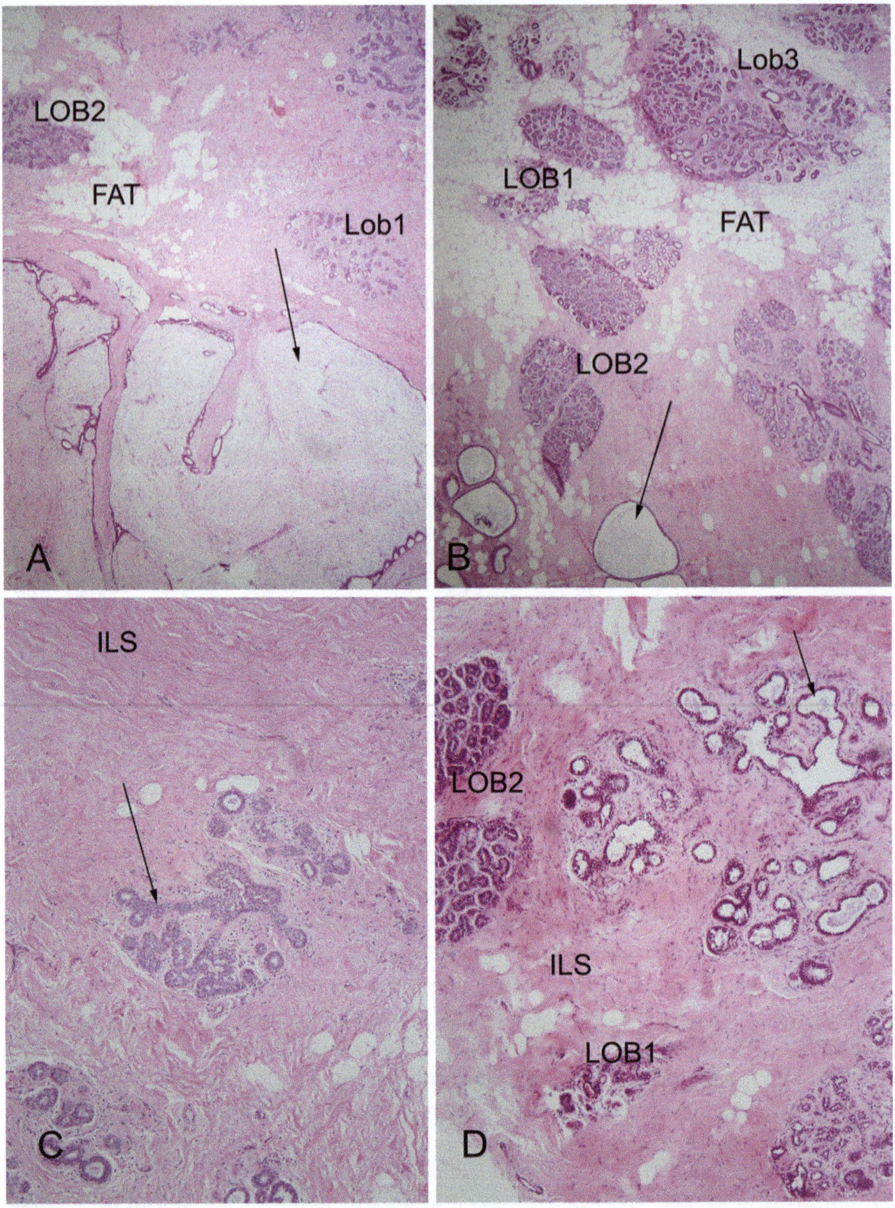

Fig. 2.2 Histological sections of tissue obtained from surgical specimens containing cancerous or benign diseases of the breast. Stained with hematoxylin and eosin, ×2. (**a**) Breast tissue from a premenopausal parous woman. Fat tissue (FAT) is separating areas of interlobular stroma composed of fibro connective tissue. The glandular or parenchyma is formed by Lobule Types 1 and 2 (LOB1 and Lob2). The *arrow points* toward a fibroadenomatous lesion. (**b**) Breast tissue from a premenopausal parous woman. The fat (FAT) and interlobular stroma are separating Lobule Types 1, 2, and 3 (LOB1, LOB2, and LOB3). The *arrow points* toward cystic structures. (**c**) Breast tissue from premenopausal nulliparous woman. The interlobular stroma (ILS) is composed of fibro connective tissue. The *arrow* is pointing an intraductal proliferation that was found 2 cm from the margin of resection of an invasive ductal carcinoma. (**d**) Breast tissue from premenopausal parous woman. The interlobular stroma (ILS) is composed of fibro connective tissue and separating LOB1 and LOB2. The *arrow* is pointing an LOB1 with cystic changes

Table 2.4 Characteristics of the patient population[a]

Patient group	Sample No.	Age	Patient's race[b]	Parity history	Breast side	Cancer	Tumor dimensions (cm)[c]	Tumor location
I	5	20	C	N	Left	No		
	13	21	B	N	Left	No		
	6	30	C	N	Left	No		
	16	61	C	N	Right	No		
II	14	39	C	N	Left	Yes	3.0	UOQ
	15	48	C	N	Right	Yes	4.0	UOQ
III	1	20	C	P	Left	No		
	2	25	C	P	Left	No		
	3	28	C	P	Left	No		
	4	30	C	P	Left	No		
	7	37	C	P	Right	No		
	9	46	C	P	Left	No		
	12	63	C	P	Left	No		
IV	8	39	C	P	Right	Yes	4.0	UOQ
	10	51	B	P	Left	Yes	3.0	LOQ
	11	58	C	P	Left	Yes	5.5	LOQ

[a]Group I, nulliparous females free of breast pathology; Group II, nulliparous females with infiltrating ductal carcinoma; Group III, parous females free of mammary pathology; Group IV, parous females with infiltrating ductal carcinoma

[b]*C* Caucasian, *B* black, *N* nulliparous, *P* parous

[c]Tumor's largest diameter

with USA HIPAA regulations. All breast core biopsies and corresponding participant data must be de-identified following the safe harbor method recommended by the institution's De-Identification of Protected Health Information policies [7–9]. In the protocol used in our studies, four core biopsies with a 10-gauge biopsy needle were taken from the upper outer quadrant (UOQ) of the right or the left breast, preferentially sampling dense or gritty areas in the breast tissue which may be appreciated on the mammogram for each of the 23 women who entered in the study. From the four core biopsies collected, the first one is preserved in 70 % ethanol for histopathological evaluation; the second, third, and fourth passes are separately preserved in an RNA-preserving fluid (RNAlater®, Ambion) for genomic analysis (see Chaps. 7–9).

For histopathological analysis the first pass is expelled from the needle into a prelabeled 15 mL tube containing 10 mL of 70 % ethanol. The needle is washed in isotonic sodium chloride solution before the next use. Each specimen is labeled with a number and stored for up to 4 weeks at 4° centigrade in the original collection tube until ready for processing. Cell clusters or tissue fragments are tightly positioned in a tissue cassette, which is then dehydrated and embedded in paraffin and sectioned at a thickness of 4 μm and stained with H&E, or processed for immunocytochemical studies utilizing an automated cell stainer (Optimax Plus Automatic Consolidated Cell Stainer, Biogenex, San Ramon, CA) following standard procedures. Figure 2.4a, b depicts the core breast biopsies of postmenopausal nulliparous women with prominent interlobular stroma composed of fibro connective tissue and

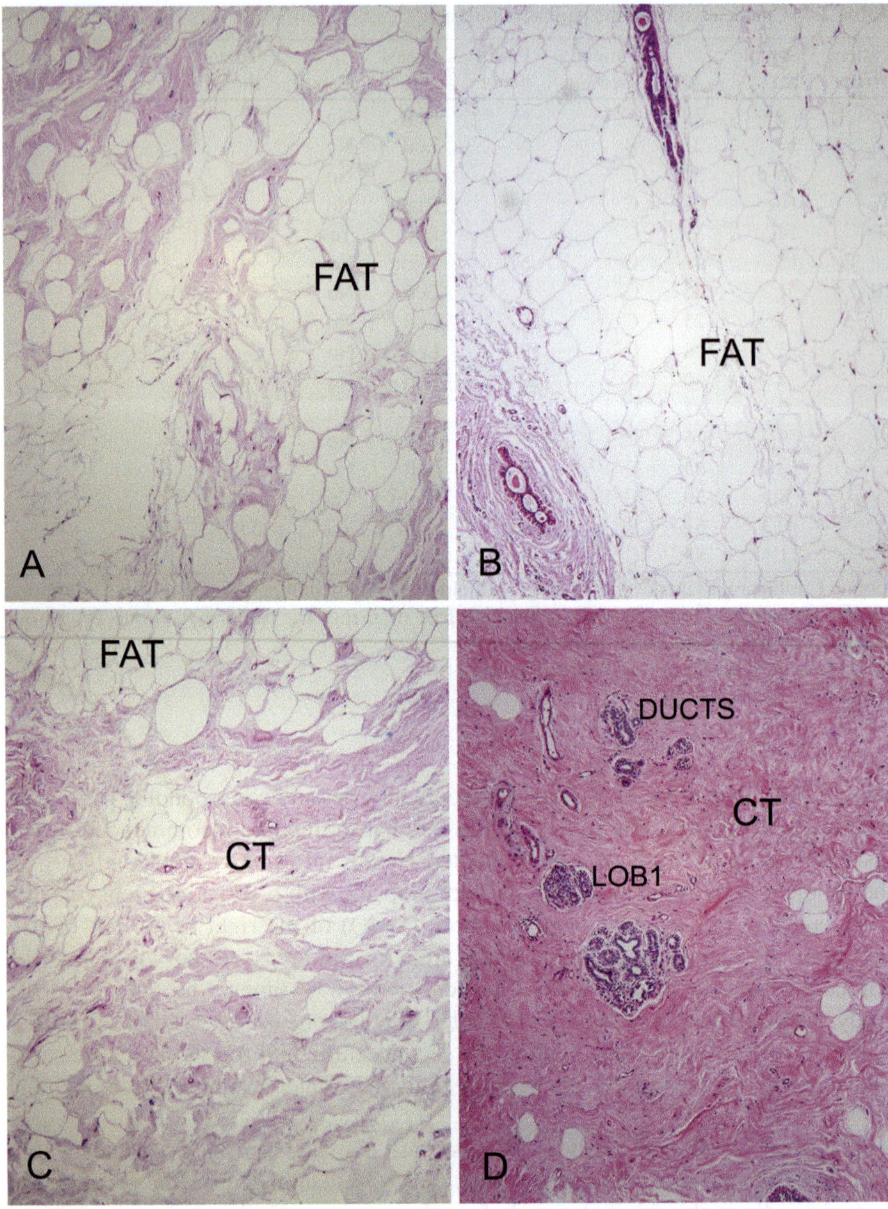

Fig. 2.3 Histological sections of postmortem specimens. Stained with hematoxylin and eosin, ×2. (**a–c**) Breast tissue from a premenopausal nulliparous woman. Fat tissue (FAT) is separating areas of fibro connective tissue (CT). (**d**) Breast tissue from a premenopausal nulliparous woman. Interlobular stroma composed of fibro connective tissue (CT) is the main component of the breast tissue and isolated LOB1 and ducts (DUCTS)

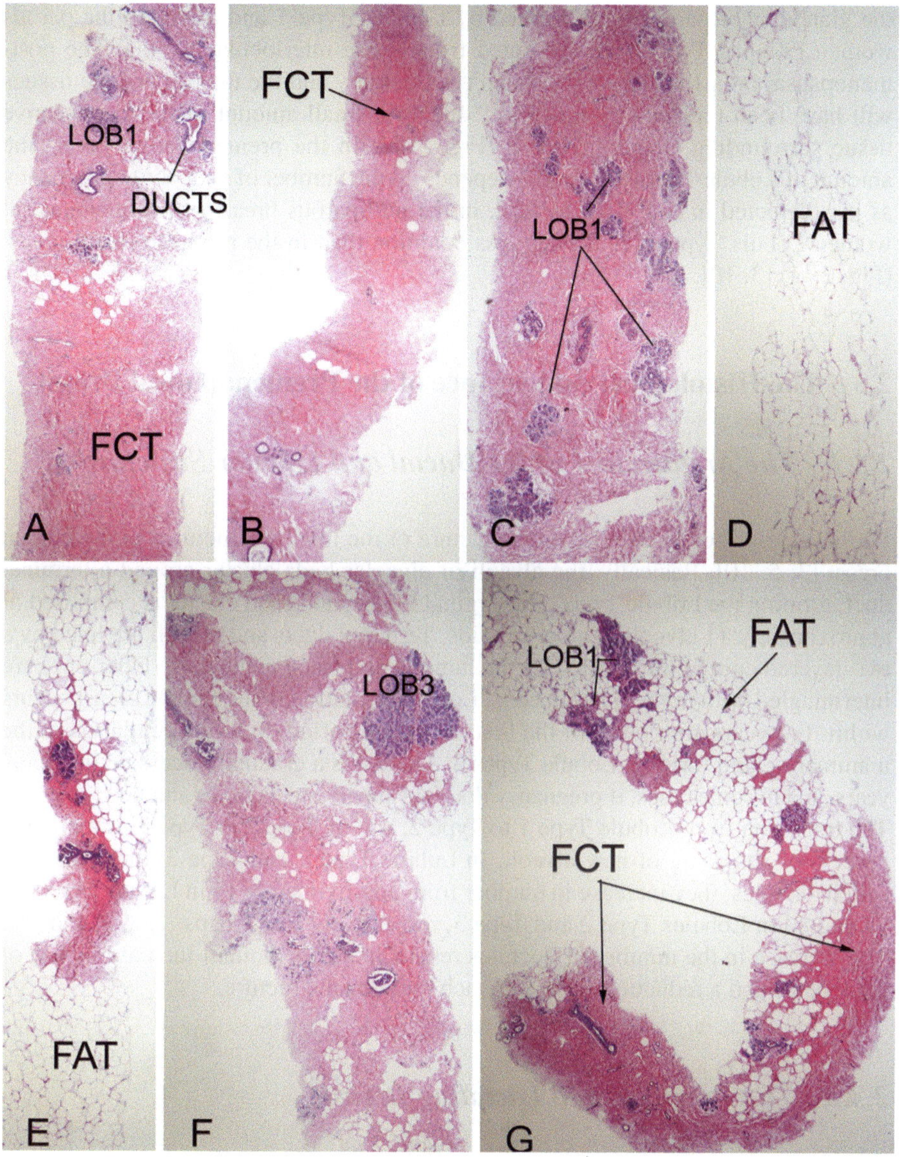

Fig. 2.4 Histological sections of core biopsies. Stained with hematoxylin and eosin, ×2. (**a**, **b**) Breast tissue from postmenopausal nulliparous women. Interlobular stroma composed of fibro connective tissue (FCT). The glandular or parenchyma is formed by Lobules Type 1 and ducts (DUCTS). (**c**) Breast tissue from a postmenopausal parous woman. The interlobular stroma composed of fibro connective tissue separating Lobules Type 1. (**d**) Breast tissue from a premenopausal nulliparous woman containing small amount of glandular tissue and surrounded by fat tissue (FAT). (**f**) Breast tissue from premenopausal parous women. Interlobular stroma composed of fibro connective tissue surrounds a well-preserved Lobule Type 3 (LOB3) and Lobules Type 3 in involution. (**g**) Breast tissue from a premenopausal nulliparous woman containing Lobules Type 1 and surrounded by fat tissue (FAT) and fibro connective tissue (FCT)

the glandular or parenchyma formed by Lobules Type 1 and ducts. In the parous women Lobules Type 1 are separated from dense interlobular stroma. The post-menopausal breast will have involuted considerably and the majority of the tissues will have been replaced by fat (Fig. 2.4d), or a small amount of fibro connective tissue surrounding Lobules Type 1 (Fig. 2.4e). In the premenopausal breast the amount of Lobule Types 1, 2, and 3 depends on the number of years after pregnancy as it is depicted in Fig. 2.4f. Instead, in the nulliparous breast of postmenopausal women Lobule Types 2 and 3 are less common than in the premenopausal breast (Fig. 2.4g) [3, 10].

2.3 The Histological Appearance of the Premenopausal Breast

2.3.1 The Architecture of the Ductal and Lobular Structure

We have described in detail the architecture of the lobular structures of the human breast [3, 5, 10]. Basically, the alveoli or alveolar buds cluster around a terminal duct, forming the Lobule Type 1, or virginal lobule, and each cluster is composed of approximately 11 alveoli or alveolar buds. Terminal ducts and alveoli are lined by a two-layered epithelium formed by the luminal and basal cells; these latter ones are intermingled with myoepithelial cells. Lobule formation in the female breast occurs within 1–2 years after onset of the first menstrual period. Full differentiation of the mammary gland forming Lobule Types 2, 3, and 4 is a gradual process taking many years, and in some cases, if pregnancy does not supervene, is never attained [2, 3, 10]. The transition from Lobule Type 1 to Type 2, and of Type 2 to Type 3, is a gradual process of sprouting of new alveoli. In Lobule Type 2 and Type 3, these are now called ductules; they increase in number from approximately 11 in Lobule Type 1 to 47 and 80 in Lobules Type 2 and Type 3, respectively (see Chaps. 1, 2, 3, and 10). The increase in the number of ductules results in a concomitant increase in size of the lobules and a reduction in size of each individual structure.

2.3.2 The Pregnant and Lactating Breast

During pregnancy, the breast attains its maximum development. This occurs in two distinctly dominant phases characteristic of the early and late stages of pregnancy [2, 3, 10]. The early stage is characterized by the proliferation of the distal elements of the ductal tree, resulting in the formation of ductules that at this stage can be called acini, thus progressing a Lobule Type 3 into a Lobule Type 4. The intensity of budding and degree of lobule formation go beyond what have been observed in the virginal breast. By the third month of pregnancy, the number of well-formed lobules exceeds the number of primitive buds; however, primitive buds are still found. In newly formed lobules, the epithelial cells composing each acinus not only increase greatly in number due to active cell division but also increase in size mainly because of cytoplasm enlargement [2]. In the middle of pregnancy, the lobules are further enlarged and increase in number. They surround the duct from which their central branch proceeds so thickly that the chief duct, the terminal or intralobular terminal duct, can no longer be recognized. The transition between the terminal ducts and the budding acini is gradual, making the histological distinction between the two difficult, since both show evidence of early secretory activity. The definitive structure of the ductal tree is essentially settled by the end of the first half of pregnancy; the mammary changes that characterize the second half of pregnancy are chiefly a continuation and accentuation of the secretory activity. Further progressive branching continues with less prominent bud formation. At this time, the formation of secreting units or acini, the differentiated structures, becomes increasingly evident. Proliferation of new acini is reduced to a minimum, and the luminae of those already formed become distended by accumulation of secretory material or colostrum [4]. The epithelium is vacuolated due to the accumulation of lipids. Under the electron microscope the mammary epithelia show numerous lipid droplets and proteinaceous material. In the Lobule Type 4, or lactating breast, the reactivity against milk fat globule protein is highly expressed [3]. The secretory acinus formed during pregnancy is a terminal outgrowth that marks the end of glandular differentiation.

During lactation, the process of growth and differentiation may be observed in the same lobule type, side by side with the process of milk secretion [1, 2, 11, 12]. No major morphological changes of the mammary gland are observed during lactation. The mammary lobules are enlarged and the acini have a dilated lumen filled with granular, slightly basophilic material admixed with fat. There is a significant variation in lobule size throughout the gland, suggestive of a variation in lactogenic activity from lobule to lobule. Milk is synthesized and released into the mammary acini and ductal system, although it can be stored for up to 48 h before the rate of milk synthesis and secretion begins to decrease. As long as milk is removed regularly from the mammary gland, the alveolar cells continue to secrete milk [2, 12, 13].

The accumulation of milk in the ductoacinar lumina and within the cytoplasm of the lactogenic epithelial cells that occurs after weaning has an inhibitory effect on further milk synthesis. This effect is followed by a series of involutional changes in the mammary gland consisting of a multifocal asynchronous process of reduction in volume of the secretory epithelial cells and further inhibition of their secretory activity.

It is considered that post-lactational regression is due to two complementary mechanisms: cell autolysis, with collapse of acinar structures and narrowing of the tubules accompanied by the appearance of round cell infiltration and phagocytes in and about the disintegrating lobules; and finally, regeneration of the periductal and perilobular connective tissue with renewed budding and proliferation in the terminal tubules. Until menopausal involution sets in, the parous organ shows more glandular tissue than if pregnancy or pregnancy and lactation had never occurred [2, 12, 13].

2.3.3 The Stroma of the Breast

Both the intralobular and the interlobular stroma are affected simultaneously during development, pregnancy, lactation, and involution [3]. These processes occur, in turn, in a synchronous manner in response to specific hormonal and growth factor stimuli [14]. Two major mechanisms are considered involved in the interaction of the stroma and epithelial cells, the production of soluble growth factors and a modification of the composition of the extracellular matrix. This interaction seems to be bi-directional such that epithelial cells are also capable of influencing stromal cell behavior and governing gene expression [15, 16]. Several hypotheses and molecular pathways have been postulated for explaining the mechanism of organogenesis of the terminal structures and its relation with the stroma [17]. However, there is not significant information available as to how this process is applied or takes place in the human breast. The study of the stroma-parenchyma ratio in the breasts of pubertal, postpubertal, parous, and pregnant women shows that the relationship between parenchyma and stroma is a dynamic process. At puberty, almost 90 % of the mammary gland is made up of stroma: the intralobular stroma, that represents 17 % of the total, consists of the connective tissue that surrounds each individual alveolar bud, and the interlobular stroma, composed of fat and connective tissue, which separates one lobule from another. The parenchyma of these glands, representing 10 % of the mammary area, is made up almost exclusively of Lobule Type 1 and ductal structures. In the glands of postpubertal and young nulliparous women, the parenchyma increases from 10 to 30 % of the total area of the gland (0-10 % is composed of Lob 1, 10-18 % of Lob 2, and 1-3 % of Lob 3) [2, 3, 10]. The intralobular stroma of these breasts represents about 28 % of the total [2, 10]. Parity induces significant differences in mammary gland development. The breasts of parous women are mostly composed of Lobule Type 3 with a markedly reduced proportion of Lobule Type 1. During the involution of the breast of parous women at menopause it is frequently observed that the intralobular stroma is composed of connective tissue, blood vessels, and moderate amount of lymphocytes intermingled with ductules of a Lobule Type 2 in regression (Fig. 2.5a, b). However, it is very unusual to see, even in postmenopausal parous women, the intralobular stroma infiltrated with lymphocytes intermingled with ductules of Lobule Type 2 in regression to Lobules Type 1 (Fig. 2.5c, d).

2.4 The Histological Appearance of the Postmenopausal Breast

Menopause supervenes as the consequence of the atresia of more than 99 % of the 400,000 follicles that are present in the ovaries of a female fetus at a gestational age of 5 months. Gonadotropin-releasing hormone secretion is also implicated in this

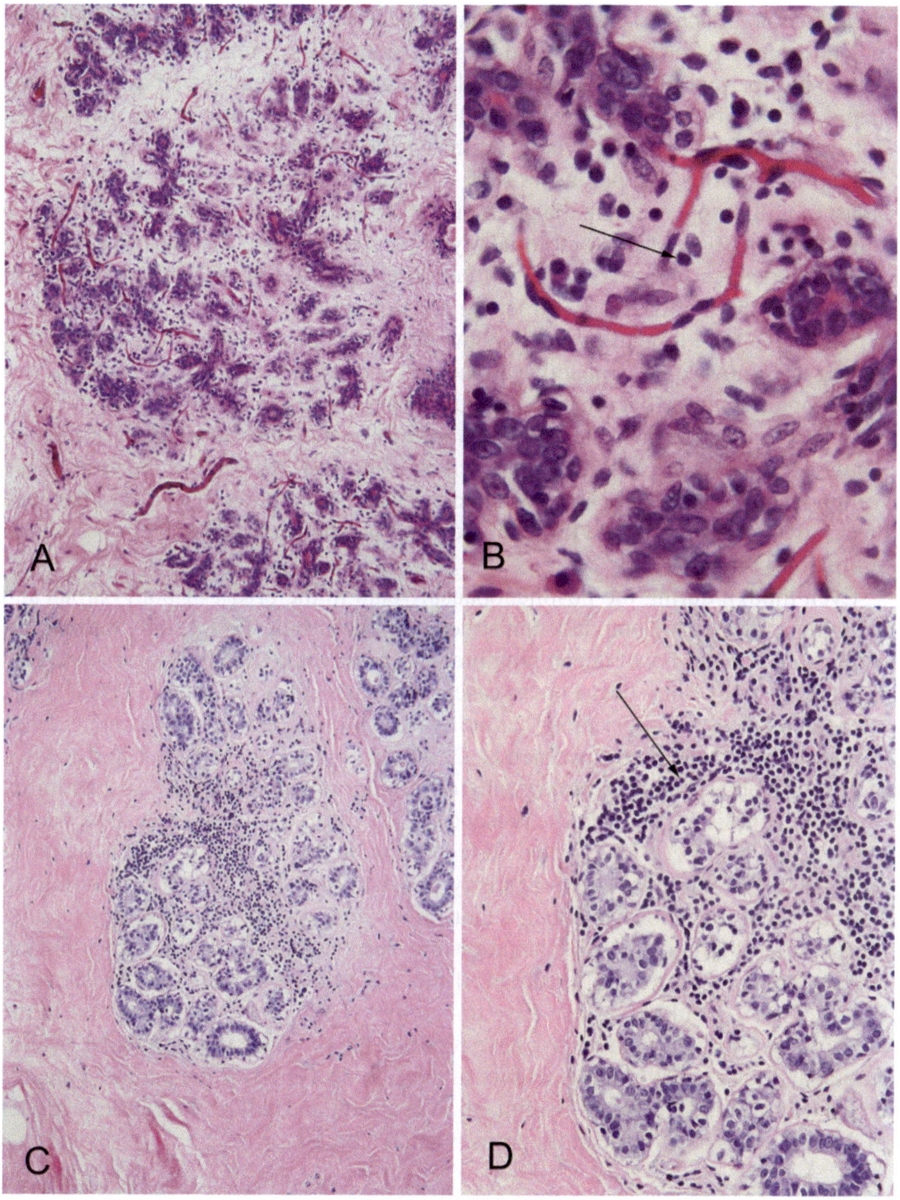

Fig. 2.5 Histological sections of core biopsies. Stained with hematoxylin and eosin. (**a**, **b**) Breast tissue from postmenopausal parous women. Intralobular stroma composed of connective tissue, blood vessels, and moderate amount of lymphocytes (*arrow*) intermingled with ductules of a Lobule Type 2 in regression. Magnification ×10 and ×40, respectively. (**c**, **d**) Breast tissue from postmenopausal parous women. Intralobular stroma composed of connective tissue, and significant amount of lymphocytes (*arrow*) intermingled with ductules of a Lobule Type 2 in regression. This type of lymphocytic infiltration is extremely unusual. Magnification ×10 and ×40, respectively

phenomenon, indicating that a hypothalamic process is involved in the development of menopause. After menopause, the breast undergoes a regressive phenomenon both in nulliparous and parous women. This regression is manifested as an increase in the number of Lobules Type 1, and a concomitant decline in the number of Lobules Type 2 and Lobule Type 3. At the end of the fifth decade of life, the breast of both nulliparous and parous women contains Lobule Type 1 [2, 3, 10]. In the breast of nulliparous women, the predominant structure is the Lobule Type 1, which comprises 65–80 % of the total lobule type components and their relative percentage is independent of age. Second in frequency is Lobule Type 2, which represents 10–35 % of the total. The least frequent are Lobules Type 3, which represent only 0–5 % of the total lobular population. As depicted in Fig. 2.6a, b, the intralobular stroma is composed of connective tissue and a moderate amount of lymphocytes intermingled with the ductules of a Lobule Type 1 in regression.

In the breast of premenopausal parous women on the other hand, the predominant lobular structure is the Lobule Type 3, which comprises 70–90 % of the total lobule component (see Chap. 1). Only after menopause does it decline in number, and the relative proportion of the three lobule types present approaches that observed in nulliparous women.

Even though during the postmenopausal years in the breast of both parous and nulliparous women the preponderant structure is the Lobule Type 1, only nulliparous women are at high risk of developing breast cancer, whereas parous women remain protected [10, 18]. Since ductal breast cancer originates in Lobule Type 1 (TDLU) [18], the epidemiological observation that nulliparous women exhibit a higher incidence of breast cancer than parous women [1, 2] indicates that Lobule Type 1 in these two groups of women might be biologically different, or exhibit different susceptibility to carcinogenesis [4, 19–21]. The presence of Lobule Type 1 in the breasts of parous women has also been interpreted as a failure of the mammary parenchyma to respond to the influences of pregnancy and lactation [10, 13]. It is possible to postulate that unresponsive lobules that fail to undergo full differentiation under the stimuli of pregnancy and lactation are responsible of cancer development, despite the parity history of a woman. If this were the case, then this unresponsive Lobule Type 1 would be as sensitive to carcinogenesis as the lobules found in the breasts of nulliparous women. We have reported the presence of intralobular hyalinization and lower proliferative activity in the Lobule Type 1 of the parous woman's breast (Fig. 2.6c, d), whereas hyalinization is absent and cell proliferation is higher in the Lobule Type 1 of the nulliparous woman's breast.

2.5 Defining the Normal Breast Histology

In summary, the human breast undergoes a complete series of changes from intrauterine life to senescence. These changes can be divided into two distinct phases; the developmental phase and the differentiation phase [3]. The developmental phase

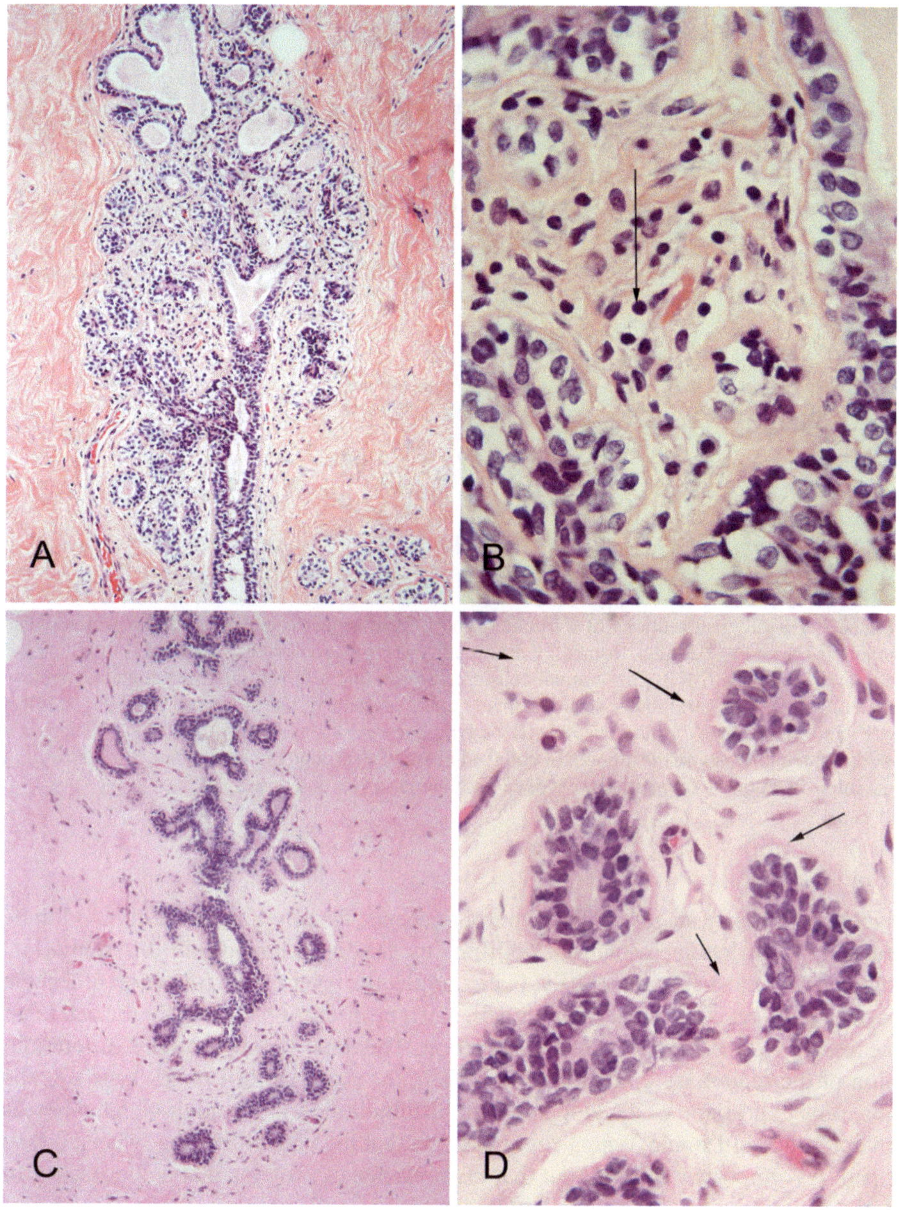

Fig. 2.6 Histological sections of core biopsies. Stained with hematoxylin and eosin. (**a**, **b**) Breast tissue from postmenopausal nulliparous women. Intralobular stroma composed of connective tissue and moderate amount of lymphocytes (*arrow*) intermingled with ductules of a Lobule Type 1 in regression. Magnification ×10 and ×40, respectively. (**c**, **d**) Breast tissue from postmenopausal parous women. Intralobular stroma composed of hyalinized connective tissue (*arrows*) in a Lobule Type 1. Magnification ×10 and ×40, respectively

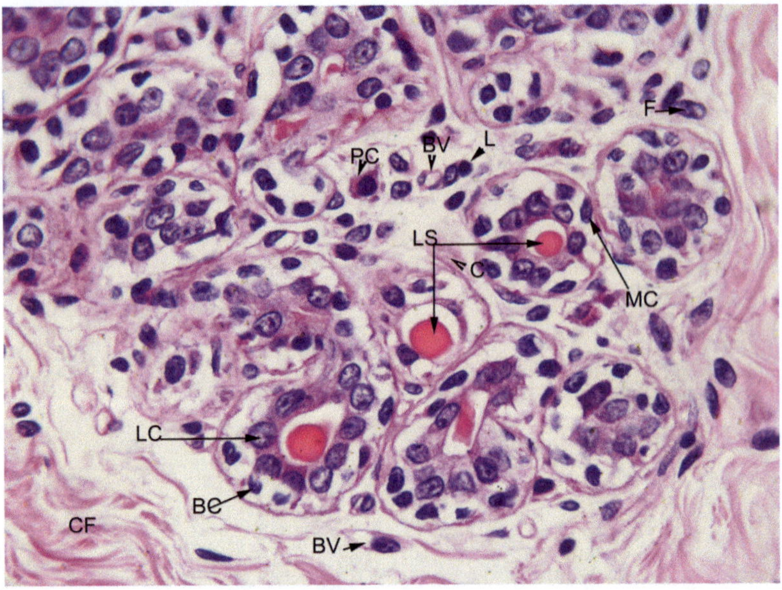

Fig. 2.7 Histological sections of a Lobule Type 2 from a postmenopausal parous woman. The lobule is formed by ductules formed by three cell types: basal cells (BC), luminal cells (LC), and myoepithelial cells (MC). The luminal cells are surrounding a lumen that may contain secretory material (LS). The intralobular stroma is made up of connective tissue formed by collagen fibers (CF) and fibroblasts (F), blood vessels (BV), lymphocytes (L), and plasma cells (PC). There is a well-defined intensity in the amount of collagen fibers (CF) around the lobular structure separating the intra- from interlobular stroma. Stained with hematoxylin and eosin, ×40

includes the early stages of gland morphogenesis, from nipple epithelium to lobule formation. In lobule formation, both processes, development and differentiation, take place almost simultaneously. For example, the progressive transition of Lobule Type 1 to Types 2, 3, and 4 requires active cell proliferation, to acquire the cell mass necessary for the function of milk secretion. This later process implies differentiation of the mammary epithelium. Therefore, the presence of Lobule Type 4 is the maximal expression of development and differentiation in the adult gland, whereas the presence of Lobule Type 3 could indicate that the gland has already been developed. It is important to point out that the presence of proteins that are indicative of milk secretion, such as α-lactalbumin, casein, or milk fat globule type membrane protein, also indicates cellular differentiation of breast epithelium. However, only when all the other components of milk (such as lactose, α-lactalbumin, casein, and milk fat) are coordinately synthesized within the appropriate structure can full differentiation of the mammary gland be acknowledged [2].

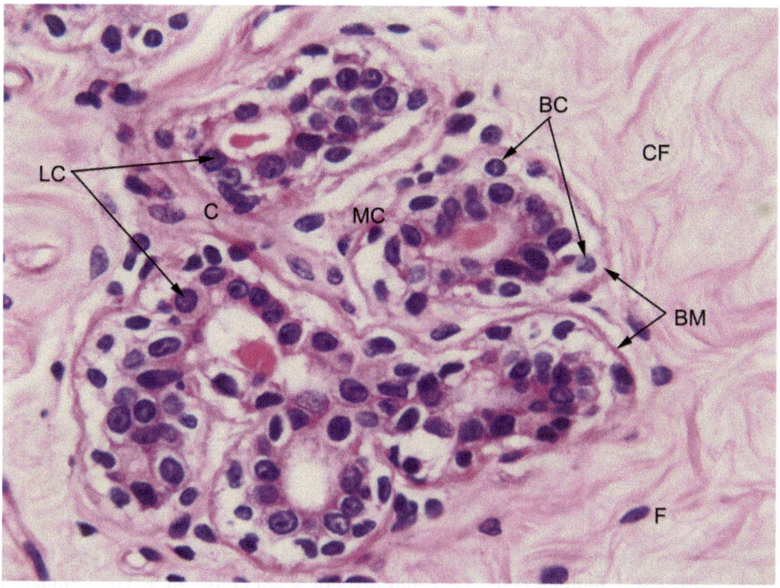

Fig. 2.8 Histological sections of a Lobule Type 1 from a postmenopausal parous woman. The lobule is formed by ductules formed by three cell types: basal cells (BC), luminal cells (LC), and myoepithelial cells (MC). The luminal cells are surrounding a lumen that may contain secretory material and lying in an eosinophilic basal membrane (BM). In the intralobular stroma it is shown connective tissue formed by collagen (C) and fibroblasts (F). The interlobular stroma is clearly demarcated and made up of abundant collagen fibers (CF) and fibroblast (F). Stained with hematoxylin and eosin, ×40

2.5.1 The Cell Types in the Normal Breast

The lobules of the human breast are formed by ductules containing three cell types: basal cells (BC), luminal cells (LC), and myoepithelial cells (MC). The luminal cells surround a lumen that may contain secretory material and lie in an eosinophilic basal membrane (BM). The intralobular stroma is made up of connective tissue formed by collagen and fibroblasts, blood vessels, lymphocytes, and plasma cells. The intensity and amount of collagen fibers around the lobular structure separating the intra- from the interlobular stroma are well defined (Figs. 2.7 and 2.8). In some cases, the basal lamina of the ductular structures is in contact with the fat cells (Fig. 2.9).

Frequently the luminal cells are separated from the basal membrane by an infranuclear vacuolization (Fig. 2.10), whereas the myoepithelial cells remain attached to basal membrane. The nucleus of the myoepithelial cells is more triangular and flatter than the basal cells (Fig. 2.11). The infranuclear vacuolization allows for an appreciation of the difference between the basal cells (BC) and myoepithelial cells (MC) (Fig. 2.12). When these histological sections react with an antibody that recognizes the estrogen receptor alpha (E2R) (Fig. 2.13), it is clear that the receptors are localized in the luminal cells of the breast epithelium whereas the basal and myoepithelial cells are negative for E2R.

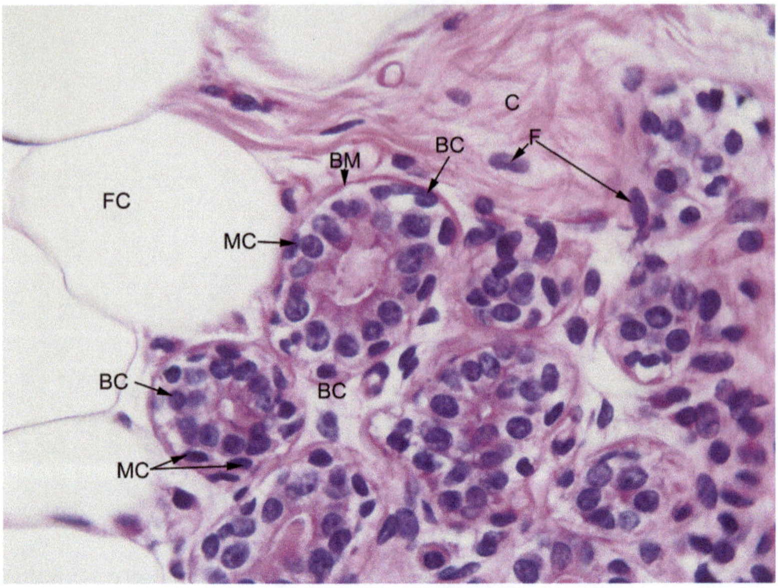

Fig. 2.9 As described in Fig. 2.8 the Lobule Type 1 is formed by ductules formed by three cell types: basal cells (BC), luminal cells (LC), and myoepithelial cells (MC). The luminal cells are surrounding a lumen that may contain secretory material and lying in an eosinophilic basal membrane (BM). In the interlobular stroma it is shown connective tissue formed by collagen (C) and fibroblasts (F), and in some cases the basal lamina of the ductular structures is in contact with the fat cells (FC). Stained with hematoxylin and eosin, ×40

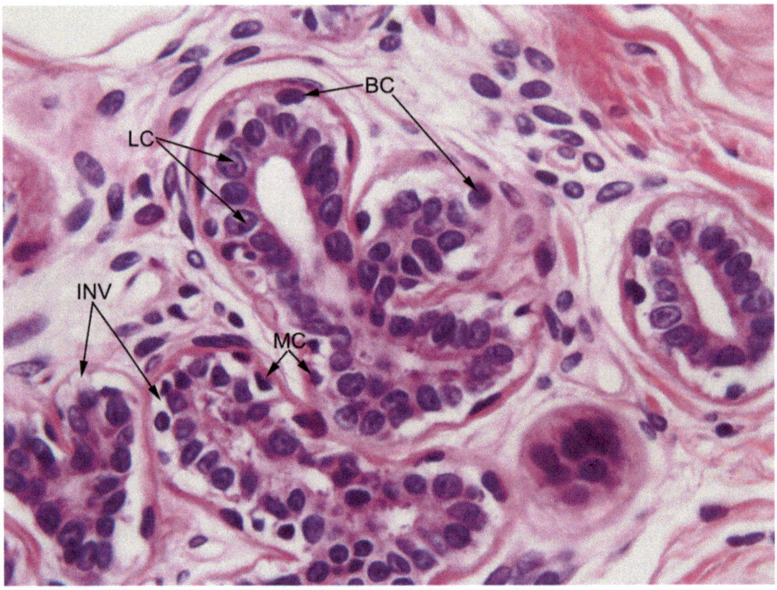

Fig. 2.10 As described in Figs. 2.8 and 2.9 the Lobule Type 1 is formed by ductules formed by three cell types: basal cells (BC), luminal cells (LC), and myoepithelial cells (MC). Frequently the luminal cells are separated from the basal membrane by an infranuclear vacuolization (INV). Stained with hematoxylin and eosin, ×40

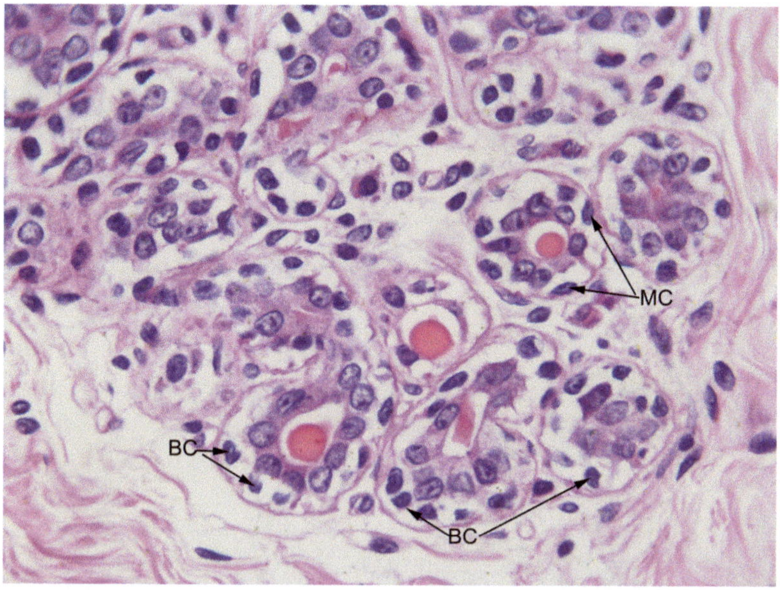

Fig. 2.11 Frequently the luminal cells are separated from the basal membrane by an infranuclear vacuolization whereas the myoepithelial cells remain attached to basal membrane (MC). The nucleus of the myoepithelial cells is more triangular and flatter than the basal cells (BC). Stained with hematoxylin and eosin, ×40

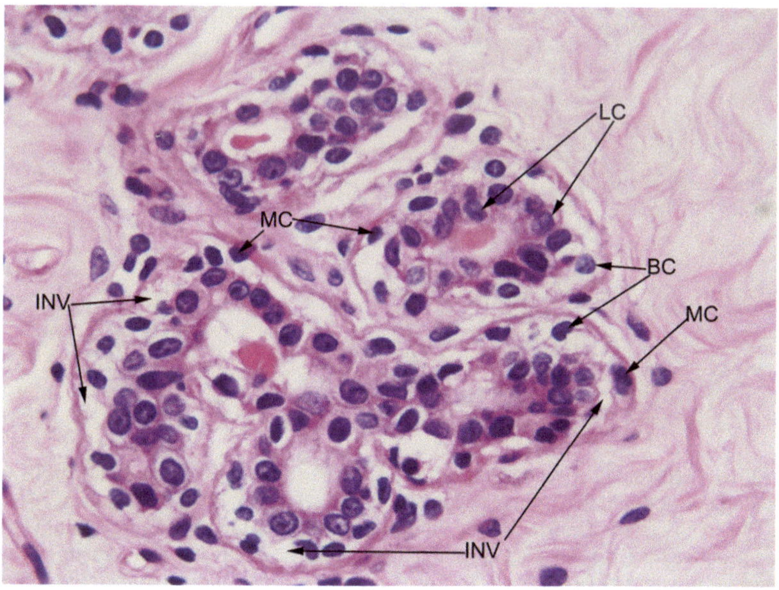

Fig. 2.12 Lobule Type 1 showing luminal cells (LC) separated from the basal membrane by an infranuclear vacuolization (INV); this separation allows appreciating the difference between the basal cells (BC) and myoepithelial cells (MC). Stained with hematoxylin and eosin, ×40

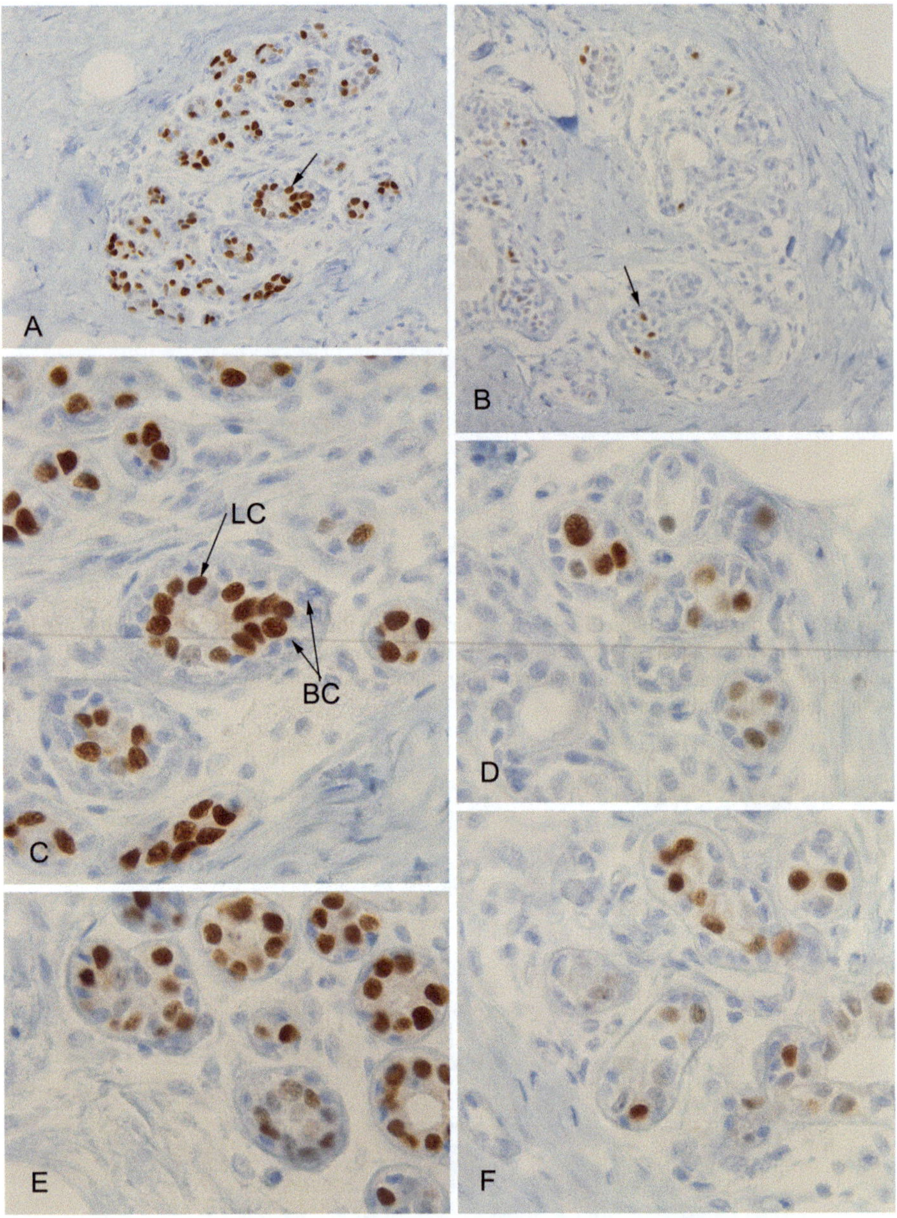

Fig. 2.13 (**a–f**) Histological sections of a Lobule Type 1 from a postmenopausal parous woman. The histological sections have been reacted with an antibody that recognizes the estrogen receptors (E2R). The receptors are localized in the luminal cells (LC) of the breast epithelium. Basal cells (BC) are negative for E2R. As it is depicted in the figures the number of cells reacting against the antibody vary in intensity (**a–c**) as well as the number of cells reacting or containing E2R vary in the same breast from lobule to lobule (compare (**a**) vs. (**b**) and (**c–e**) vs. (**d–f**)). We have used the antibody against estrogen receptor α that is a monoclonal mouse antihuman (clone 1D5) at a dilution 1:50, from DAKO an Agilent Technologies Company. Visualization of the immunocytochemical reactions was performed using the DAB from Biogenex, San Ramon, CA. Cat #: HK124-7KT. The reaction was performed with appropriate negative and positive controls; incubation and staining procedures were performed utilizing the i6000automated staining system from Biogenex

2.5.2 The Use of Immunocytochemical Markers for Differentiating the Cell Types of the Human Breast

There are several publications which use an immunocytochemical approach to define the different cell types and the role of each as a stem cell in the human breast [22–24]; however, there is a lack of agreement, mainly regarding the use of cytokeratins as a marker of stemness. We will refer to this specific topic in Chap. 6 of this book. In the present chapter, we want to discuss our own results using the same techniques used by other authors in paraffin-embedded tissue.

Using an antibody that recognizes the cytokeratin 5/6 (K5/6) we have identified that there is a difference in reactivity in the breast, mainly in the Lobule Type 1, depending on the parity history of the woman (Figs. 2.14 and 2.15). In the postmenopausal nulliparous woman, the K5/6 reacted positively in the basal cells as shown in Fig. 2.14a, b, d. K5/6 also reacts in the myoepithelial cells (Fig. 2.14c) and is also present in the luminal border of the luminal cells as well as in the lateral border of these cells (Fig. 2.14c). In the postmenopausal parous woman the K5/6 reacts in the luminal border of the luminal cells as well as in the lateral border of these cells (Fig. 2.15a–c). K5/6 is negative in the myoepithelial cells and basal cells (Fig. 2.15d, e). A more specific staining for the myoepithelial cells is the smooth muscle antigen or SMA. The SMA reacts in all the cells that contain actin filaments like in the blood vessels (Fig. 2.16a). It is difficult to separate those that are basal from myoepithelial cells using this technique (Fig. 2.16b–d).

The cytokeratin 5 (K5) has been considered a marker of basal cells as shown in Fig. 2.17a–c. K5 could also react in the myoepithelial cells but does not react with the same intensity as SMA as shown in Fig. 2.16. The luminal cells are completely negative for K5.

A double staining to differentiate the basal and luminal cells can be applied using a cocktail containing antibodies against keratin 5 (K5) and keratin 19 (K19). The K5 reacts in the basal cells as shown in Fig. 2.18a–d. This antibody against K5 seems to be more specific for basal cells and it does not react in the myoepithelial cells as compared to SMA, as shown in Fig. 2.16. The luminal cells are positive for K19 in the luminal border. This double staining procedure was performed using a primary monoclonal antibody against cytokeratin 5 in rabbit combined with a mouse monoclonal antibody against cytokeratin 19. The different colors, brown for the basal cells and purple for the luminal cells, were obtained using a secondary antibody cocktail of anti-mouse polymer-AP and anti-rabbit polymer-HRP.

Another cytokeratin that can be used for detecting luminal cells is keratin 6 (K6). An antibody against K6 reacts in the luminal cells (Fig. 2.19a–c) and is weakly reactive in the basal and myoepithelial cells in sporadic regions of the same breast as it is shown in Fig. 2.19d. Keratin 18 (K18) reacts in the luminal cells (LC) and is not reactive in the basal and myoepithelial cells (Fig. 2.20a, b).

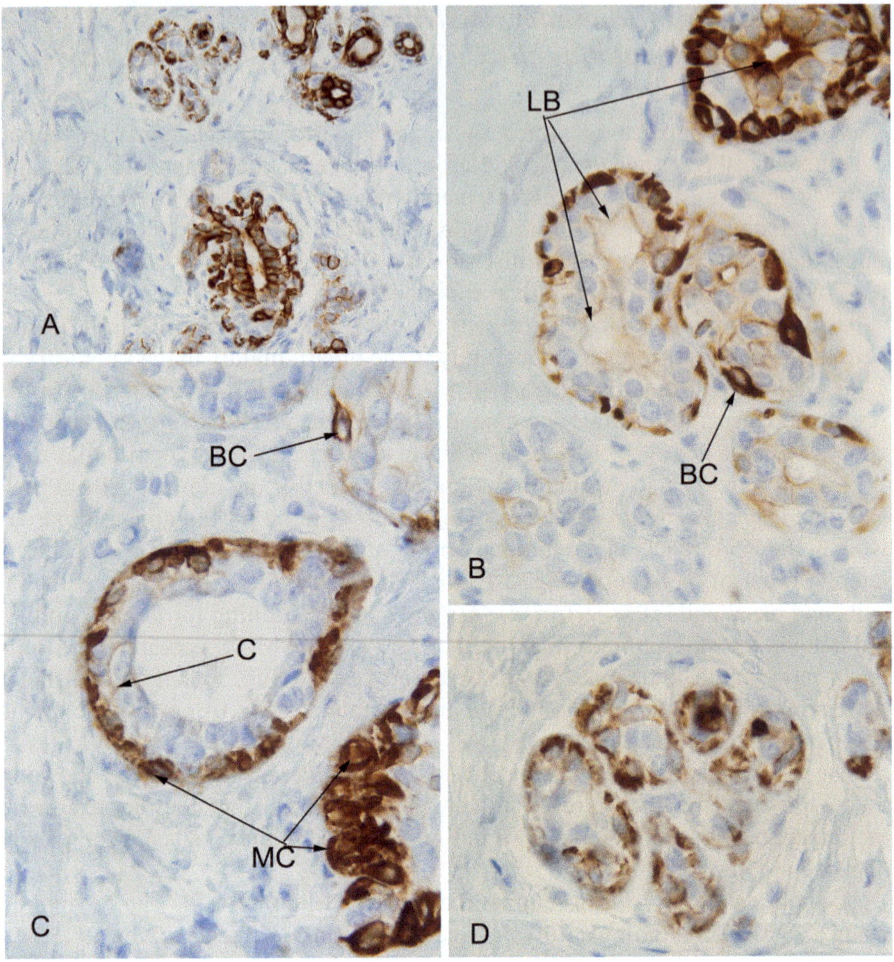

Fig. 2.14 (**a–d**) Histological sections of a Lobule Type 1 from a postmenopausal nulliparous woman. The histological sections have been reacted with an antibody that recognizes the cytokeratin 5/6 (K5/6). The K5/6 reacted positively in the basal cells as shown in (**b**). K5/6 also reacts in the myoepithelial cells (MC) as it is shown in (**c**). K5/6 also is present in the luminal border (LB) of the luminal cells as well as in the lateral border of these cells (C). We have used the monoclonal mouse antihuman cytokeratin 5/6, Clone E6/17 C5, at a dilution of 1:50 (DAKO an Agilent Technologies Company). Visualization of the immunocytochemical reactions was performed using the DAB from Biogenex, San Ramon, CA. Cat #: HK124-7KT. The reaction was performed with appropriate negative and positive controls; incubation and staining procedures were performed utilizing the i6000automated staining system from Biogenex

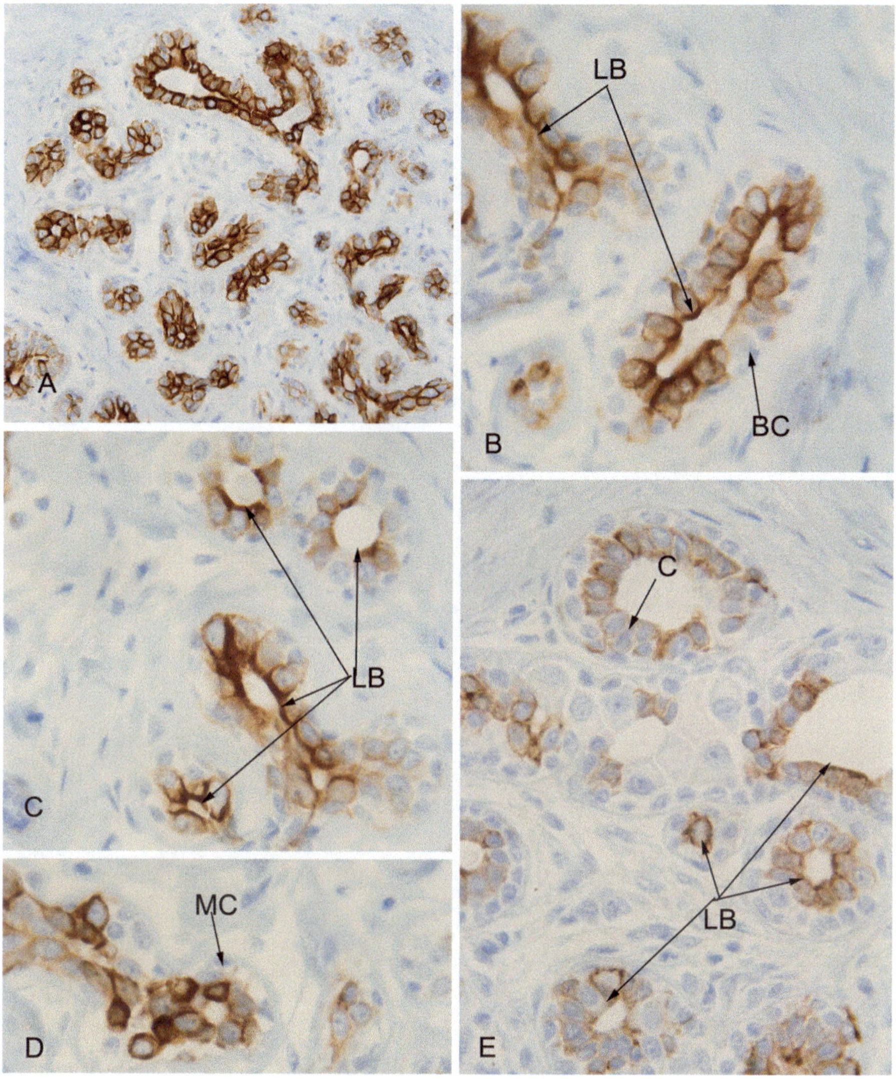

Fig. 2.15 (**a–e**) Histological sections of a Lobule Type 1 from a postmenopausal parous woman. The histological sections have been reacted with an antibody that recognizes the cytokeratin 5/6 (K5/6). The K5/6 when reacted in the parous breast tissue presents a different location (**a–e**) than the one shown in Fig. 2.14. K5/6 reacts in the luminal border (LB) of the luminal cells as well as in the lateral border of these cells (C) but is negative in the myoepithelial cells (MC) in (**d**) and basal cells (BC) are negative. We have used the monoclonal mouse antihuman cytokeratin 5/6, Clone E6/17 C5, at a dilution of 1:50 (DAKO an Agilent Technologies Company). Visualization of the immunocytochemical reactions was performed using the DAB from Biogenex, San Ramon, CA. Cat #: HK124-7KT. The reaction was performed with appropriate negative and positive controls; incubation and staining procedures were performed utilizing the i6000automated staining system from Biogenex

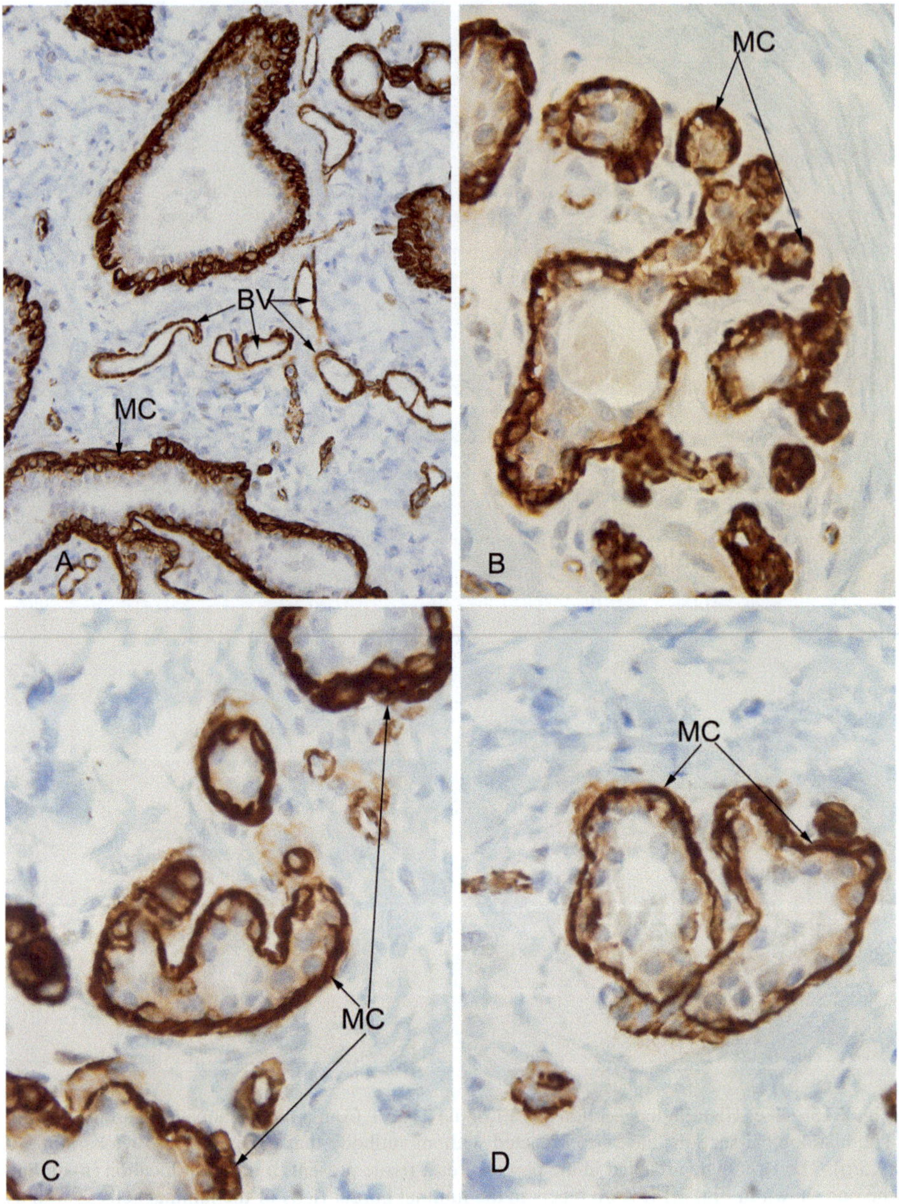

Fig. 2.16 (**a–d**) Histological sections of a Lobule Type 1 from a postmenopausal parous woman. The histological sections have been reacted with an antibody that recognizes the smooth muscle antigen or SMA. The SMA has been considered a marker of myoepithelial cells (MC). It reacts in all the cells that contain myosin like in the blood vessels (BV) in (**a**). It is difficult to separate those that are basal cells from myoepithelial cells using this technique. We have used a monoclonal mouse antihuman actin (Smooth Muscle) clone 1A4 at a dilution of 1:50 from DAKO an Agilent Technologies Company. The reaction was developed using DAB from Biogenex, San Ramon, CA. Cat #: HK124-7KT. The sections were counterstained with Mayer's hematoxylin (Biogenex, San Ramon, CA. Cat #: HK100-9K)

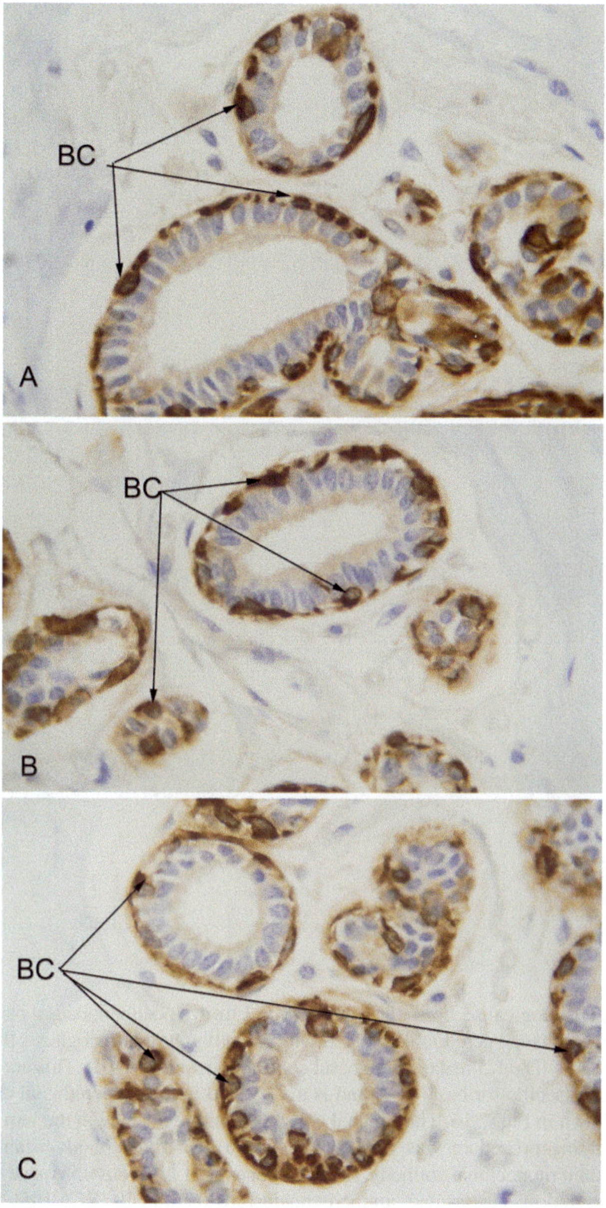

Fig. 2.17 (**a–c**) Histological sections of a Lobule Type 1 from a postmenopausal nulliparous woman. The histological sections have been reacted with an antibody that recognizes the keratin 5 (K5). The K5 has been considered a marker of basal cells as shown in (**a**)–(**c**). K5 also reacts in the myoepithelial cells (MC) but does not react with the same intensity than SMA as shown in Fig. 2.16. The luminal cells are completely negative with this antibody. The primary antibody against cytokeratin 5 [EPR1600Y], rabbit monoclonal antibody, was obtained from Biogenex, Cat #: AN494-5M, that is ready to use as prediluted concentration. The secondary antibody is also from Biogenex, Cat #: HK519-YAK, Polymer HRP. Both antibodies were incubated during 30 min at room temperature. The reaction was developed using DAB from Biogenex (Cat #: HK124-7KT). The sections were counterstained with Mayer's hematoxylin (Biogenex, Cat #: HK100-9K)

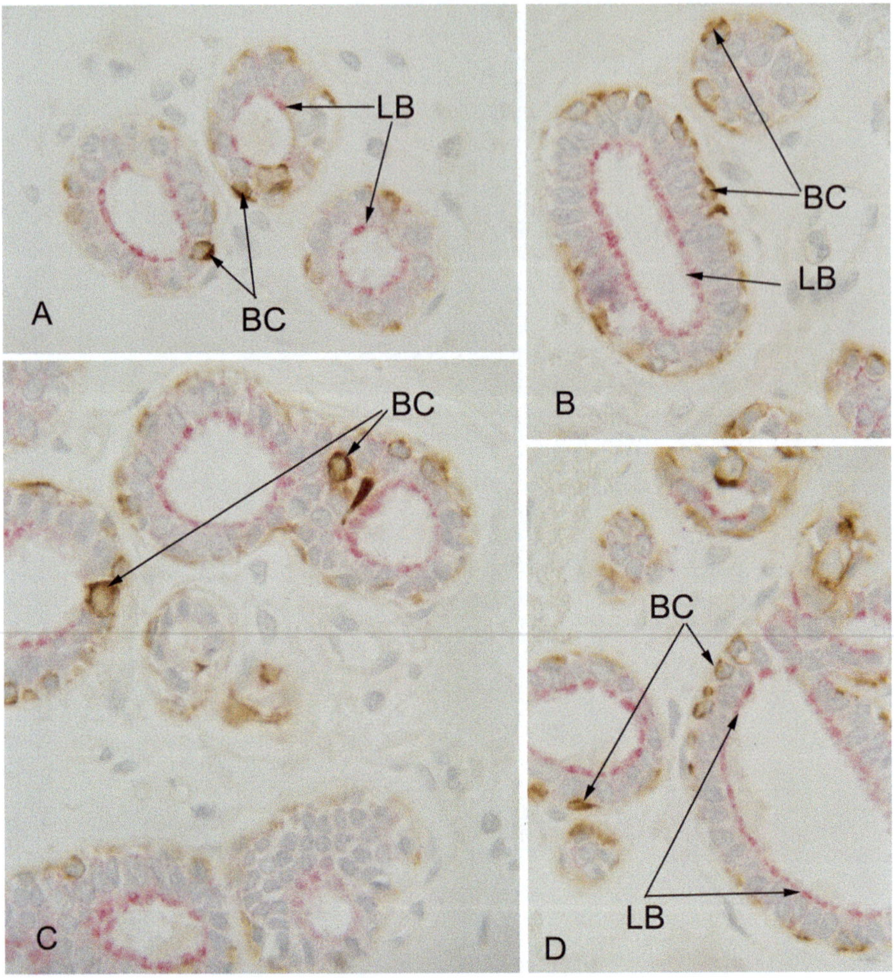

Fig. 2.18 (**a**–**d**) Histological sections of a Lobule Type 1 from a postmenopausal parous woman. The histological sections have been double reacted with an antibody that recognizes the keratin 5 (K5) and keratin 19 (K19). The K5 reacts in the basal cells as shown in (**a**)–(**d**). This antibody against K5 seems to be more specific for basal cells and is not reacted in the myoepithelial cells as compared with SMA as shown in Fig. 2.16. The luminal cells are positive for K19 in the luminal border (LB). This double staining procedure was performed using a primary antibody against cytokeratin 5 [EPR1600Y], rabbit monoclonal antibody, from Biogenex, Cat #: AN484-5M, that is ready to use as prediluted concentration. The primary antibody against cytokeratin 19 [RCK108], mouse monoclonal antibody, was from Biogenex, Cat #: AN484-10M, that is ready to use as prediluted concentration. The secondary antibody was a cocktail of anti-mouse polymer-AP and anti-rabbit polymer-HRP detection cocktail from Biogenex, Cat #: QS-200-60K. The incubation time for the first antibody was 60 min and for the secondary antibody 30 min, both at room temperature. The reaction was developed using Fast red (Biogenex, Cat #: HK 181-7K) and DAB (Biogenex, Cat #: HK124-7KT). The sections were counterstained with Mayer's hematoxylin (Biogenex, Cat #: HK100-9K)

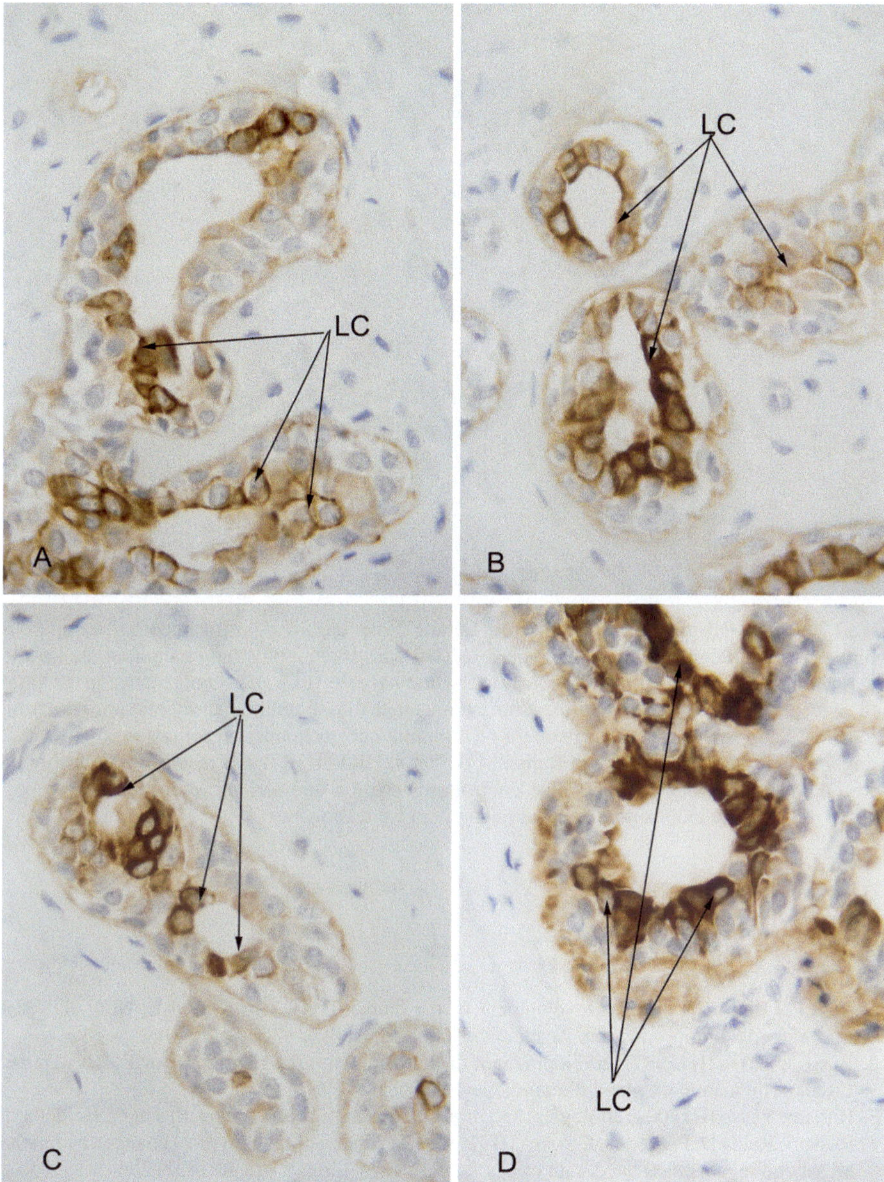

Fig. 2.19 (**a–d**) Histological sections of a Lobule Type 1 from a postmenopausal parous woman. The histological sections have been reacted with an antibody that recognizes the keratin 6 (K6). This antibody against K6 reacts in the luminal cells (LC) and is weakly reactive in the basal and myoepithelial cells in sporadic regions of the same breast as it is shown in (**d**). We have used the monoclonal mouse antihuman rabbit monoclonal antibody, Cat #: AN500-10M, from Biogenex, San Ramon, CA. Visualization of the immunocytochemical reactions was performed using the DAB from Biogenex, San Ramon, CA. Cat #: HK124-7KT. The reaction was performed with appropriate negative and positive controls; incubation and staining procedures were performed utilizing the i6000automated staining system from Biogenex

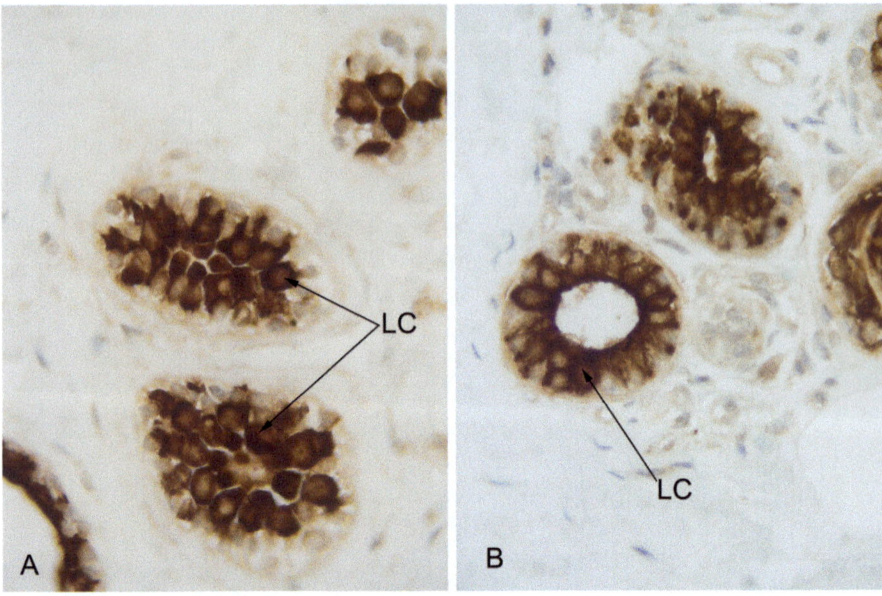

Fig. 2.20 (**a**, **b**) Histological sections of a Lobule Type 1 from a postmenopausal nulliparous woman. The histological sections have been reacted with an antibody that recognizes the keratin 18 (K18). This antibody against K18 reacts in the luminal cells (LC) and is not reactive in the basal and myoepithelial cells. We have used the monoclonal mouse antihuman cytokeratin 18, from Biogenex, San Ramon, CA. Visualization of the immunocytochemical reactions was performed using the DAB from Biogenex, San Ramon, CA. Cat #: HK124-7KT. The reaction was performed with appropriate negative and positive controls; incubation and staining procedures were performed utilizing the i6000automated staining system from Biogenex

Literature Cited

1. Russo J, Russo IH (1998) Development of the human breast. In: Knobil E, Neill JD (eds) Encyclopedia of reproduction. Academic, New York
2. Russo J, Russo IH (1987) Development of the human mammary gland. In: Neville MC, Daniel C (eds) The mammary gland. Plenum, New York, pp 67–93
3. Russo J, Russo IH (2004) Biological and molecular basis of breast cancer. Springer, Heidelberg
4. Russo J, Reina D, Frederick J, Russo IH (1988) Expression of phenotypical changes by human breast epithelial cells treated with carcinogens in vitro. Cancer Res 48:2837–2857
5. Russo J, Calaf G, Roi L, Russo IH (1987) Influence of age and gland topography on cell kinetics of normal human breast tissue. J Natl Cancer Inst 78:413–418
6. Russo J, Romero AL, Russo IH (1994) Architectural pattern of the normal and cancerous breast under the influence of parity. Cancer Epidemiol Biomarkers Prev 3:219–224
7. Peri S, de Cicco RL, Santucci-Pereira J, Slifker M, Ross EA, Russo IH, Russo PA, Arslan AA, Belitskaya-Levy I, Zeleniuch-Jacquotte A, Bordas P, Lenner P, Ahman J, Afanasyeva Y, Johansson R, Sheriff F, Hallmans G, Toniolo P, Russo J (2012) Defining the genomic signature of the parous breast. BMC Med Genomics 5:46–57

8. Russo J, Santucci-Pereira J, de Cicco RL, Sheriff F, Russo PA, Peri S, Slifker M, Ross E, Mello ML, Vidal BC, Belitskaya-Levy I, Arslan A, Zeleniuch-Jacquotte A, Bordas P, Lenner P, Ahman J, Afanasyeva Y, Hallmans G, Toniolo P, Russo IH (2011) Pregnancy-induced chromatin remodeling in the breast of postmenopausal women. Intl J Cancer 131:1059–1070
9. Belitskaya-Levy I, Zeleniuch-Jacquotte A, Russo J, Russo IH, Bordas P, Ahman J, Afanasyeva Y, Johansson R, Lenner P, Li X, de Cicco-Lopez RL, Peri S, Ross E, Russo PA, Santucci-Pereira J, Sheriff FS, Slifker M, Hallmans G, Toniolo P, Arslan AA (2011) Characterization of a genomic signature of pregnancy identified in the breast. Cancer Prev Res 4:1457–1464
10. Russo J, Rivera R, Russo IH (1992) Influence of age and parity on the development of the human breast. Breast Cancer Res Treat 23:211–218
11. Russo I, Russo J (1994) Role of hCG and inhibin in breast-cancer (review). Int J Oncol 4:297–306
12. Russo J, Tay LK, Russo IH (1982) Differentiation of the mammary gland and susceptibility to carcinogenesis. Breast Cancer Res Treat 2:5–73
13. Russo J, Hu YF, Silva ID, Russo IH (2001) Cancer risk related to mammary gland structure and development. Microsc Res Tech 52:204–223
14. Russo IH, Russo J (1996) Mammary gland neoplasia in long-term rodent studies. Environ Health Perspect 104:938–967
15. Xie J, Haslam SZ (1997) Extracellular matrix regulates ovarian hormone-dependent proliferation of mouse mammary epithelial cells. Endocrinology 138:2466–2473
16. Petersen OW, Ronnov-Jessen L, Weaver VM, Bissell MJ (1998) Differentiation and cancer in the mammary gland: shedding light on an old dichotomy. Adv Cancer Res 75:135–161
17. Sternlicht MD, Kouros-Mehr H, Lu P, Werb Z (2006) Hormonal and local control of mammary branching morphogenesis. Differentiation 74:365–381
18. Russo J, Gusterson BA, Rogers AE, Russo IH, Wellings SR, van Zwieten MJ (1990) Comparative study of human and rat mammary tumorigenesis. Lab Invest 62:244–278
19. Hu YF, Russo IH, Zalipsky U, Russo J (1996) Lack of involvement of bcl2 and cyclin D1 in the early phases of human breast epithelial cell transformation by environmental chemical carcinogens. Proc Am Assoc Cancer Res 37:1005a
20. Russo J, Russo IH (1997) Role of differentiation in the pathogenesis and prevention of breast cancer. Endocr Relat Cancer 4:7–21
21. Russo J, Hu Y-F, Yang X, Russo IH (2000) Developmental, cellular, and molecular basis of human breast cancer. J Natl Cancer Inst Monograph 27:17–38
22. Clarke CL, Sandle J, Parry SC, Reis-Filho JS, O'Hare MJ, Lakhani SR (2004) Cytokeratin 5/6 in normal human breast: lack of evidence for a stem cell phenotype. J Pathol 204:147–152
23. Bocker W, Hungermann D, Decker T (2009) Anatomy of the breast. Pathologe 30:6–12
24. Bocker W, Hungermann D, Weigel S, Tio J, Decker T (2009) Immunohistochemistry in breast pathology: differential diagnosis of epithelial breast lesions. Pathologe 30:13–19

Chapter 3
Methodology for Studying the Compartments of the Human Breast

Jose Russo, Fathima Sheriff, Ricardo Lopez de Cicco, Thomas J. Pogash, Theresa Nguyen, and Irma H. Russo

Keywords Laser capture microdissection (LCM) • DNA extraction • LCM from paraffin-embedded tissue • DNA obtained by LCM • RNA obtained by LCM • Frozen section for LCM

3.1 Introduction

A shortcoming of using human breast tissue for molecular analysis is the heterogeneous nature of the sample. To resolve this, various microdissection techniques have been employed to obtain homogeneous cells, such as manual scraping of tissue with scalpel blades, needles, or other probes to positively select cells of interest. These techniques are limited because of poor delineation of tissue and high susceptibility to contamination from dissimilar cells. Infrared laser capture microdissection (LCM), developed at the National Institute of Health, has become increasingly commercially available in the last two decades of the twentieth century [1–3]. Laser capture microdissection, also called microdissection, laser microdissection (LMD), is a method for isolating cells or specific regions of interest from cells, tissue, or organisms. This technique enables researchers to investigate DNA/RNA and proteins from specific cells or regions of tissue. LCM consists of an LMD system, camera, and software used to select and collect the areas of interest. The basic principle of LCM involves a laser which fuses the desired material onto a specialized cap which can be then close over a 0.5 mL microcentrifuge tube. Before LCM can be performed, a paraffin-embedded tissue (PET) is sectioned with a regular

J. Russo, M.D. (✉) • F. Sheriff, M.D. • R.L. de Cicco, Ph.D. • T.J. Pogash, B.S.
T. Nguyen, B.S. • I.H. Russo, M.D.
Irma H Russo MD Breast Cancer Research Laboratory, Fox Chase Cancer Center, Cottman Avenue 333, Philadelphia, PA 19111, USA
e-mail: jose.russo@fccc.edu; Irma.Russo@fccc.edu

J. Russo and I.H. Russo, *Techniques and Methodological Approaches in Breast Cancer Research*, DOI 10.1007/978-1-4939-0718-2_3,

microtome, or a cryostat is used to create histological sections from frozen tissue. The histological sections are placed onto a membrane slide which is then freshly hemotoxylin and eosin (H&E) stained. After the slide has been stained, the sample is set in 100 % EtOH followed by air drying.

3.2 The LCM Equipment

There are several prestigious LCM equipment such as Zeiss, Leica, and Arcturus. Our first experience was with the PixCell Laser Capture Microscope with an infrared diode laser (Arcturus Engineering, Inc, Mountain View, CA) [1, 2]. Cells were captured using a 30 μm pulse to focally melt a thermoplastic membrane attached to a transparent flat cap. The amplitude and pulse duration of the laser were adjusted to 40 mW and 40 ms, respectively. After LCM, the cap containing the captured tissue was placed on a 0.5 mL standard Eppendorf microfuge tube. Recently, we have been using the LMD6500 model which includes software that allows the user to control the microscope with different objectives (1.25, 5, 10, 20, 40, 63×) for optimum viewing to select the specific area desired. This model marks a significant improvement in the quality of images obtained and in the precision of the cuts (Fig. 3.1).

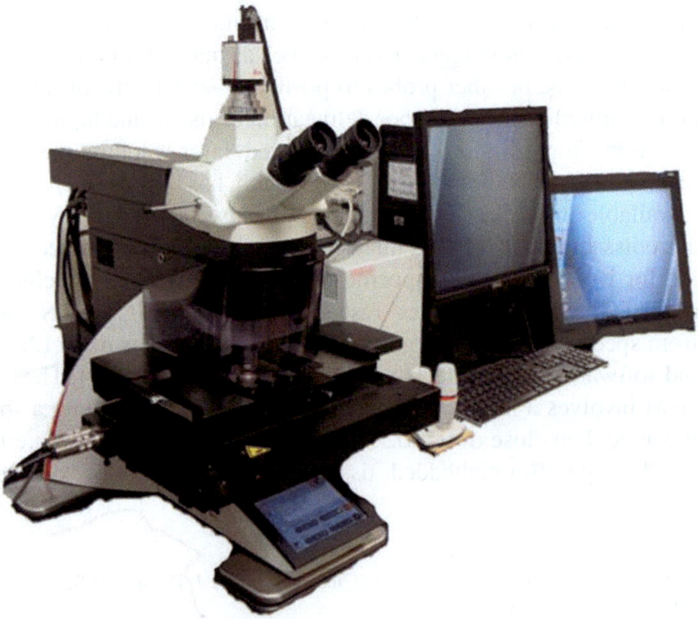

Fig. 3.1 Photograph of the LMD6500

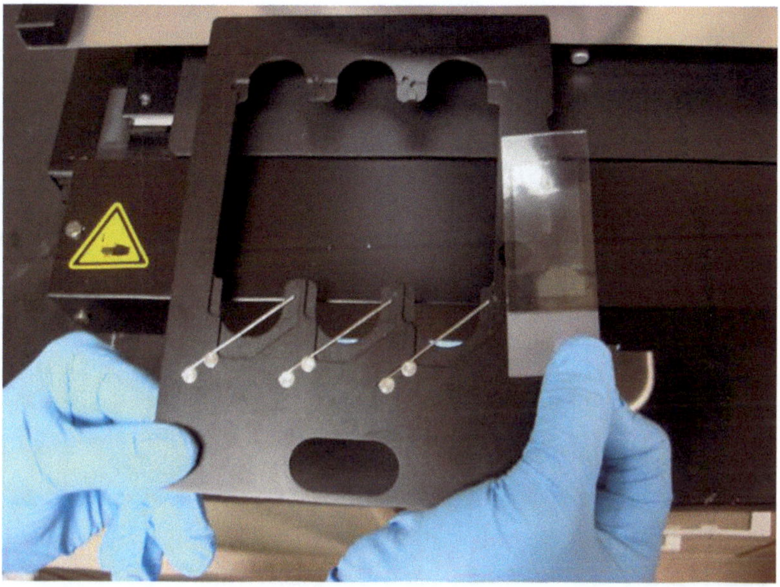

Fig. 3.2 In the LMD6500 up to 3 slides can be mounted and used at the same time in an upside down position

An important component of these instruments is the laser control which allows the user to adjust various power levels. This means the ability to adjust cutting laser's power, speed, aperture, etc. depending on, for example, the thickness of the section or tissue type. As shown in Fig. 3.2, up to 3 slides can be mounted and use consecutively in an upside down position. When this takes place, a tube holder slides out automatically (Fig. 3.3). Up to four 0.5 mL tubes can be inserted into the tube holder with spaces designated A,B,C,D. The amount of time the instrument is in operation becomes crucial once the samples are placed in the machine. The laser capture must be completed in 45 min or less per each set of tubes, in part so the buffer does not dry out, but more importantly so good quality for DNA/RNA can be obtained. At this stage, buffer must be added to caps as appropriate 20–30 μL into cap of 0.5 mL tube. At this stage, the laser cut and separates the collect cell from the tissue. The precut images are saved and can be taken at any magnification preferable with 20× or 40× objectives.

It is advisable to capture a test line somewhere on an unneeded portion of the specimen in order to figure out the optimal settings for the laser. Once the line has been drawn using the imaging software, the laser will then cut on the drawn line. If collecting lots of little areas, it is best to compile a list of the areas and total cut area so the collection tubes can contain same amount of tissue. The software has a function which enables this (Fig. 3.4a). Figure 3.4b shows the image after the specimen has been cut; this result can usually be obtained with a pulse of 35 ms and power 8 mW.

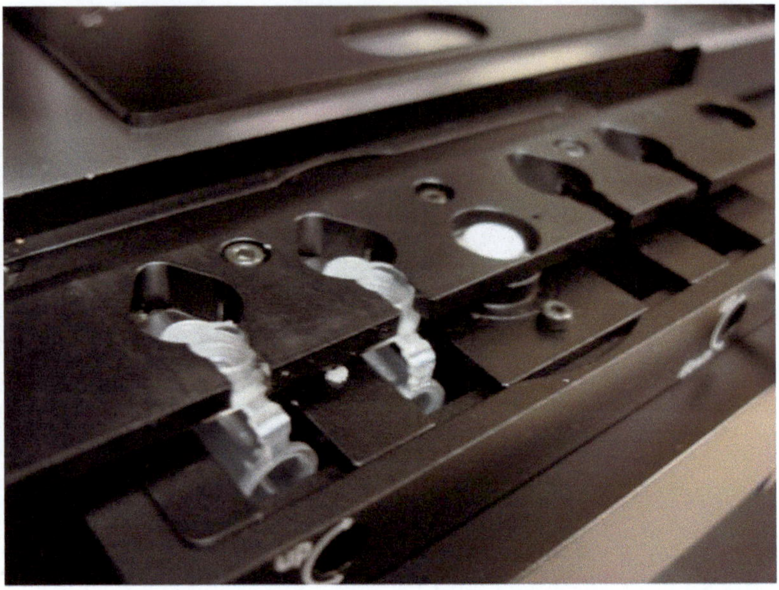

Fig. 3.3 In the LMD6500 a tube holder contains up to 0.5 mL 4 tubes designated A, B, C, D

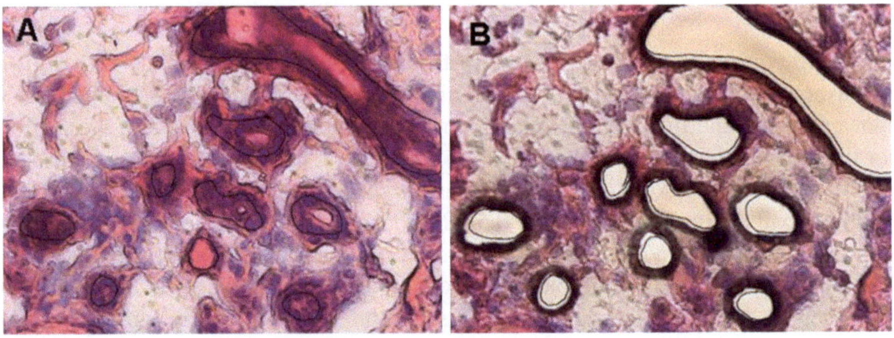

Fig. 3.4 (**a**) The laser will cut on the drawn line. (**b**) Shown is the image after the cut, which can be obtained usually with a pulse of 35 ms and power 8 mW

3.3 DNA Extraction in LCM from Paraffin-Embedded Tissue

For decades, PET has been used by pathologists to examine microscopic sections of pathologic conditions. Pathology and molecular biology merged when it became possible to obtain DNA from PET, rendering these tissues valuable material for molecular analysis. LCM yields homogeneous populations of targeted cells from specific microscopic regions of breast tissue sections [4]. Using LCM on PET, it is possible to obtain a homogeneous supply of nuclei from which DNA can be extracted. DNA from PET can serve as substrate for in vitro amplification using polymerase chain reaction (PCR) [5]. The efficiency of PCR amplification is related to the purity and to the quantity of DNA in the sample [6]. Lack of amplification can be attributed to insufficient DNA or the presence of PCR inhibitors in the DNA extract. Knowing the quantity of DNA in the sample can help distinguish between these two possible causes [6]. Knowledge about the quantity of DNA also allows the researcher to maximize the number of PCR tests performed on a sample by consuming only as much DNA as necessary [6]. We have been able to determine the amount of DNA that can be obtained by LCM, as well as information on the number of cells captured and the DNA concentration per capture. One of the methods for quantifying DNA that we have used is to fill each microfuge tube with 50 µL of DNA digestion buffer (0.05 % proteinase K, 10 mM Tris-HCI (pH 8.0), 1 mM EDTA, *1* % Tween 20). The cap (containing the captured cells) was fit onto the microfuge tube using a cap spacer. Each tube was maintained in an upright position and pre-heated in a 42 °C oven for 5 min. Tubes were then turned upside down to contact the digestion buffer with the tissue on the cap and incubated overnight at 42 °C. Following incubation, the tubes were centrifuged for 5 min and the cap was removed. Each tube was heated to 95 °C for 8 min to inactivate the proteinase K. The DNA in solution was then used directly as a template for PCR.

3.3.1 Yield of DNA Obtained by LCM in PET

Our first goal when we began using this methodology was to determine the number of nuclei per capture that could be obtained from a homogeneous population of cells isolated via LCM. For this reason, we used breast carcinoma in situ [7] to determine the number of nuclei per field. 32 captures were obtained from the epithelial component of the tumor. Positive images of each capture were made and the number of nuclei present was counted [7]. On average, each capture contained 21 ± 5.4 nuclei, within a diameter of 33 ± 6.4 µm. Our next goal was to determine the concentration of DNA within the nuclei obtained from our captures. UV spectrophotometry is the principal method for quantifying nucleic acids [8] and has a level of sensitivity in the nanograms range. Our initial attempts to measure the absorption of DNA at 260 Å failed using traditional UV spectrophotometry. We concluded that our samples contained amounts of DNA below the threshold (nanograms) required by UV spectrophotometry. A solution was to utilize fluorescent dyes with subsequent measurement by fluorometry. We utilized a fluorochrome, PicoGreen (Molecular Probes, Inc., Eugene, OR) [9], which has the sensitivity to detect from 25 pg/mL to 1 µg/mL DNA. An aqueous working solution 200-fold dilution of the concentrated reagent in TE solution (10 mM Tris-HCI, 1 mM EDTA, pH 7.5) that was freshly prepared in a sterile plastic container and covered in foil to protect from light was added to the sample. The relatively new intercalating dye PicoGreen was chosen to quantify the extracted DNA. PicoGreén provides a linear correlation between double-stranded (ds) DNA concentration and fluorescence with a detection range

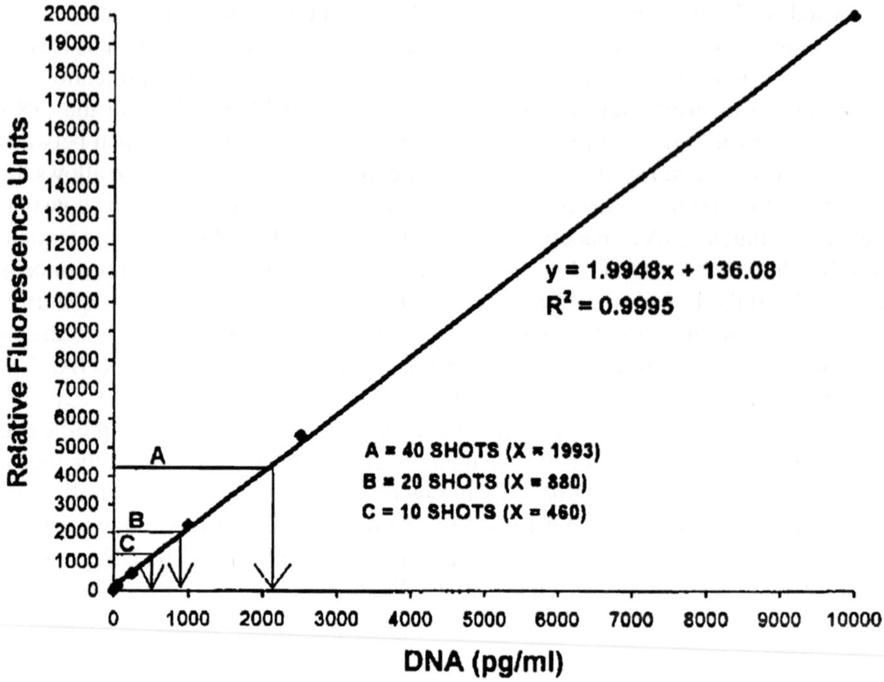

Fig. 3.5 Dynamic range and sensitivity of the PicoGreen dsDNA quantification assay. Lambda DNA was serially diluted and assayed for linearity. The total amount of DNA (pg) was plotted against relative fluorescent units. The fluorescence was linear over the range of 25 pg to 10 ng with a correlation coefficient of 0.9995. Added to the linear regression line is the placement of the relative fluorescent units obtained from the DNA contained within 40, 20, and 10 captures (Reprinted from: Balogh, G.A., Heulings, R., Mailo, D, Wang, R., Li, YZ., Hardy, R., and Russo J. International J. of Oncology 31: 253–260, 2007)

from 25 pg/mL to 1 μg/mL dsDNA [6]. The assay is highly selective for dsDNA in solution with minimal interference by RNA, single-stranded DNA, and nucleotides. Figure 3.5 confirms that PicoGreen detection of the lambda DNA standard (Molecular Probes, Inc.) was linear over the range of 25 pg to 10 ng; the equation for the linear regression line was calculated as $y = 1.9948x \pm 136.08$ with a correlation coefficient of 0.9995 (R^2) [7].

To determine the amount of DNA found in each capture, we obtained three groups of five samples with each group containing 40, 20, and 10 captures. This method of grouping also enabled us to determine whether a relationship existed between the number of captures and concentration of DNA. DNA was extracted from each sample using a digestion buffer and its relative fluorescence was measured by a fluorometer. Fluorescence (235 RFU) of negative controls (digestion buffer + PicoGreen) was subtracted from the average RFU of each sample. DNA concentration was determined using the linear regression equation of the DNA standard ($y = 1.9948x \pm 136.08$), where y represents the average fluorescence of each

Table 3.1 Conversion of total DNA concentration obtained from the standard regression line to total DNA per cell

No.of captures	Total no. of cells	Mean relative fluorescent units	DNA standard plot (1:50 dilution) pg/mL	DNA prior to (1:50 dilution) pg/mL	Total DNA in 50 μL of digestion buffer pg	Total DNA/ capture pg	Total DNA/ cell pg
40	840	3,992	1,933	96,650	4,833	120.8	5.75
20	420	1,992	880	44,000	2,200	110.0	5.24
10	210	1,055	460	23,000	1,150	115.0	5.48

Reprinted from: Blumenstein R., Dias M., Russo I.H., Tahin Q. and Russo, J. DNA content and cell number determination in microdissected samples of breast carcinoma in situ. Int. J. of Oncology 21:447–450, 2002

sample and x the DNA concentration. The concentration of each group was also demonstrated graphically in Fig. 3.5 by drawing a perpendicular line from the intersection of the fluorescence (y) with the regression line to the x-axis. Table 3.1 summarizes our results and indicates the concentration of DNA per capture and per cell, approximately 115.3 and 5.5 pg, respectively. This is lower than the 400 pg/capture suggested previously [3]; however, no information on the diameter, amplitude, and pulse duration of the microbeam was available for comparison.

To demonstrate graphically the relationship between the number of captures and the concentration of DNA, each group containing 40, 20, and 10 captures was plotted relative to its respective fluorescence. These results are demonstrated in Fig. 3.6, which shows a linear relationship with a regression line having a correlation coefficient of 0.9995 (R^2). This is in agreement with others [10] that have found a high degree of linearity between the amount of DNA and the number of cells captured.

In summary this was the first study [7] to quantify the number of nuclei and the concentration of DNA obtained per laser capture. We have also demonstrated a high degree of linearity between the number of captures and the concentration of DNA. Variation in concentration from the results reported here should be expected since differences in section thickness, age of tissue, type of cells, adherence to slide, number of cells captured, size of capture, processing of tissue, and completeness of digestion will influence results. This protocol utilizing PicoGreen overcame the difficulties encountered when using UV spectrophotometry for detection of subnanogram amounts of DNA.

3.3.2 *Yield of RNA Obtained by LCM from Paraffin-Embedded Tissue*

We have used LCM to selectively obtain pure cell populations from the breast epithelial ductal structures and from the stroma for RNA isolation [9–12]. The technique allows the precise extraction of pure cell populations from fixed tissue sections

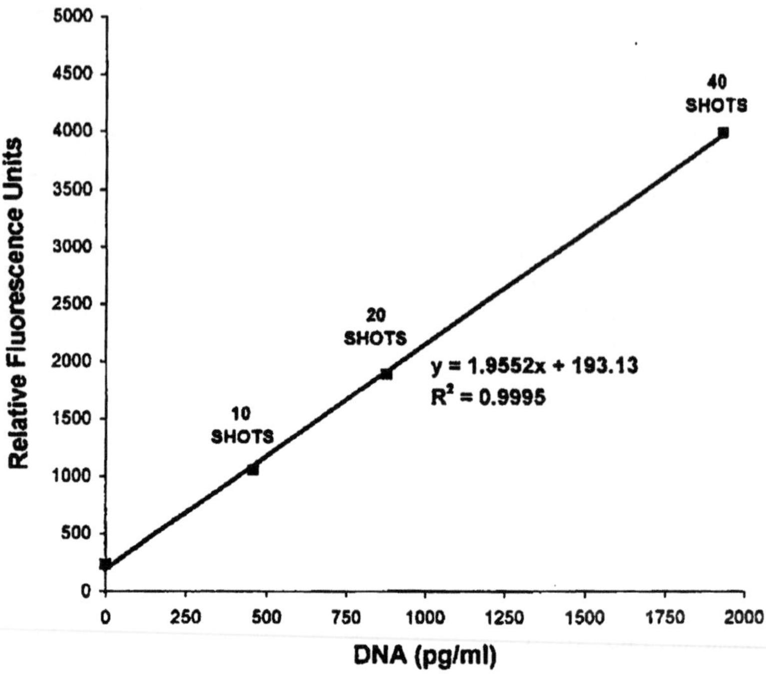

Fig. 3.6 Relationship between the number of captures and the concentration of DNA. A linear relationship between the concentration of DNA and the number of captures was observed with a regression line having a correlation coefficient of 0.9995 (R^2) (Reprinted from: Balogh, G.A., Heulings, R., Mailo, D, Wang, R., Li, YZ., Hardy, R., and Russo J. International J. of Oncology 31: 253–260, 2007)

by adhering the selected cells to a thermoplastic film with an infrared laser pulse [2]. When the film is lifted, the "captured" cells can be transferred directly into a digestion buffer for extraction of nucleic acids for subsequent analysis [1]. Several studies have shown the genomic expression from isolated tumor breast cells using LCM [13–19]; however, very few publications [4, 16–19] show the successful gene expression study using microarray technologies from laser captured breast epithelial cells from paraffin-embedded normal breast tissue. In addition we aimed to compare two different methods of RNA amplification, the linear one already described in the literature [20, 21] and a PCR-based method developed in our laboratory. Altogether, the final objective is the optimization of the cells obtained by LCM from ethanol-fixed and paraffin-embedded breast tissue for obtaining good quality and high yield of RNA to perform genomic studies using the cDNA microarray

The normal breast tissues were obtained from reduction mammoplasties of postmenopausal women after respective informed consent forms had been signed by the patients, according to the regulations of the Human Subjects Protection Committee. Fragments of breast tissue were fixed in ethanol 70 % for 8 h at room temperature, and then embedded in paraffin. 5 μm sections were cut and stained with

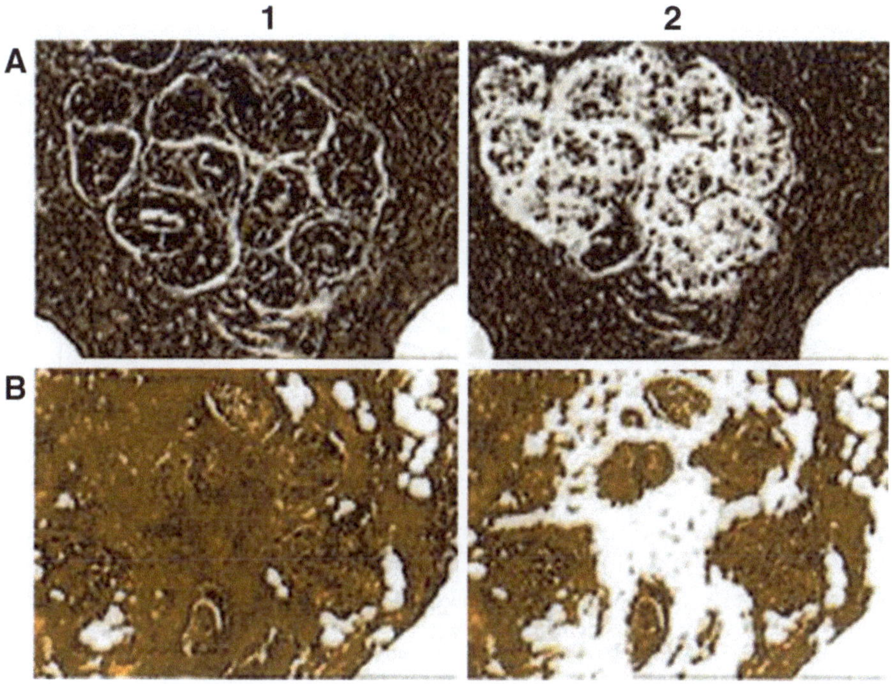

Fig. 3.7 Normal breast epithelial and interlobular stroma cells dissected by LCM. Pure breast epithelial cells from Lobule Type 1 were captured, separating them from the surrounding cells of the stroma. (**a**) Lobule Type 1 from the normal breast, section of 5-μm ethanol-fixed paraffin-embedded, before (1) and after LCM (2). (**b**) Interlobular stroma cells before (1) and after the LCM (2). LCM was performed using the PixCell II LCM system (Arcturus Engineering, Mountain View, CA) (Reprinted from: Balogh, G.A., Heulings, R., Mailo, D, Wang, R., Li, YZ., Hardy, R., and Russo J. International J. of Oncology 31: 253–260, 2007)

hematoxylin and eosin, at which point LCM was performed. Figure 3.7 shows the representative image of the epithelium and stroma cells before and after LCM. In order to investigate how many cells were extracted using LCM and the correlation between the number of the cells and the RNA isolated, we performed a curve using the pool of the RNA isolated in each sample and the number of the cells calculated using the following formula: *Number of cells* $= [(s/t) \times (m/r) \times (n)] \times N$. In this formula, s = size of the probe used for LCM (in μm); t = size of the cells (in μm); m = thickness of the cell captured (μm); r = thickness of the cell (μm); n = number of captures observed in the capsule after LCM; and N = number of the replicates. For example, if in the capsule 100 captures were observed, we used a 30-μm probe size to perform LCM, and if the size of the epithelial cells was 12 μm, the number of the cells would be $= [(30/12\ \mu m) \times (6/12\ \mu m) \times 100] \times 3 = 375$ cells. Figure 3.8 shows the curve of total RNA concentration isolated from LCM of breast epithelial cells and interlobular stromal cells versus the number of captured cells [4].

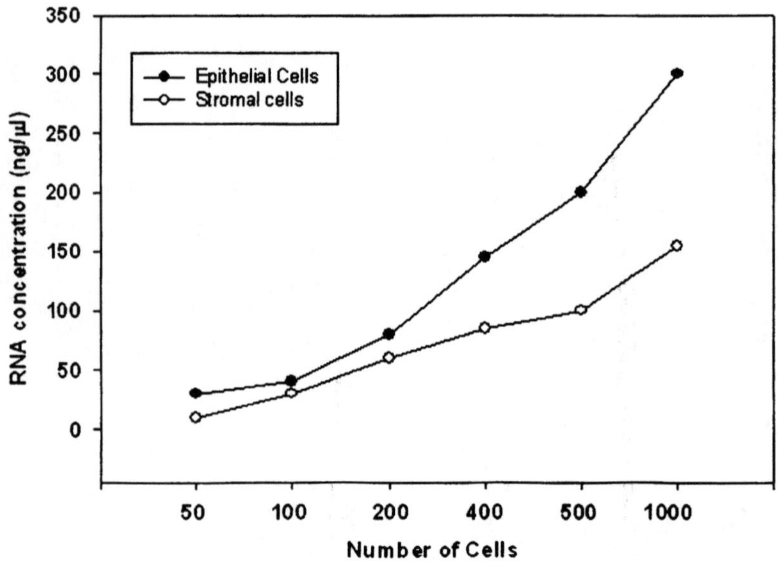

Fig. 3.8 Total RNA concentration versus number of epithelial and stroma cells captured by LCM from normal breast tissue (Reprinted from: Balogh, G.A., Heulings, R., Mailo, D, Wang, R., Li, YZ., Hardy, R., and Russo J. International J. of Oncology 31: 253–260, 200731: 253–260, 2007)

From the LCM capsules, the RNA was isolated using TRIzol (Invitrogen, Inc.). The concentration and the quality of total isolated RNA were measured in the nanodrop (Nanodrop Technologies, Inc.) and in the Bioanalyzer (Agilent technologies). We used the RNA obtained from the laser capture microdissected epithelia and stroma of the normal breast for amplification using the PCR-RNA amplification developed in our laboratory (Fig. 3.9) and the established linear amplification method (Fig. 3.10). The RNA was isolated by placing the LCM cap in a 50-μL of TRIzol reagent. The cells were in contact with the reagent and placed on ice for 30 min (it can be stored overnight at −20 °C). The following step was incubation at room temperature for 5 min, followed by centrifugation for 15 min at 13,000 rpm at 4 °C, and addition of 20 μL of chloroform for every 50 μL of TRIzol. The mixture was added to PLGT Heavy Gel tubes, previously centrifuged for 2 min at 13,000 rpm, and centrifuged at 13,000 rpm for 15 min at 4 °C. The aqueous phase was collected and 200 μL of 100 % EtOH, 2 μL of linear acrylamide, and 20 μL of AcNaOH were added followed by incubation for 2 h at −80 °C. At the end of the incubation, the sample was centrifuged at 13,000 rpm at 4 °C for 30 min. The pellet was dried and washed with 200 μL of 70 % EtOH. After mixing, it was spun at 13,000 rpm at 4 °C for 15 min. The pellet was dried by removing EtOH in vacufuge for 2–3 min. The pellet was dissolved in 11 μL of DEPC water and measured in Nanodrop [4].

The total RNA was amplified using a PCR method based on dT-T7-RNA amplification. The amplification procedure involves four steps (Fig. 3.9): (1) cDNA synthesis using the RT enzyme (SuperScript from Invitrogen), (2) amplification

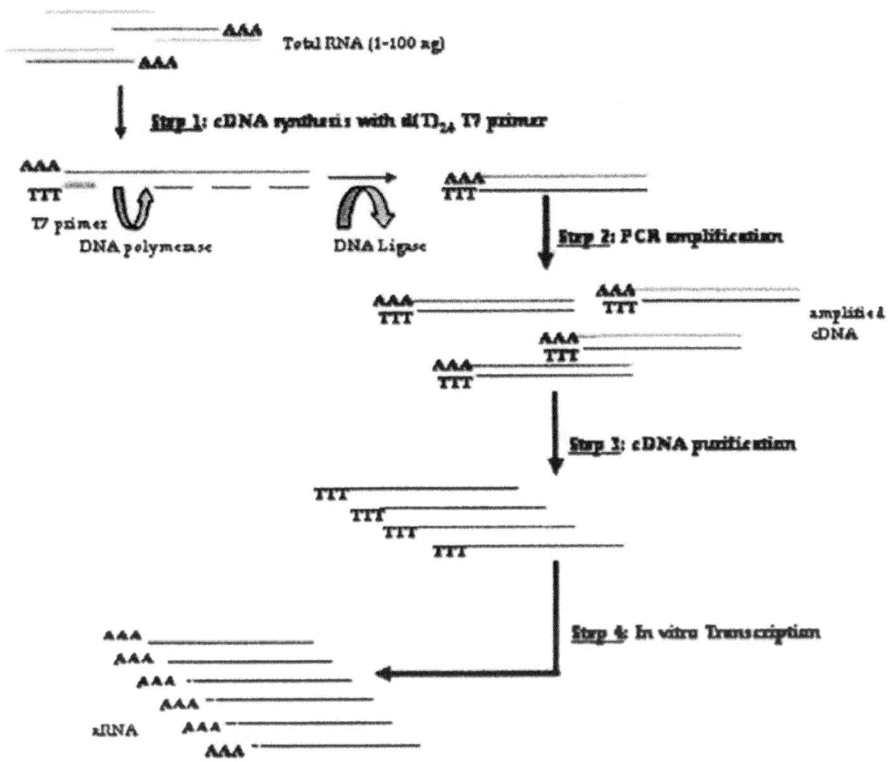

Fig. 3.9 Diagram of PCR methodology (Reprinted from: Balogh, G.A., Heulings, R., Mailo, D, Wang, R., Li, YZ., Hardy, R., and Russo J. International J. of Oncology 31: 253–260, 2007)

based on dT-T7 and switch T7 primers (Table 3.2), (3) in vitro transcription reaction, and (4) RNA purification and quantification. We performed two rounds of amplifications. We have used a human universal reference RNA from Stratagene as reference that was amplified identically to the sample under study. The RNA amplification procedures are described schematically in Fig. 3.9 [4].

For the linear amplification procedure, we employed a modification of the Eberwine amplification procedure [20] modified by Upson-Bellacosa (UB) [22] (Fig. 3.10). In this procedure a double-stranded cDNA (ds-cDNA) was synthesized from 1 μg of total RNA using the SuperScript Double-Stranded cDNA Synthesis Custom Kit (Invitrogen, Carlsbad, CA), then mixed with 2 μL of 20 μm oligo-(dT)24-T7 primer [5′-AAACGACGGCCAGTGAATTGTAATACGACTCACTATAGGCGC-(dT)24-3′] in a total reaction volume of 11 μL to initiate first-strand cDNA synthesis. The oligo-(dT)24-T7 primer and RNA were heat-denatured at 70 °C for 10 min, followed by annealing at 42 °C for 2 min. Four microliters of 5× first-strand reaction buffer (Invitrogen), 1 μL of 10 mM dNTPs, 2 μL of 0.1 M DTT, 0.8 μL of T4gp32 (5.0 mg/mL) (USB), 1 μL of RNasin (40 U/μL) (Promega, Madison, WI), and 1 μL of

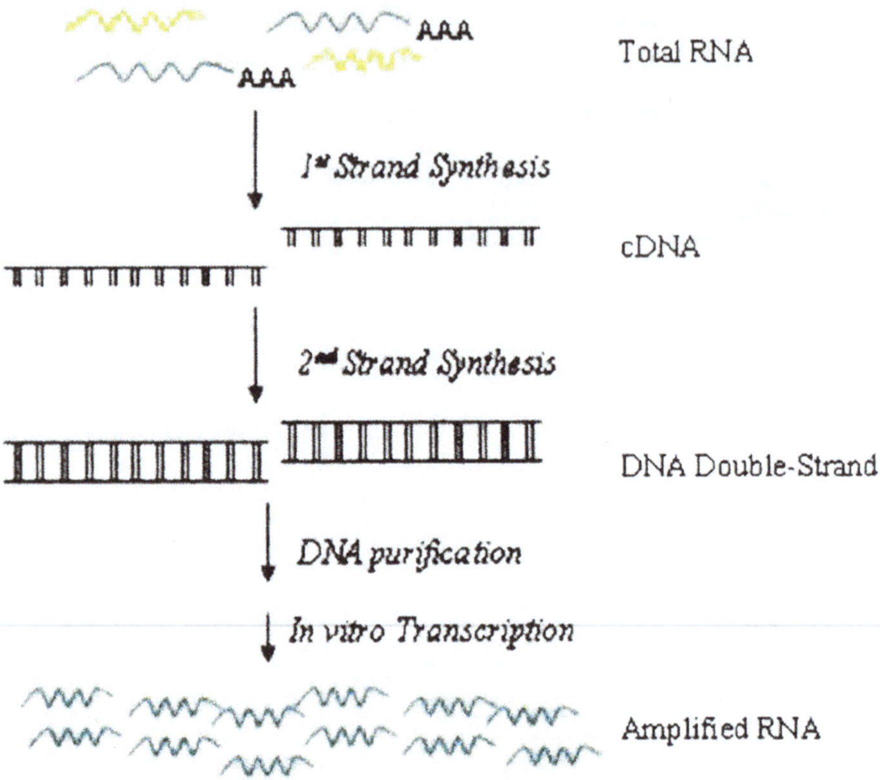

Fig. 3.10 Diagram of linear methodology (Reprinted from: Balogh, G.A., Heulings, R., Mailo, D, Wang, R., Li, YZ., Hardy, R., and Russo J. International Journal of Oncology 31: 253–260, 2007)

Table 3.2 Description of the primers for cDNA amplification

Primers used	Description	Sequence
Primer 1	Oligo Dt-T7 primer	5′-gca tta gcg gcc gcg aaa tta ata cga ctc act ata ggg aga ttt ttt ttt ttt ttt ttt ttt-3′
Primer 2	Switch primer	5′-aaa cga cgg cca gtg aat tgt aat acg act cac tat agg cgc ggg-3′

Reprinted from: Balogh, G.A., Heulings, R., Mailo, D, Wang, R., Li, YZ., Hardy, R., and Russo J. International J of Oncology 31: 253–260, 2007

SuperScript II (200 U/μL) were added to the reaction and incubated at 42 °C for 2 h. The reaction was heat-inactivated at 65 °C for 15 min and chilled on ice for 5 min. For second-strand synthesis, 30 μL of 5× second-strand reaction buffer, 3 μL of 10 mM dNTPs, 4 μL of DNA Polymerase I (10 U/μL), 1 μL of *E. coli* RNase H

(2 U/μL), 1 μL of *E. coli* DNA ligase (1 U/μL), and 41 μL of RNase-free water were added to the reaction and incubated at 16 °C for 2 h. Two microliters of T4 DNA Polymerase (5 U/μL) were added to the reaction and incubated at 16 °C for an additional 15 min. The reaction was heat-inactivated at 70 °C for 10 min. The resulting ds-cDNA was extracted with phenol/chloroform and then again with chloroform. The purified ds-cDNA was washed three times with 500 μL RNase-free water in Microcon YM-100 spin columns (Millipore, Bedford, MA), centrifuged at 500 × *g* for 18 min. The ds-cDNA was eluted and adjusted to a final volume of 9 μL for amplification by T7 RNA Polymerase [4]. The Ampliscribe T7 transcription kit (Epicentre Technologies, Madison, WI) was used for two consecutive rounds of in vitro RNA transcription. Nine microliters of the ds-cDNA, 2 μL of 10× reaction buffer (Epicentre), 1 μL each of 100 mM ATP, GTP, CTP, and UTP, 2 μL of 100 mM DTT, 1 μL of RNasin (40 U/μL) (Promega), and 2 μL of T7 RNA Polymerase (Epicentre) were incubated at 37 °C for 5 h. The subsequent amplified, complementary RNA (aRNA) was extracted and a second round of RNA amplification was conducted, using the aRNA as a template to synthesize ds-cDNA. The resulting aRNA was extracted and adjusted to a final volume of 20 μL, the quality of aaRNA, and the concentration was measured in the nanodrop and in the bioanalzyer (Agilent Technologies, CA) [4].

For real-time PCR (RT-PCR), the cDNA was synthesized for the control genes 18S, milk fat gene, whey acidic protein, inhibin alpha and beta gene using Taqman primer and probe set sequences. A ready to use primer and probe set pre-designed by Applied Biosystems (Assay-on demand Gene Expression Product from ABI) was used for the detection of the gene expression [4]. All RT-PCR reactions were performed on the ABI PRISM 7000 Sequence Detection System (Applied Biosystems) using the fluorescent Taqman methodology (TaqMan One Step RT-PCR Master Mix Reagents, Applied Biosystems). The PCR cycle at which the fluorescence arises above the background signal is called the cycle threshold (Ct). We used 100 ng of amplified RNA for each RT-PCR reaction in a total volume of 50 μL. Primer and probe concentration for the target gene were optimized according to the manufacturer's procedure. The thermal cycling conditions comprised 30 min at 48 °C, 10 min at 95 °C, and 40 cycles of 15 s denaturalization at 95 °C and 60 s annealing at 60 °C. The results were expressed as relative gene expression (RGE) using the ΔCt method, as described by Livak and Schmittgen [23] and Benoy et al. [24].

As we expected, the concentration of total RNA in the microcaptured epithelial cells was greater than the one obtained in the interlobular stroma cells (Fig. 3.8) [4]. This is explained by the fact that the number of nuclei in the interlobular stroma is significantly lower compared to the number of nuclei captured in the epithelial cells. The total RNA isolated was split in order to be amplified in parallel using both the PCR-based (Fig. 3.9) and the linear RNA amplification methodologies (Fig. 3.10). Table 3.3 shows the concentration of the initial starting RNA, the concentration of the aRNA in μg/μL, the volume in μl, the ratio 260/280, and the times of amplification of mRNA. The amplified RNA yield increased nearly 1,200-fold

Table 3.3 RNA amplification by PCR and linear methodology

	RNA (ng/μL)	RNA volume (μL)	aRNA (ng/μL)	aaRNA	aaRNA volume (ng/μL)	Ratio	Times of amplification
BLR-PCR amplification method							
Epithelial cells	4.30	10	2.360	498.2	100	2.00	1,158
Stroma cells	2.31	10	1.849	314.3	100	2.00	1,359
Linear amplification method							
Epithelial cells	4.30	10	1.18	23.60	100	1.80	54
Stroma cells	2.31	10	0.43	18.49	100	1.78	80

Reprinted from: Balogh, G.A., Heulings, R., Mailo, D, Wang, R., Li, YZ., Hardy, R., and Russo J. International J of Oncology 31: 253–260, 2007

with 20 cycles in the PCR methodology after two rounds of amplification. As expected, the yield with the linear amplification procedure after two rounds of amplification was checked by the Bioanalyzer (Agilent Technologies) (Figs. 3.11 and 3.12) [4]. The fluorescent probes were performed in triplicate and after checking the quality; the replicate from the same sample was combined and re-distributed into three separate tubes in order to have an identical replicate. Equal amount of fluorescent probe was used to hybridize to cDNA microarrays containing 15,000 genes, placing in the green channel (Cy3) the human universal reference RNA and in the red channel (Cy5) the amplified RNA samples. We have performed the data analysis considering replicates; we combined the replicates from each condition such as aaRNA linear method and aaRNA by our PCR method. We employed the GeneSight software version 4.2 to analyze the gene expression profile. Typically, local background correction takes place spot by spot before replicate combination. GeneSight determines which values to combine by comparing the Gene IDs and combining all spots with the same ID. The combined replicate transformations also eliminate the outlier features compared with the other values for the same gene ID. For the remaining genes, the software computes the coefficient of variance, as a measure of confidence, which is available later for queries and the report. Lowess method was used to perform the data normalization, selecting a quadratic degree of fitness. We used a sub-grid normalization scope, omitting all the outliers and flagged spots from 1 to 7. The scatter plot of Fig. 3.13 represents the gene expression of 12,565 genes out of 15,000 genes, showing a comparison between amplified RNA using linear methodologies versus amplified RNA using PCR methodology. The correlation coefficient from both methodologies was 0.87 demonstrating that differences in gene expression for samples amplified using linear and PCR protocols are minor [4].

We have further verified the correlation between both methods by processing frozen tissue of rat mammary gland, mouse liver, and human breast tissue (Fig. 3.14). The quality of the amplified RNA using the PCR methodology was checked by the Bioanalyzer and in all the cases the amount of the aaRNA was higher using our PCR

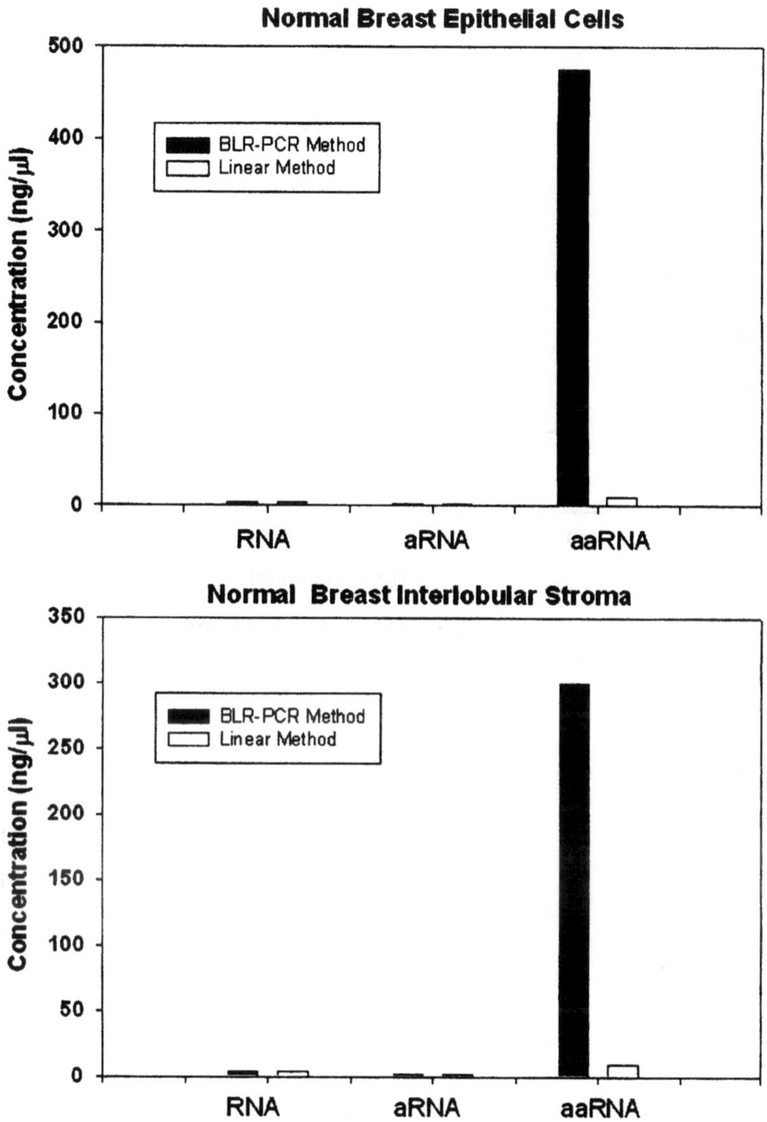

Fig. 3.11 Normal breast epithelium and interlobular stroma RNA amplification of the breast epithelia and interlobular stroma using our methodology (BLR) and the linear amplification procedures (Reprinted from: Balogh, G.A., Heulings, R., Mailo, D, Wang, R., Li, YZ., Hardy, R., and Russo J. International J. of Oncology 31: 253–260, 2007)

methodology than the linear one. We hybridized cDNA microarrays containing 15,000 genes, placing in the green channel (Cy3) the human universal reference RNA and in the red channel (Cy5) the RNA samples. The quality of the array was

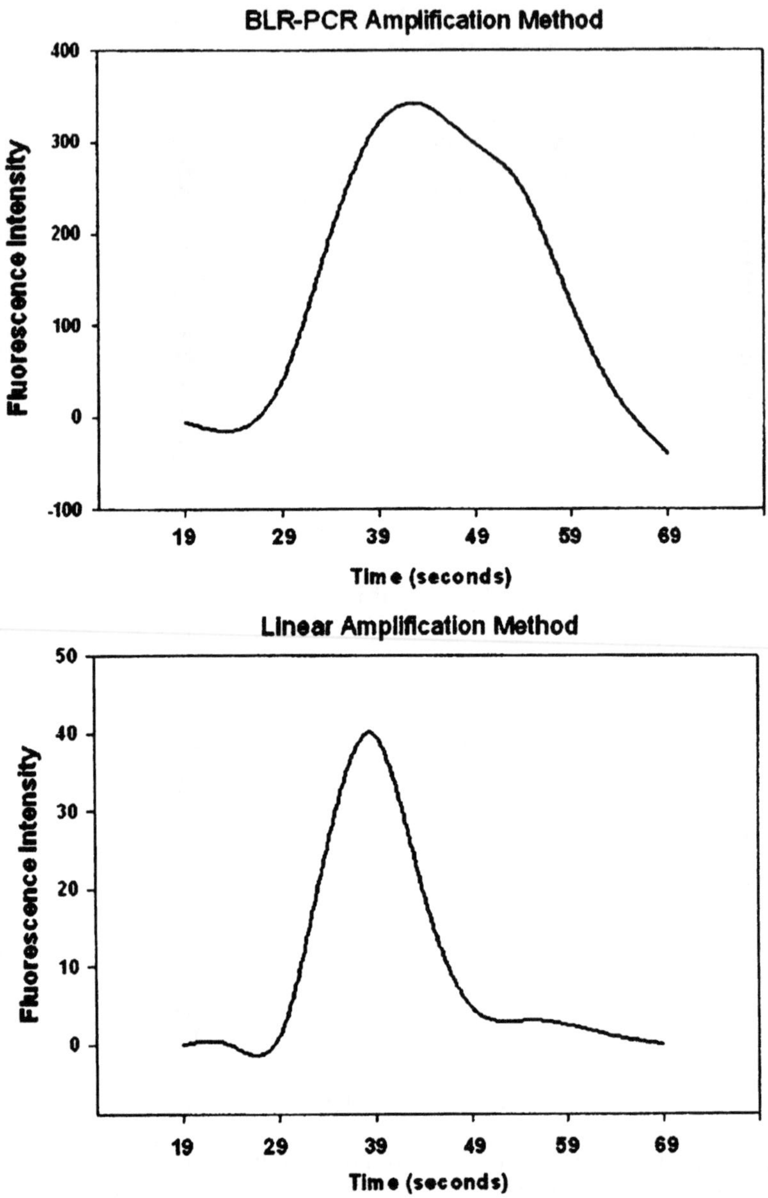

Fig. 3.12 Electropherogram of aaRNA quality using the Bioanalyzer (Agilent Biotechnologies). The *Y* represents the intensity of fluorescence (signifying the quantity of total RNA) and *X* represents the time of elution in seconds (signifying the molecular weight of the RNA measured). The upper graphic represents the aaRNA of breast epithelial cells using the BLR-PCR method and the lower graphic represents the aaRNA of breast epithelial cells using the linear methodologies. In both cases the same starting amount of RNA was used (Reprinted from: Balogh, G.A., Heulings, R., Mailo, D, Wang, R., Li, YZ., Hardy, R., and Russo J. International J. of Oncology 31: 253–260, 2007)

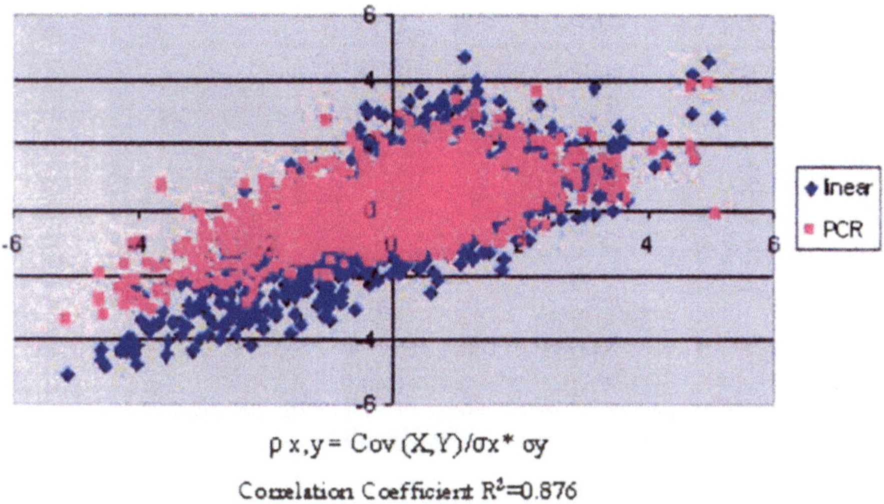

Fig. 3.13 Correlation coefficient of gene expression analysis from linear and BLR-PCR amplification methodologies (Reprinted from: Balogh, G.A., Heulings, R., Mailo, D, Wang, R., Li, YZ., Hardy, R., and Russo J. International J. of Oncology 31: 253–260, 2007)

highly comparable as were the scatter plots of gene expression comparison between the RNA amplified using linear methodology versus RNA amplified using PCR methodology. The correlation coefficient from both methodologies was 0.87 for the rat mammary gland, 0.72 for the mouse liver, and 0.89 for the human breast (Fig. 3.14) [4]. As another quality control, we have validated by RT-PCR, the genes 18S, milk fat, whey acid protein, inhibin alpha and beta genes using Taqman primer and probe set sequences. The RGE was not significantly different from the total RNA versus the amplified RNA by both procedures. Altogether, our data demonstrated the proof of principle that paraffin sections of ethanol breast fixed tissue can be used for LCM and sufficient quality of amplified RNA for high-density cDNA microarray analysis can be produced. The quality and reproducibility of PCR-RNA amplification procedures are quite comparable with the traditional linear methodology. We show that the correlation between the gene expression analyses from both methods is high—the amplification procedure maintains the gene expression profile of the original sample. Our PCR method of RNA amplification is fast and economical, because it needs fewer reagents compared to the linear amplification procedure. It is accurate; keeping intact the integrity of the original RNA and most importantly it gives a very high yield, enough to perform genomic studies from LCM samples from normal breast tissue [4].

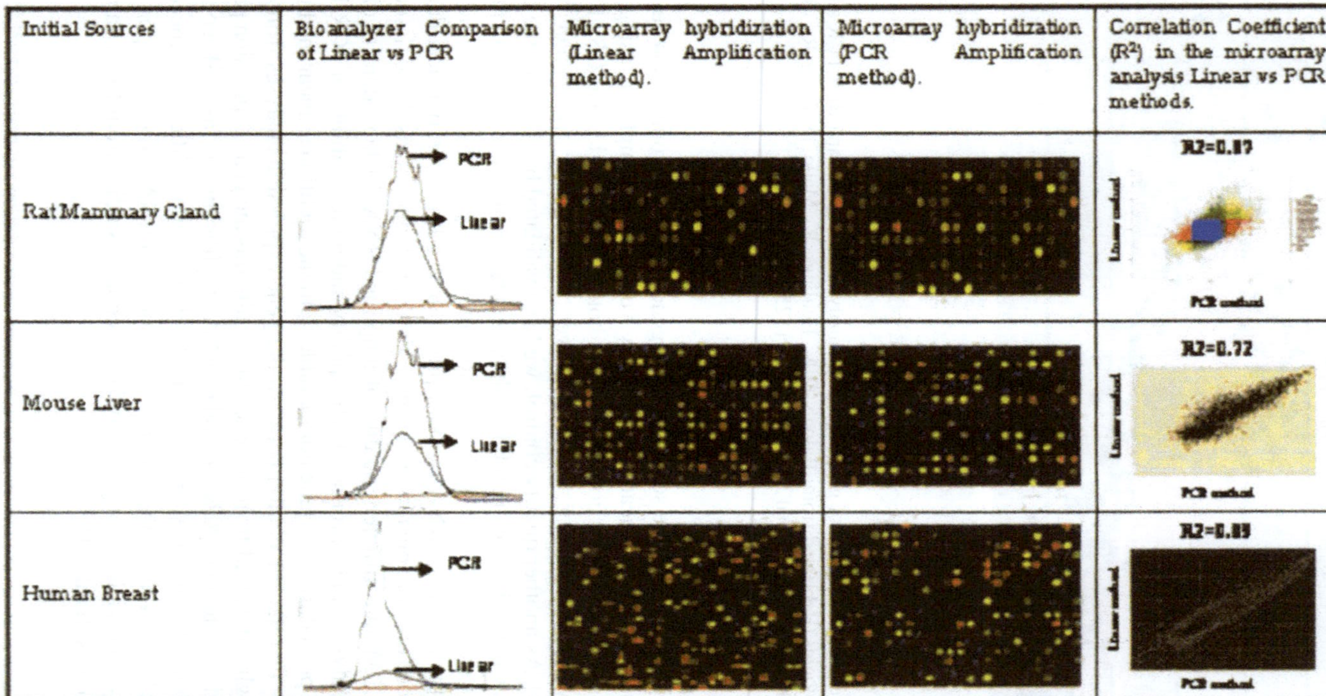

Fig. 3.14 Microarray hybridization using RNA obtained from linear and PCR amplification. The source of RNA was from rat mammary gland, mouse liver, and human breast (Reprinted from: Balogh, G.A., Heulings, R., Mailo, D, Wang, R., Li, YZ., Hardy, R., and Russo J. International J. of Oncology 31: 253–260, 2007)

3.4 The Use of Frozen Section for LCM

To perform LCM on frozen sections, the section should be cut 1 h prior to staining. Slides are placed on dry ice immediately after sectioning where they remain until staining commences. Slides are then fixed in 70 % ethanol for 30 s and dipped in nuclease-free water for 1 min, followed by hematoxylin treated with ProtectRNA Rnase Inhibitor (Sigma, Cat#R7397) and in bluing reagent for 30 s. Once stained, they are dehydrated in 70 % ethanol, 95 % ethanol, 100 % ethanol, and xylene. The slides are allowed to air dry for about 1 min before beginning the LCM (Fig. 3.15).

3.4.1 Yield of DNA Obtained by LCM in Frozen Sections

DNA can be isolated from the captured cells with the PicoPure DNA Extraction Kit (Arcturus cat# KIT0103) using the manufacturer's protocols. For example, to obtain a concentration of 57–62 μg/mL of DNA around 1,400 to 3,000 laser pulses are needed. This large variation is due to the varied cellularity of the human breast

3.4.2 LCM on Frozen Rat Mammary Tumor for RNA Extraction

To assess how many captures would be needed to yield sufficient RNA quantities for Microarray Analysis and Real-Time RT-PCR from frozen rat mammary tumors, we performed LCM on two frozen sections, 10 μm in thickness, from a rat mammary cribriform adenocarcinoma (Fig. 3.16). An initial frozen section was stained with H&E to verify the areas for capture. The PixCell II Manuel Laser Capture Microdissection System was used with a laser spot size of 15 μm and 70 mW laser power setting. For each slide, three caps were collected, totaling six caps. The caps

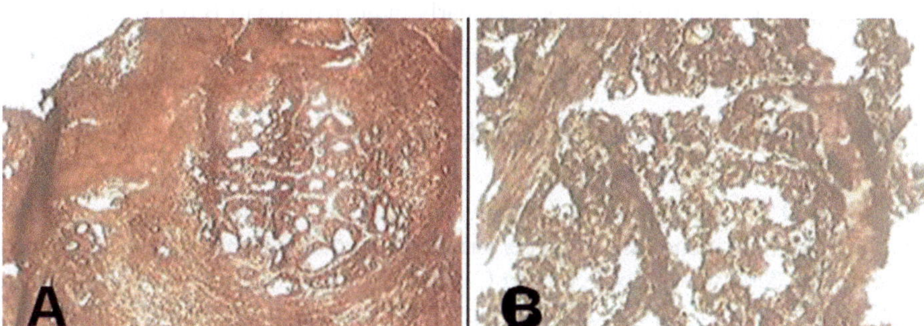

Fig. 3.15 LCM of frozen tissue of a rat mammary carcinoma cribriform type. (**a**) Histological section stained with H&E to show the area to be dissected and (**b**), after LCM

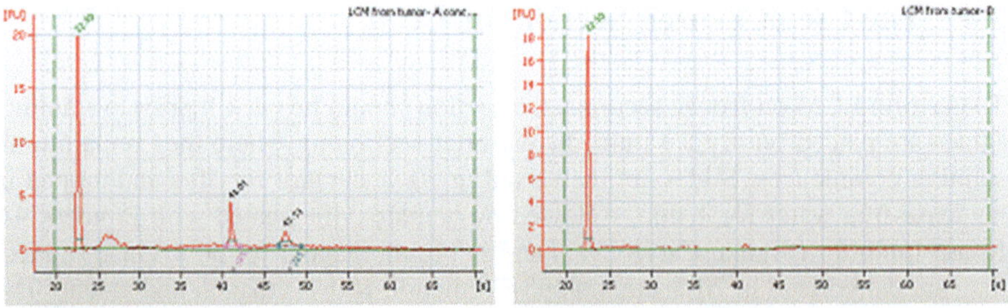

36ng/µl (volume = 20µl)	
# of caps	3
Average # of laser pulses per cap	6085

11ng/µl (volume= 50µl)	
# of caps	3
Average # of laser pulses per cap	5175

Fig. 3.16 Electropherogram images from Agilent Bioanalyzer 6000 showing the quality of the RNA obtained from LCM. Sample A (*left*) showed a greater concentration of RNA than sample B (*right*). The presence of peaks in A is indicative of this higher concentration. While the quantity of RNA is not vast, our study showed that it is of sufficient quality to perform cDNA microarrays

were immediately placed on dry ice in Trizol Reagent. RNA extraction was performed within 2 h of collection. After initial centrifugation, caps were pooled into two groups, A and B. The groups yielded approximately 36 and 11 ng/μL of RNA, respectively. As seen in Fig. 3.16, sample A shows a greater concentration of RNA than sample B. The presence of peaks in A is indicative of this higher concentration. In this case, the quantity of RNA is of sufficient quality to perform cDNA microarrays. It is important to note that these samples were captured from tumor tissue; in doing LCM from epithelial cells and stroma it is expected a much smaller yield due to the reduced number of cells.

Using the same material described above, we proceeded to determine how much RNA could be obtained by an individual cap of LCM material. Slides were processed for LCM, and RNA was extracted using Trizol, then quantified using the Nanodrop after which they were measured in the Bioanaylzer to assess quality. Samples 1LCM, 2LCM, and 3LCM each represent one captured cap, each processed individually (Fig. 3.17a). The quality of the RNA is clearly indicated in part B of Fig. 3.17.

3.4.3 LCM in Frozen Human Breast Tissue

The slides of human breast tissue were stained using Arcturus HistoGene Frozen section Staining Kit and LCM was performed using the Arcturus Veritas-automated system. The cells were captured onto CapSure/E macro LCM Caps and the RNA was isolated from the captured cells using the Pico pure RNA isolation kit and DNA extracted with the Pico pure DNA extraction kit according to manufacturer's protocols. The amount of laser captures needed depends of the methodology to be used. For example, the amount of captures needed to study the epithelium of the normal

A **B**

1LCM (1 cap)	• Concentration: **2.01ng/ul** • Total amount of RNA: **20ng**	260/280 ratio: 2.68	RIN*: 7.9	
2LCM (1 cap)	• Concentration: **1.89ng/ul** • Total amount of RNA: **19 ng**	260/280 ratio: 0.87	RIN:7.8	
3LCM (1 cap)	• Concentration: **6.21ng/ul** • Total amount of RNA: **60 ng**	260/280 ratio: 1.68	RIN:8.4	

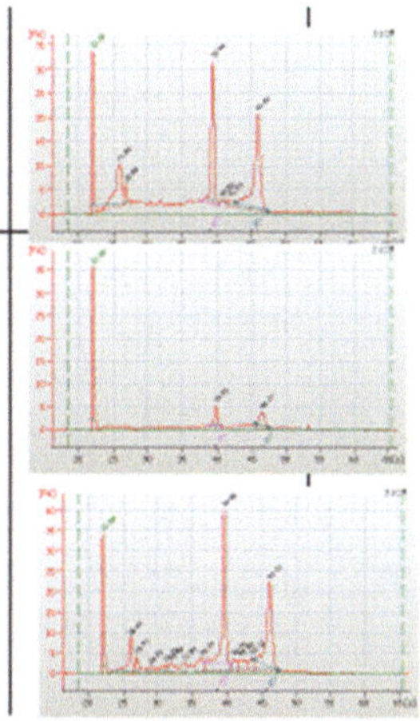

* RNA Integrity Number

Fig. 3.17 Samples 1LCM, 2LCM, and 3LCM each represent one captured cap, each processed individually. (**a**) Depicts the amount of RNA obtained, and (**b**) the quality of the RNA using the Agilent Bioanalyzer

breast using the Illumina system requires 63 caps per sample whereas if Affymetrix U133 plus 2 is used, only 5 caps per sample or core biopsy are needed. The amount of RNA per cells is around 10–30 pg total RNA and each diploid human cell contains ~6 pg genomic DNA. This data can provide a baseline for determining the amount needed for each methodology. When LCM is to be used to extract protein, the total protein levels per cell vary greatly depending on the cell types being studied. Cell size is probably one of the biggest factors. A simple experiment can be done to determine the total protein amount per cell type; we use a modified Lowry Protein Assay or a similar simple protein assay. Using BSA as standard, simply plot various samples with known numbers of cells to a standard of known BSA. It should work out to be a linear relationship. Once standard a line is obtained, it is possible to estimate the protein amount per 1,000 cells and lower.

We also performed studies utilizing breast core biopsies preserved in RNAlater® and the total RNA was extracted using All prep RNA/DNA mini kit (QIAGEN). The concentration of the samples was measured using Nanodrop ND1000 and quality was assessed using Bioanalyzer Pico Chips (Agilent Technologies). If the quality and amount of total RNA obtained was not enough to perform subsequent experiments, and there was at least one core biopsy available, the extraction was repeated. The range of concentration was 11.7–346.1 ng/μL in a volume of approximately 35 μL. These samples were amplified and labeled using 3′ IVT Express Kit (Affymetrix) following the manufacturer's protocol we performed arrays. The arrays were subjected to QC analysis using standard Affymetrix quality measures (Scale Factor, Percent Present, and Average Background). In addition, quality

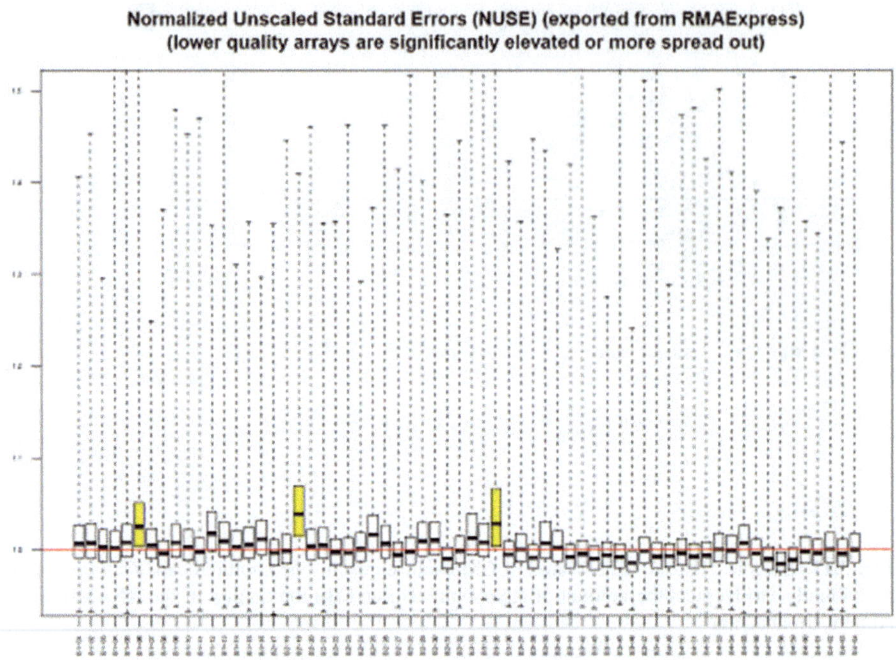

Fig. 3.18 Normalized unscaled standard error (NUSE) plot of the discovery set. *Boxes* in *yellow* correspond to arrays that failed the QC analysis

was assessed using graphical tools based on Affymetrix probe-level models (PLM). The Normalized Unscaled Standard Error (NUSE) plot, in particular (Fig. 3.18), was used to disqualify lower quality arrays.

3.5 Use of LCM for Separating Epithelium and Stroma in Normal Human Breast

For this specific work, the core biopsies fixed in RNAlater were frozen and the cryostat was used to perform histological section of the frozen tissue. The frozen sections were then stained with hemotoxylin and eosin (H&E) specially prepared to avoid RNA degradation. Once each slide was stained, the VERITAS Microdissection Instrument (Arcturus, CA, USA) was used to select and capture all the epithelial cells present in each section. The tissue left on the slide was then scraped into a different tube and classified as stroma. The collected cells were maintained in the extraction buffer for RNA extraction, which was performed using the Arcturus®PicoPure®RNA Isolation Kit (Life Technologies) following the manufacturer's protocol. Also, total RNA was extracted from a second core biopsy, in which no LCM was performed. All the extracted RNA had the concentration measured using Nanodrop ND1000 and quality checked using Bioanalyzer Pico Chips

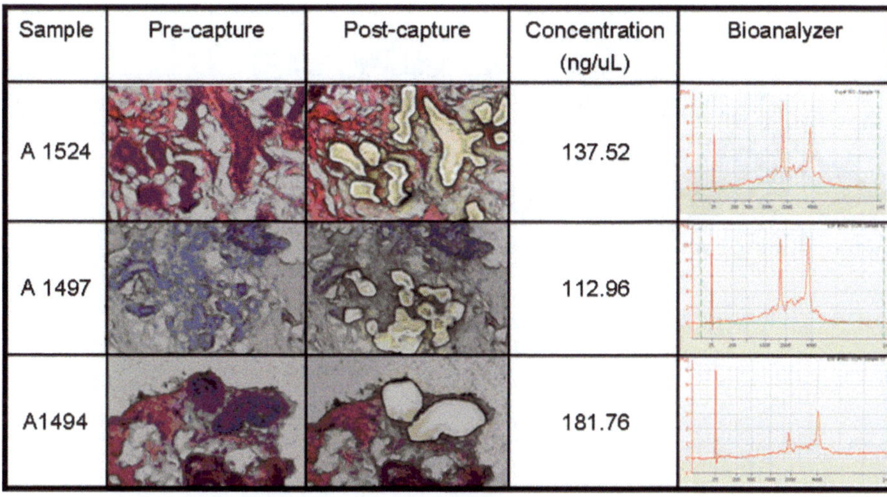

Fig. 3.19 Example of samples pre-capture and post-capture. RNA concentration obtained and bioanalyzer images showing the quality of the samples

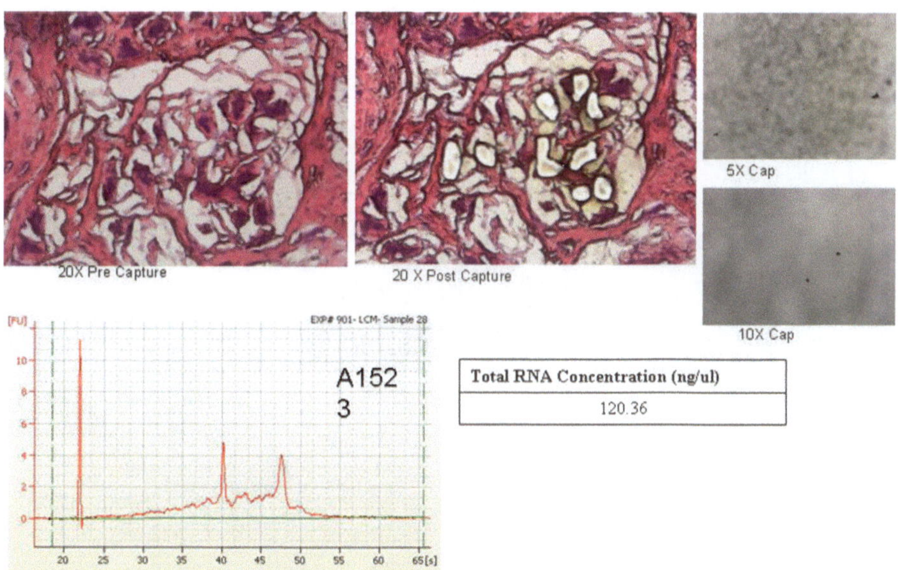

Fig. 3.20 Example of a sample pre-capture and post-capture, RNA concentration obtained and bioanalyzer images showing the quality of the sample

(Agilent Technologies). If the quality or quantity of RNA acquired was insufficient to perform the arrays, the LCM was repeated. The RNA concentration for the epithelial component ranged from sample to sample (Figs. 3.19, 3.20, 3.21, 3.22, 3.23, 3.24, and 3.25), and the same for the stroma (Fig. 3.26). The RNA extracted was

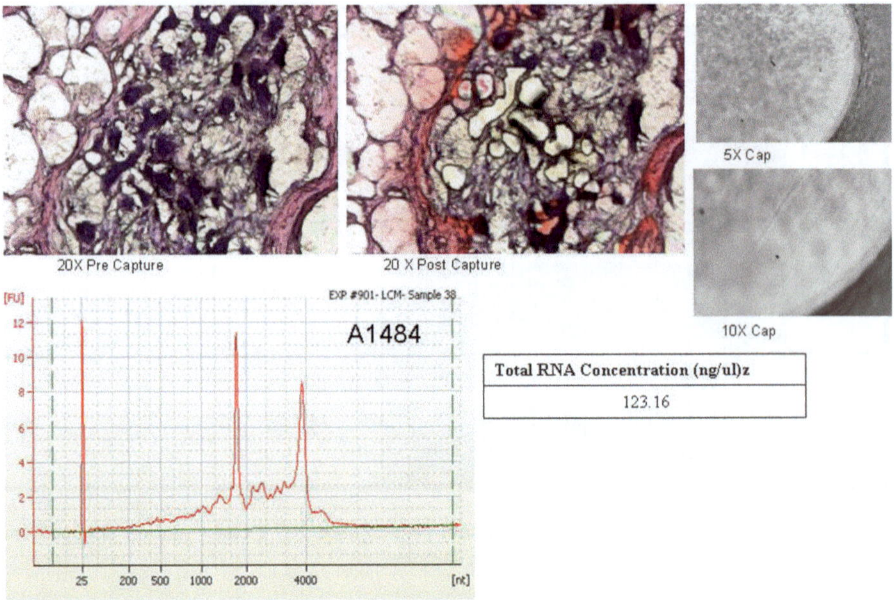

Total RNA Concentration (ng/ul)z
123.16

Fig. 3.21 Example of a sample pre-capture and post-capture, RNA concentration obtained and bioanalyzer images showing the quality of the sample

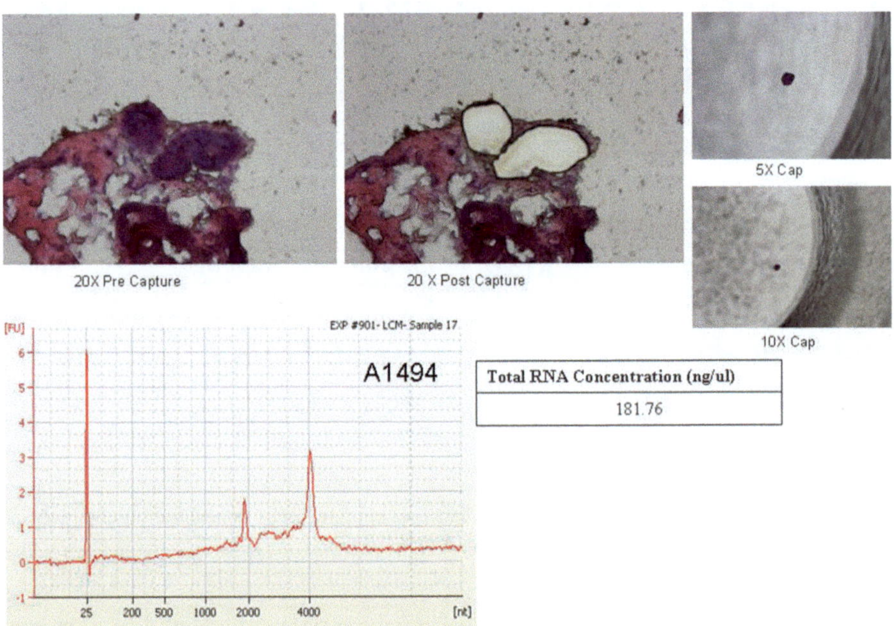

Total RNA Concentration (ng/ul)
181.76

Fig. 3.22 Example of a sample pre-capture and post-capture, RNA concentration obtained and bioanalyzer images showing the quality of the sample

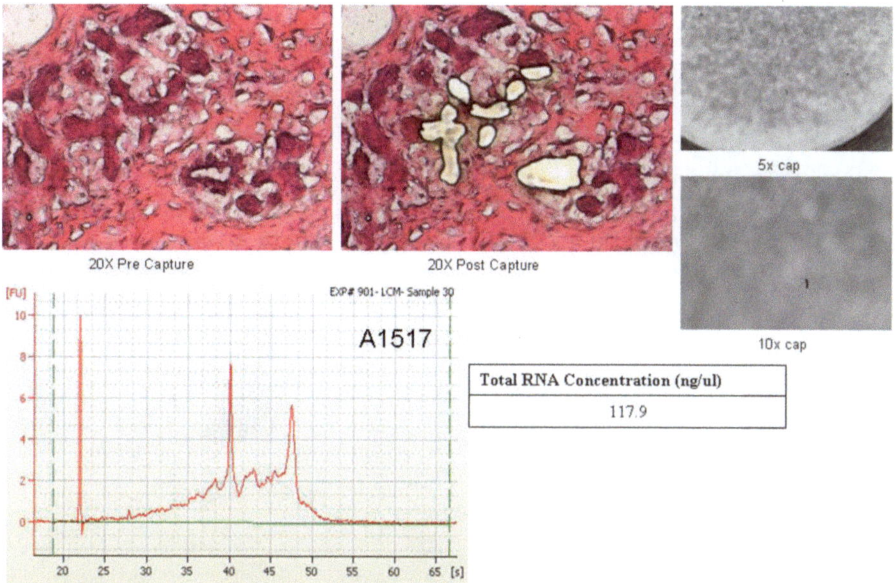

Fig. 3.23 Example of a sample pre-capture and post-capture, RNA concentration obtained and bioanalyzer images showing the quality of the sample

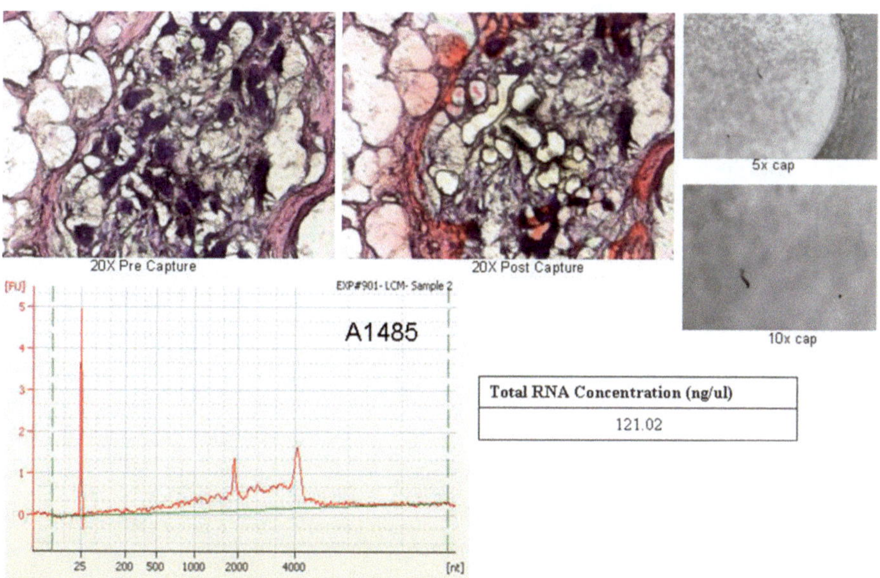

Fig. 3.24 Example of a sample pre-capture and post-capture, RNA concentration obtained and bioanalyzer images showing the quality of the sample

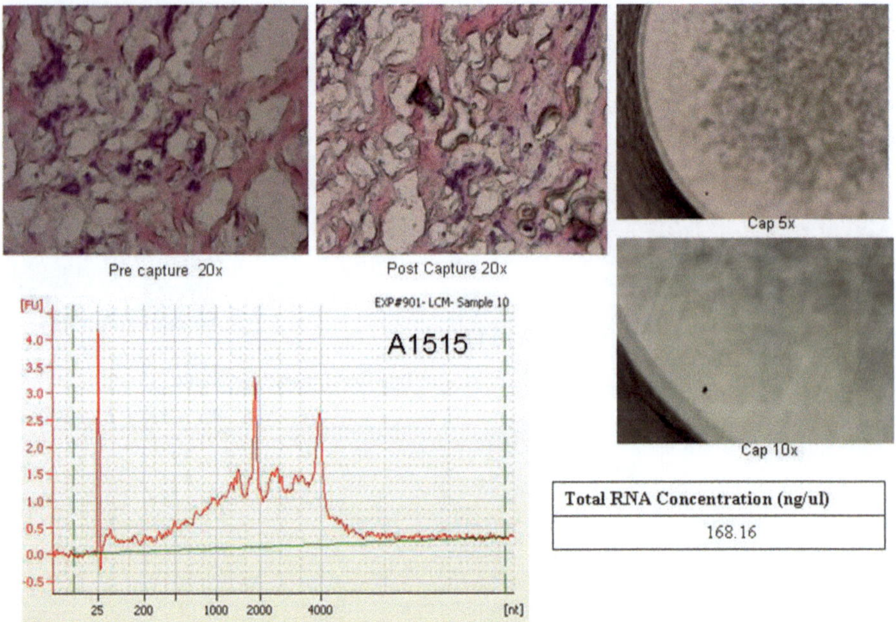

Total RNA Concentration (ng/ul)
168.16

Fig. 3.25 Example of a sample pre-capture and post-capture, RNA concentration obtained and bioanalyzer images showing the quality of the sample

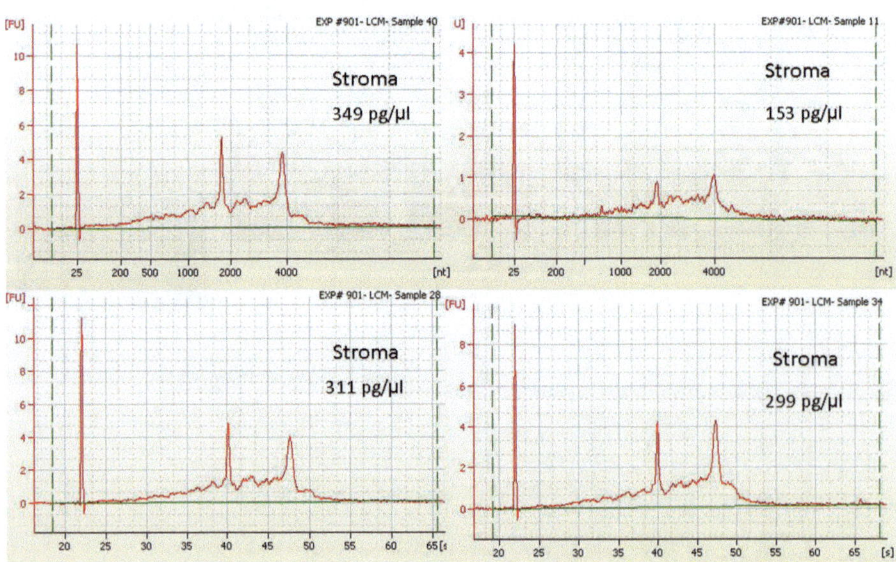

Fig. 3.26 Quality and amount of RNA obtained from LCM of the normal stroma of the human breast

used to perform microarray analysis. All arrays were subjected to QC analysis which ensured that they were in the acceptable ranges for standard Affymetrix quality measures (Scale Factor, Percent Present, and Average Background). In addition, quality was assessed using graphical tools based on Affymetrix PLM. The NUSE plot, in particular, was used to disqualify lower quality arrays.

3.6 Conclusions

The use of LCM in both PETs and frozen sections are suitable for genomic studies of normal breast tissue. There is a significant difference in the amount of RNA and DNA obtained from the epithelium and from the stroma, as well as from sample to sample. The quality analysis of the RNA and DNA obtained is crucial for molecular studies.

Literature Cited

1. Emmert-Buck MR, Bonner RF, Smith PD, Chuaqui RF, Zhuang Z, Goldstein SR, Weiss RA, Liotta LA (1996) Laser capture microdissection. Science 274:998–1001
2. Bonner RF, Emmert-Buck MR, Cole K, Pohida T, Chuaqui RF, Goldstein SR, Liotta LA (1997) Laser capture microdissection: molecular analysis of tissue. Science 278:1481–1483
3. Simone NL, Bonner RF, Gillespie JW, Emmert-Buck MR, Liotta LA (1998) Laser-capture microdissection: opening the microscopic frontier to molecular analysis. Trends Genet 14:272–276
4. Balogh GA, Heulings R, Mailo DA, Russo PA, Sheriff F, Russo IH, Moral R, Russo J (2006) Genomic signature induced by pregnancy in the human breast. Int J Oncol 28:399–410
5. Mies C (1994) Molecular biological analysis of paraffin-embedded tissues. Hum Pathol 25:555–560
6. Walsh PS, Varlaro J, Reynolds R (1992) A rapid chemiluminescent method for quantitation of human DNA. Nucleic Acids Res 20:5061–5065
7. Blumenstein R, Dias M, Russo IH, Tahin Q, Russo J (2002) DNA content and cell number determination in microdissected samples of breast carcinoma in situ. Int J Oncol 21:447–450
8. Singer VL, Jones IJ, Yue ST, Haugland RP (1997) Characterization of PicoGreen reagent and development of a fluorescence-based solution assay for double-stranded DNA quantification. Anal Biochem 249:228–238
9. Zhuang Z, Bertheau P, Emmert-Buck MR et al (1995) A microdissection technique for archival DNA analysis of specific cell populations in lesions <1 mm in size. Am J Pathol 146:620–625
10. Shibata D, Hawes D, Li ZH, Hernandez AM, Spruck CH, Nichols PW (1992) Specific genetic analysis of microscopic tissue after selective ultraviolet radiation fractionation and the polymerase chain reaction. Am J Pathol 141:539–543
11. Whetsell L, Maw G, Nadon N, Ringer D, Schaefer FVL (1992) Polymerase chain reaction microanalysis of tumors from stained histological slides. Oncogene 7:2355–2362
12. Kuppers R, Rajewsky K, Zhao M et al (1994) Hodgkin disease: Hodgkin and Reed-Sternberg cells picked from histological sections show clonal immunoglobulin gene rearrangements and appear to be derived from B cells at various stages of development. Proc Natl Acad Sci U S A 91:10962–10966

13. Schuetz CS, Bonin M, Clare SE et al (2006) Progression-specific genes identified by expression profiling of matched ductal carcinomas *in situ* and invasive breast tumors, combining laser capture microdissection and oligonucleotide microarray analysis. Cancer Res 66:5278–5286
14. Yang F, Foekens JA, Yu J et al (2006) Laser microdissection and microarray analysis of breast tumors reveal ER-alpha related genes and pathways. Oncogene 25:1413–1419
15. Nishidate T, Katagiri T, Lin ML et al (2004) Genome-wide gene expression profiles of breast-cancer cells purified with laser microbeam microdissection: identification of genes associated with progression and metastasis. Int J Oncol 25:797–819
16. Sgroi DC, Teng S, Robinson G, Le Vangie R, Hudson JR Jr, Elkahloun AG (1999) *In vivo* gene expression profile analysis of human breast cancer progression. Cancer Res 59:5656–5661
17. Luzzi V, Mahadevappa M, Raja R, Warrington JA, Watson MA (2003) Accurate and reproducible gene expression profiles from laser capture microdissection, transcript amplification, and high density oligonucleotide microarray analysis. J Mol Diagn 5:9–14
18. King C, Guo N, Frampton GM, Gerry NP, Lenburg ME, Rosenberg CL (2005) Reliability and reproducibility of gene expression measurements using amplified RNA from laser microdissected primary breast tissue with oligonucleotide arrays. J Mol Diagn 7:57–64
19. Luzzi V, Holtschlag V, Watson MA (2001) Expression profiling of ductal carcinoma *in situ* by laser capture microdissection and high-density oligonucleotide arrays. Am J Pathol 158: 2005–2010
20. Eberwine J, Yeh H, Miyashiro K, Cao Y, Nair S, Finnell R, Zettel M, Coleman P (1992) Analysis of gene expression in single live neurons. Proc Natl Acad Sci U S A 89:3010–3014
21. Phillips J, Eberwine JH (1996) Antisense RNA amplification: a linear amplification method for analyzing the mRNA population from single living cells. Methods 10:283–288
22. Upson JJ, Stoyanova R, Cooper HS et al (2004) Optimized procedures for microarray analysis of histological specimens processed by laser capture Microdissection. J Cell Physiol 201: 366–373
23. Livak KJ, Schmittgen TD (2001) Analysis of relative gene expression data using real-time quantitative PCR and the 2(-Delta Delta C(T)) Method. Methods 25:402–408
24. Benoy IH, Elst H, van der Auwera I et al (2004) Real-time RT-PCR correlates with immunocytochemistry for the detection of disseminated epithelial cells in bone marrow aspirates of patients with breast cancer. Br J Cancer 91:1813–1820

Chapter 4
Methodological Approach to Tissue Microarray for Studying the Normal and Cancerous Human Breast

Jose Russo, Fathima Sheriff, Thomas J. Pogash, Theresa Nguyen, Julia Santucci-Pereira, and Irma H. Russo

Keywords Tissue microarray (TMA) • Making the TMA • Quality control of the TMA • TMA of normal breast tissue

4.1 Introduction

Tissue microarray (TMA) technology is a tool that allows for the rapid evaluation of a large number of tissue samples on a single section of a microscope slide [1]. Cores of paraffin embedded tissue are removed from the original donor block and re-embedded into a recipient block in an ordered pattern of hundreds of cores. The cores can be from a variety of normal and abnormal types of tissue or stages of tumor and can be used to survey marker expression. This is more expedient than staining and scoring full sections from each donor block; it conserves the reagents used in the assay (e.g., immunohistochemistry or fluorescent in situ hybridization) [1–9], and extends the use of the original donor blocks. Many techniques for preparing the TMAs are in use with variable performance characteristics depending upon the tissue type, the coring procedure, and instrumentation (Fig. 4.1). The advantage of TMA is that it allows for examination of markers with clinical features, associations to be made between different markers and of markers with response to therapy or prognosis, and assessment of markers for the purpose of guiding treatment.

J. Russo, M.D. (✉) • F. Sheriff, M.D. • T.J. Pogash, B.S. • T. Nguyen, B.S.
J. Santucci-Pereira, Ph.D. • I.H. Russo, M.D.
Irma H Russo MD Breast Cancer Research Laboratory, Fox Chase Cancer Center, Cottman Avenue 333, Philadelphia, PA 19111, USA
e-mail: jose.russo@fccc.edu; julia.pereira@fccc.edu; Irma.Russo@fccc.edu

J. Russo and I.H. Russo, *Techniques and Methodological Approaches in Breast Cancer Research*, DOI 10.1007/978-1-4939-0718-2_4,

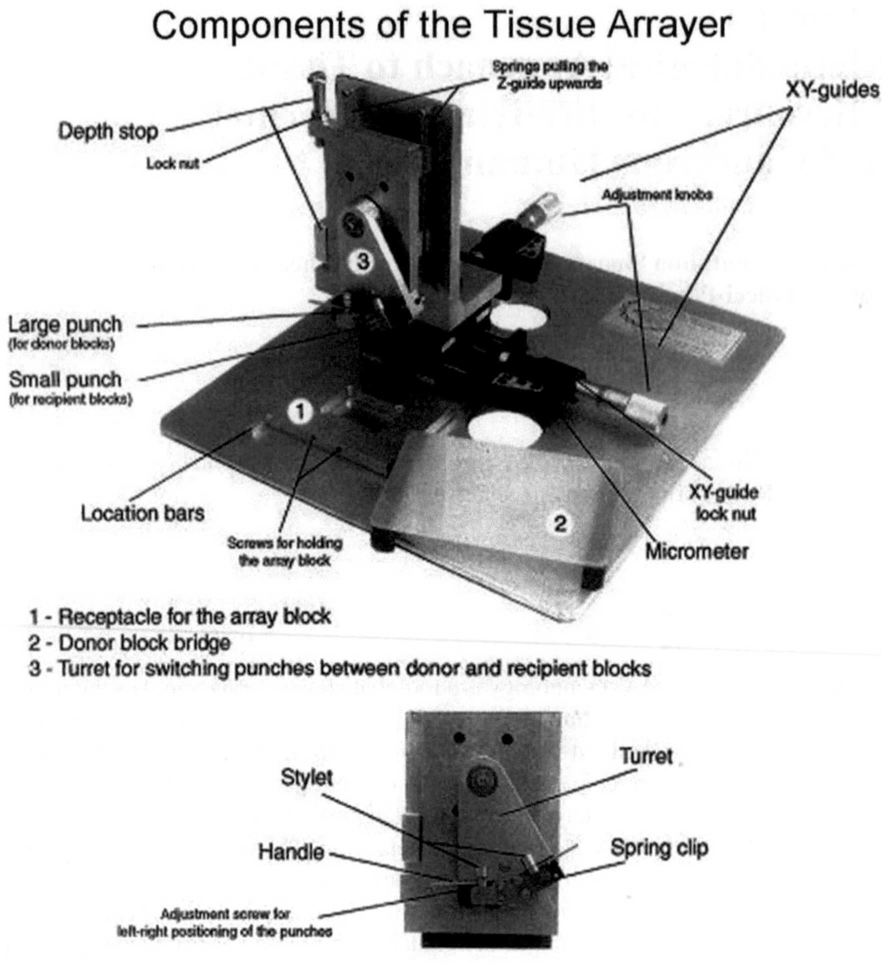

Fig. 4.1 Beecher Instruments, Inc. MTA1 Manual Tissue Arrayer Serial Number 2003527 (*Source*: http://www.diagnostictechnology.com.au/persistent/catalogue_files/products/BI-MTA1-p.pdf)

4.2 TMA Design and Processing

In designing the TMA, several considerations must be kept in mind. A basic guideline is depicted in Fig. 4.2. Among those specific to the TMA format are tumor and marker heterogeneity, the limited number of sections obtainable from an array block, the use of adequate control cores and the arrangement of cores on the array. Finally, it is important to consider the general sample size for marker studies. Another important step is the selection of the specimen donor blocks, which most of the time are derived from archival paraffin-embedded tissue. We have constructed TMA for breast cancer at different stages of the disease (I, II, and III) and also TMA

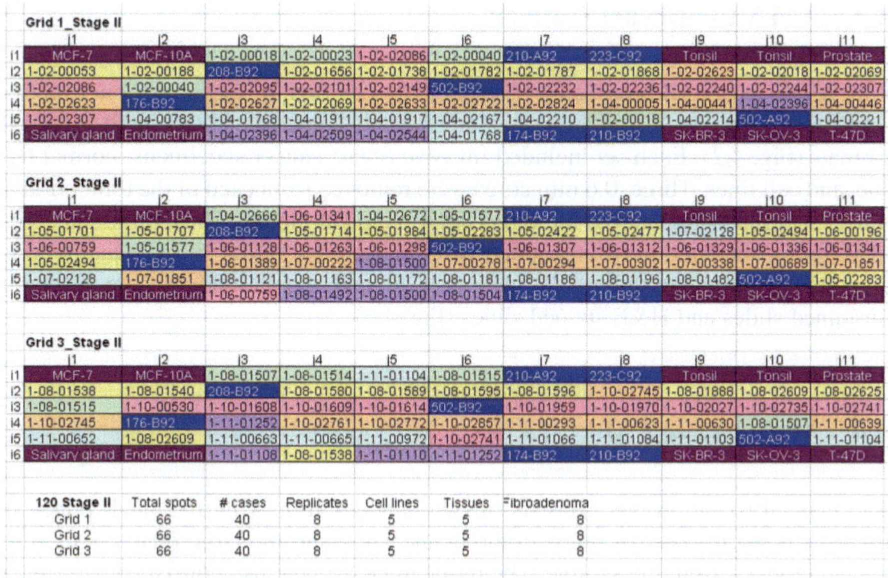

Grid 1_Stage II

	j1	j2	j3	j4	j5	j6	j7	j8	j9	j10	j11
i1	MCF-7	MCF-10A	1-02-00018	1-02-00023	1-02-02086	1-02-00040	210-A92	223-C92	Tonsil	Tonsil	Prostate
i2	1-02-00053	1-02-00188	208-B92	1-02-01656	1-02-01738	1-02-01782	1-02-01787	1-02-01868	1-02-02623	1-02-02018	1-02-02069
i3	1-02-02086	1-02-00040	1-02-02095	1-02-02101	1-02-02149	502-B92	1-02-02232	1-02-02236	1-02-02240	1-02-02244	1-02-02307
i4	1-02-02623	176-B92	1-02-02627	1-02-02069	1-02-02633	1-02-02722	1-02-02824	1-04-00005	1-04-00441	1-04-02396	1-04-00446
i5	1-02-02307	1-04-00783	1-04-01768	1-04-01911	1-04-01917	1-04-02167	1-04-02210	1-02-00018	1-04-02214	502-A92	1-04-02221
i6	Salivary gland	Endometrium	1-04-02396	1-04-02509	1-04-02544	1-04-01768	174-B92	210-B92	SK-BR-3	SK-OV-3	T-47D

Grid 2_Stage II

	j1	j2	j3	j4	j5	j6	j7	j8	j9	j10	j11
i1	MCF-7	MCF-10A	1-04-02666	1-06-01341	1-04-02672	1-05-01577	210-A92	223-C92	Tonsil	Tonsil	Prostate
i2	1-05-01701	1-05-01707	208-B92	1-05-01714	1-05-01984	1-05-02283	1-05-02422	1-05-02477	1-07-02128	1-05-02494	1-06-00196
i3	1-06-00759	1-05-01577	1-06-01128	1-06-01263	1-06-01298	502-B92	1-06-01307	1-06-01312	1-06-01329	1-06-01336	1-06-01341
i4	1-05-02494	176-B92	1-06-01389	1-07-00222	1-08-01500	1-07-00278	1-07-00294	1-07-00302	1-07-00338	1-07-00689	1-07-01851
i5	1-07-02128	1-07-01851	1-08-01121	1-08-01163	1-08-01172	1-08-01181	1-08-01186	1-08-01196	1-08-01482	502-A92	1-05-02283
i6	Salivary gland	Endometrium	1-06-00759	1-08-01492	1-08-01500	1-08-01504	174-B92	210-B92	SK-BR-3	SK-OV-3	T-47D

Grid 3_Stage II

	j1	j2	j3	j4	j5	j6	j7	j8	j9	j10	j11
i1	MCF-7	MCF-10A	1-08-01507	1-08-01514	1-11-01104	1-08-01515	210-A92	223-C92	Tonsil	Tonsil	Prostate
i2	1-08-01538	1-08-01540	208-B92	1-08-01580	1-08-01589	1-08-01595	1-08-01596	1-10-02745	1-08-01888	1-08-02609	1-08-02625
i3	1-08-01515	1-10-00530	1-10-01608	1-10-01609	1-10-01614	502-B92	1-10-01959	1-10-01970	1-10-02027	1-10-02735	1-10-02741
i4	1-10-02745	176-B92	1-11-01252	1-10-02761	1-10-02772	1-10-02857	1-11-00293	1-11-00623	1-11-00630	1-08-01507	1-11-00639
i5	1-11-00652	1-08-02609	1-11-00663	1-11-00665	1-11-00972	1-10-02741	1-11-01066	1-11-01084	1-11-01103	502-A92	1-11-01104
i6	Salivary gland	Endometrium	1-11-01108	1-08-01538	1-11-01110	1-11-01252	174-B92	210-B92	SK-BR-3	SK-OV-3	T-47D

120 Stage II	Total spots	# cases	Replicates	Cell lines	Tissues	Fibroadenoma
Grid 1	66	40	8	5	5	8
Grid 2	66	40	8	5	5	8
Grid 3	66	40	8	5	5	8

Fig. 4.2 Organization of TMA design and construction

for normal breast tissue. We will first address the breast cancer TMA followed by the TMA of normal breast tissue, which presents special considerations.

The TMA is constructed in a basic grid-like format, as depicted in Fig. 4.2, and each core must be taken from a morphologically representative region of the tissue to be examined. The size of the core will depend upon that region. For example, 0.6 mm cores are adequate for breast cancers, but cores of 1 or 2 mm in diameter are more suitable for normal breast tissue. In designing the array it is important to include controls, such as normal or benign lesions from same tissue type, cell lines known as positive or negative for some "standard" markers, and also different tissues such as spleen, tonsil, liver, and appendix in order to determine the specificity of an immunoreaction.

Sample size requirements may differ markedly depending on the goal of the study and the assistance of a biostatistician is highly encouraged in order to determine the statistical power needed. Whereas it is not in the scope of this book to discuss statistics, we want to offer one example in the event the number of specimens which must be tested in order to have sufficient power to detect an association between a marker and a clinical feature or breast cancer stage needs to be determined. If type I error (α) is used, it is probable that the statistical test will find a significant association when in reality there is none. On the other hand, if power ($1-\beta$) is used, it is probable that the statistical test will find a significant association where there truly is one. When Markers for Guiding Treatment are tested, a simple rule is that the number of events required for sufficient power to test an interaction effect may be as high as four times the number required to test a main effect of the same magnitude.

4.3 TMA for Breast Cancer

For stage I, II, and III breast cancer a set of 88, 120, and 40 specimen donor blocks respectively were selected from our archives to construct TMAs with identical formats (Fig. 4.2). Each set included invasive breast cancer specimens, normal tissue, and cell lines. Three, 0.6 mm cores were removed from each of the tumor donor blocks and placed in each of the three TMA recipient blocks. Thus, 40 tissue cores of 0.6 mm diameter were embedded in each recipient TMA block. Quality was assessed at both the individual section level and the individual core level on unstained slides and H&E stained slides (Fig. 4.3).

4.4 Making the TMA

The manual tissue arrayer consists of two punches (Fig. 4.1); the small punch is used to create an array of holes in the blank block and the large punch is used to collect biopsies from the sample. The tissue microarrayer also consists of a turret, a depth stop, an array bracket, a z-guide, and X and Y micrometers (Fig. 4.1). Fisher Brand Histosette tissue cassettes (Catalogue No. 15182701A) and Fisher Brand TissuePrep, superior embedding compound paraffin (Catalogue No. T565), are used to create both the donor and array block. In order to construct an array, a blank paraffin block, paraffin embedded tissue, and a manual tissue arrayer are required. The blank paraffin block serves as the recipient, or array, block that will receive core

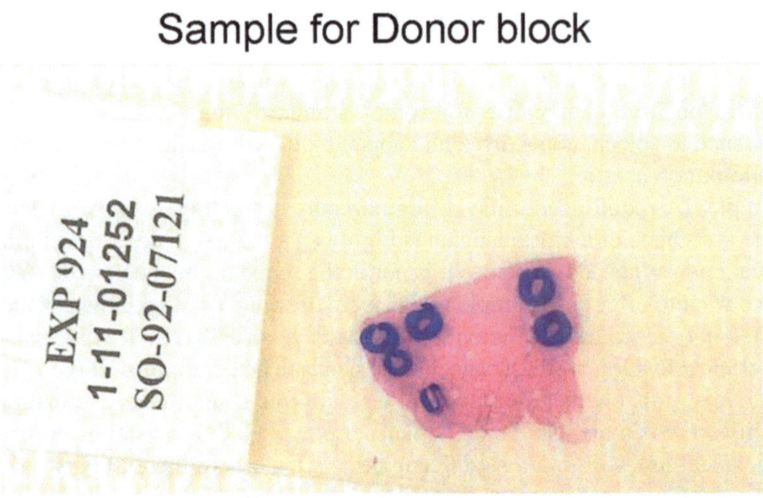

Fig. 4.3 Sample from a donor block; freshly stained slide with H&E contains selected areas marked with ink

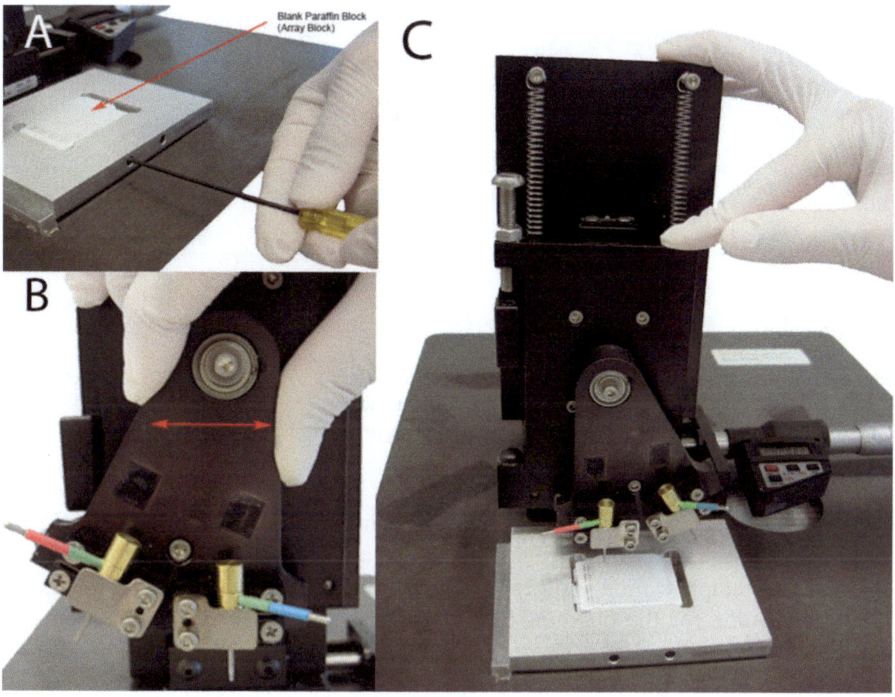

Fig. 4.4 (**a**) Array block secured into array bracket. (**b**) Turret is rotated to the left to use small punch (*red pin*), and turret is rotated to the right to use large punch (*blue pin*). (**c**) An array hole is created by the operator pressing down the z-guide until the depth stop reaches its set limit. The core will be filled with a core biopsy collected by the larger punch from a donor block

biopsies collected by the tissue microarrayer. This blank paraffin block is placed in the array bracket found on the tissue arrayer's stage and is secured by tightening the two bolts found on the front of the array bracket (Fig. 4.4a). This ensures that the recipient block does not move during the process. The depth stop is then adjusted to create a 1-mm deep array into the blank paraffin block (see Fig. 4.1 for location of depth stop). This hole, or array, is created by first rotating the turret (Fig. 4.4b, c) so the small punch will be in position to create an array in the array block. The operator will then push down on the z-guide until the depth stop reaches its set limit (Fig. 4.4d). Once the punch reaches its limit, the user then raises the punch from the array block and the hole created is exposed. This hole will be filled with the biopsy collected from the donor block.

The paraffin embedded sample desired to be studied serves as the donor block. This donor block must first be sectioned with a microtome and a fresh hematoxylin and eosin (H&E) slide of the section must be prepared to serve as a template for the tissue microarrayer (Fig. 4.5a). The template slide is viewed under a microscope and several areas of interest can be chosen for collection by the tissue microarrayer

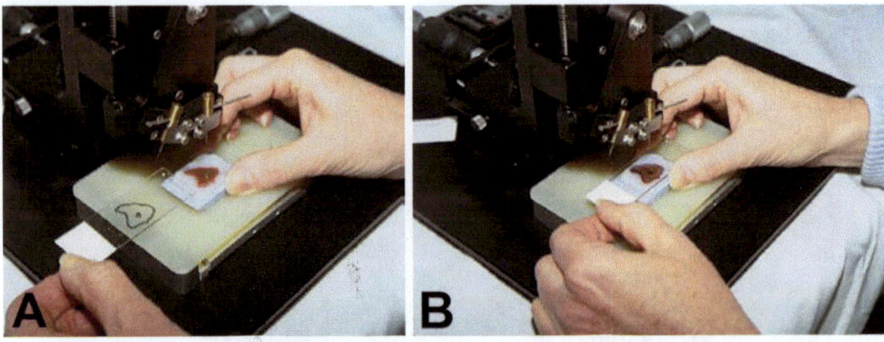

Fig. 4.5 (**a**) It is important to align the fresh H&E stained slide with the donor block and (**b**) large punch to ensure the correct biopsy is collected from the desired location (*Source*: http://www.diagnostictechnology.com.au/persistent/catalogue_files/products/BI-MTA1-p.pdf)

(Fig. 4.3). The areas of interest are selected using an ink marker on the template slide before the tissue microarrayer is used to collect a biopsy. After the area is selected, the template slide is placed over the donor block so that the template can serve as a guide for the operator to collect the correct area of interest from the embedded tissue (Fig. 4.5a). The donor block with the template slide is placed on the donor block bridge and is slid into position so that the large punch can collect the desired area (Fig. 4.5b). Once the area of interest is lined up with the tissue microarrayer, the operator removes the template slide and presses down on the z-guide in order to draw a biopsy from the area of interest using the large punch located on the turret. Since the donor block rests on the bridge it must be secured by the hand of the operator. In addition, the donor block bridge raises the height of the depth stop so the operator must use their best judgment on how deep to press down on the z-guide in order to collect a good core. The donor block bridge is then removed and the biopsy is placed into the array block manually by the operator. To do this, the operator lowers the turret down towards the array block until the top of the large punch is inside the recipient hole created earlier in the process. Next, the operator presses the top of the large punch so that the biopsy obtained is forced out of the punch. It is important to make sure that the biopsy is securely placed in the array block. If the biopsy is not secure then the operator must press down on the biopsy with a slide until it fully enters the recipient array. After the biopsy has been placed into the array block the X and Y micrometers can be used to move the position of the next biopsy (Fig. 4.6a, b). This process is repeated until all the samples are embedded into the array block.

It is important to emphasize the importance of preparing a fresh Hematoxylin and eosin (H&E) stained histologic section ("guide slide") (Fig. 4.5a) from each donor block, taking care to prevent distortion of the section while mounting on the slide. Each "guide slide" is examined by a pathologist who uses an indelible marker to draw tightly circumscribed outlines around the areas corresponding to the desired target tissue elements [invasive tumor, DCIS, and normal]. More target tissue than

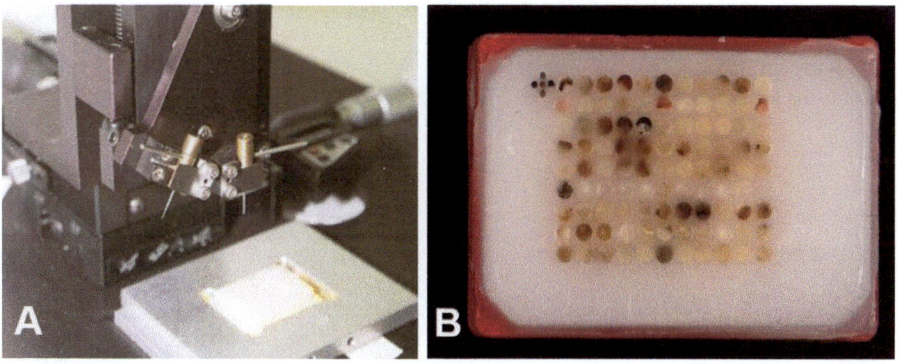

Fig. 4.6 (**a**) The recipient paraffin block (*Source*: http://www.diagnostictechnology.com.au/persistent/catalogue_files/products/BI-MTA1-p.pdf). (**b**) A finished array block (*Source*: http://en.wikipedia.org/wiki/Tissue_microarray)

needed is circled to complete the arrays. These markings are used to direct the technologist performing the coring procedure to withdraw the required cores of specific histology. After construction of the complete array, a glass slide is applied to the surface of the recipient block, and gentle pressure is applied to uniformly fix the donor cores in position within the block. The TMA block is then placed face down on a glass slide and incubated over night in a 37 °C incubator. The next day, the TMA block and glass slide are incubated in a 60 °C oven for 10 min. The block is allowed to slowly come to room temperature, at which time the glass slide is removed and the block is stored at ambient temperature until histologic sectioning (Fig. 4.7).

4.4.1 Histological Processing of the TMA Blocks

After the TMA blocks are constructed, they can be stored or immediately used. In general, we cut batches of ten sections and number them from 1 to 10 and continue to number them consecutively until the block is fully used. Each section is mounted on Fisher brand Superfrost/PLUS pre-cleaned slides (Fisher Scientific, Pittsburgh, PA). Basically, TMA blocks are cooled face down on an ice tray, and placed on a microtome (Model 2135, Leica Microsystems, Bannockburn, IL). The surface of the block is leveled to the microtome blade by eye inspection or by using a collimator (Richard-Allan Scientific). Histologic sections are cut at 4 μm thickness. Ribbons of histologic sections are removed from the microtome using blunt forceps and floated in a water bath prior to capture on the surface of charged glass slides. Sections are stored refrigerated at 4 °C in histologic boxes which are then sealed in a plastic bag with Nitrogen. The slides containing the cut sections are labeled sequentially using the convention (array block)—(section number). For example, if three blocks are generated (blocks A, B, and C), the fifth section from array block B is labeled B-5.

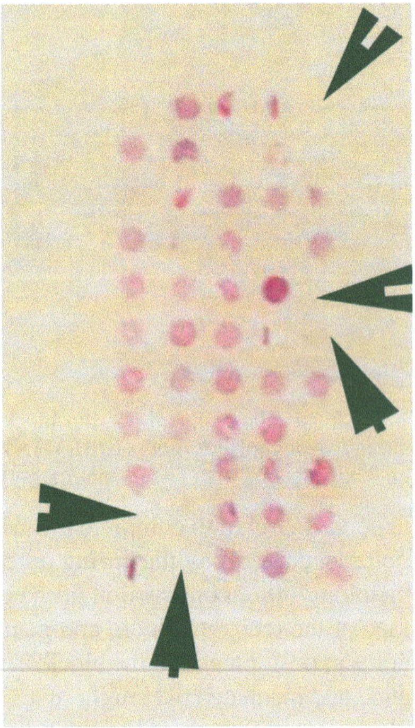

Fig. 4.7 Histological section of the A TMA block. Stained with H&E, ×2. The *arrows* point to the missing cores

4.4.2 Quality Control of the TMA

Every ten sections cut from each block are evaluated for overall section quality by a pathologist. The quality of each tissue core section is scored by microscopic examination as "Acceptable," "Damaged," or "Missing" from each section (Fig. 4.7). Acceptable cores contain representative tissue of the designated type. Damaged cores contain tears or wrinkles in the tissue section usually from the sectioning procedure. Missing cores are assigned when either no tissue is present at that position of the TMA or the designated tissue type is absent. This usually results from drop off of the tissue section from the slide, exhaustion of the core depth on successive sections due to the shallow depth of breast tissue in the original donor block, or the variable extent of normal and DCIS breast tissue in the original specimen blocks. However, TMA sections with poor core quality can still produce high proportions of cores that are suitable for immunohistochemical assays. Many of the sections could appear to be "dirty" or have defects such as distortions, wrinkles, or tears. The sources of these aberrations can include structural defects in the recipient block, contamination of the mounting media, and problems introduced by the use of tape for sectioning. Although many of the sections may not "look pretty," the fraction of acceptable cores that could be evaluated successfully in these sections is often high.

4.5 TMA of Normal Breast Tissue

The most important concept to bear in mind when TMA of normal breast tissue is constructed is having a strict quality of normality, as has been summarized in Table 2.2. Breast tissue that contains hyperplastic epithelium in the ductal structures, lobular hyperplasia, cystic formation, apocrine metaplasia, or other metaplastic changes, or another morphological indication of derangement of the normal structure as described above must be deleted from the study of normal breast tissue. In constructing a TMA of the normal breast, the different lobular structures and their size should also be accounted for, as depicted in Figs. 4.8 and 4.9.

A Lobule Type 1 measures approximately 500 μm (Fig. 4.10), therefore, a core measuring 0.6 cm in diameter will be able to incorporate one of these lobular structures and will allow for approximately 80 good 5 μm sections (Fig. 4.11). For Lobule Type 2, which is around 2 mm in diameter (Fig. 4.12), or Lobule Type 3, which is around 3 mm in diameter (Fig. 4.13), the number of possible sections containing these structures is around 300 at 5 μm thickness (Fig. 4.14). The same principle applies to Lobule Type 4 (Fig. 4.15). The amount of stroma also varies depending of the reproductive cycle of the human breast (Fig. 4.16). In the selection process for adequate lobular structures is very important to keep in mind the criteria of normality described in Table 2.2. For example, in the section depicted in Figs. 4.17, 4.18, and 4.19, only the circled areas are adequate for TMA. In the section depicted in Fig. 4.20 only encircled areas B and C are adequate for constructing a TMA of normal breast tissue, while in part A Lobule Type 1 contains fribrous intralobular stroma or mazoplasia that disqualify it according to the criteria listed in Table 2.2.

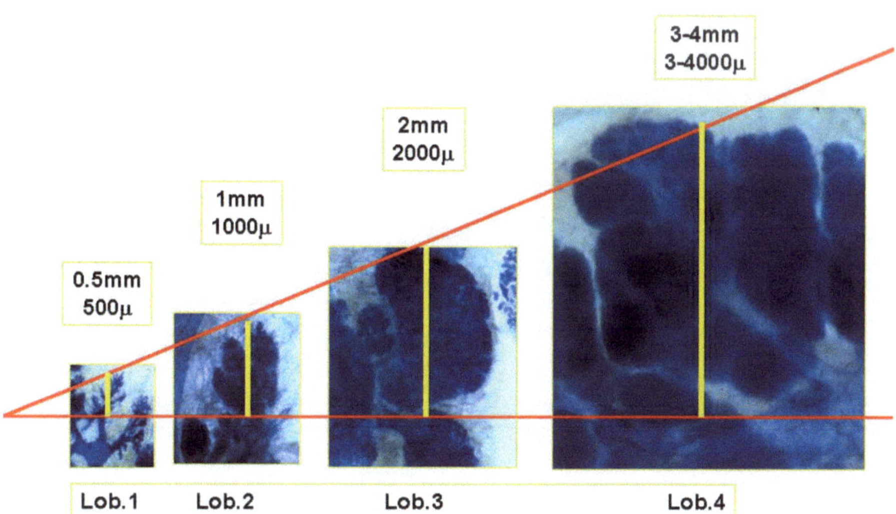

Fig. 4.8 Whole mount of the lobular structures of the human breast

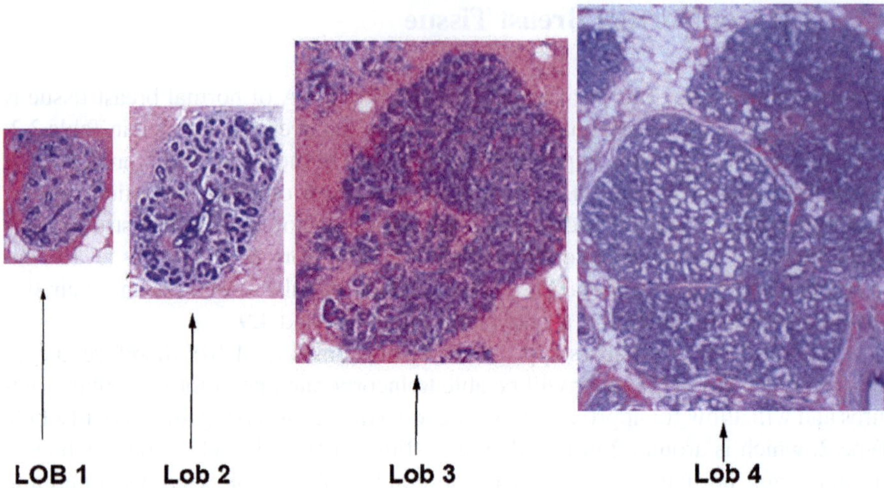

Fig. 4.9 Histological sections of the lobular structures of the human breast. Stained with H&E, ×2

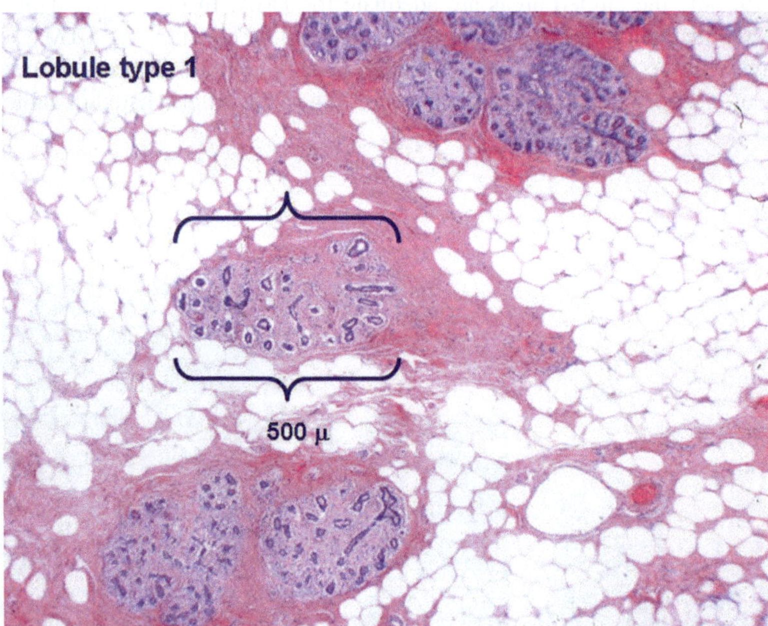

Fig. 4.10 Histological sections of the lobular structure Type 1 measuring approximately 500 μm in its larger diameter

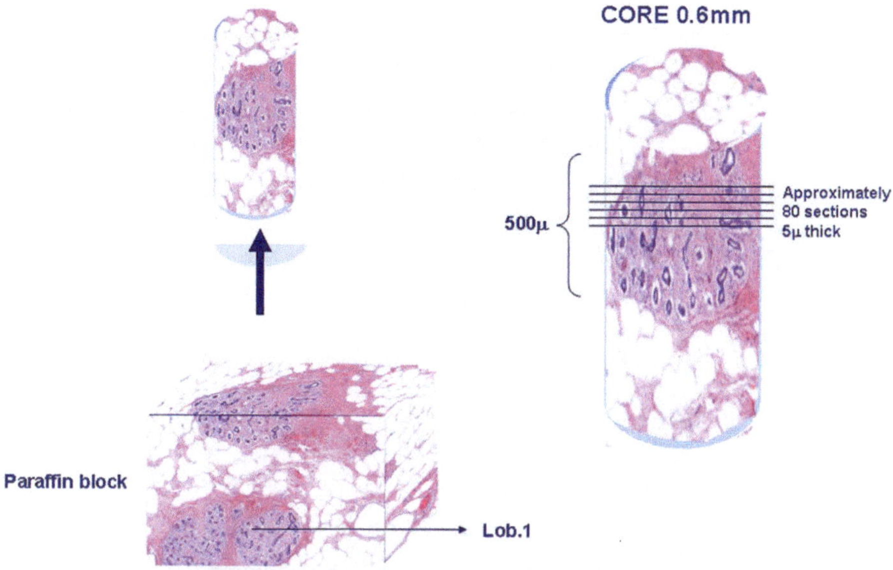

Fig. 4.11 A histological section of a core containing a lob 1. It is calculated that based on the size of this lobule approximately 80 sections of 5 µm each can be obtained

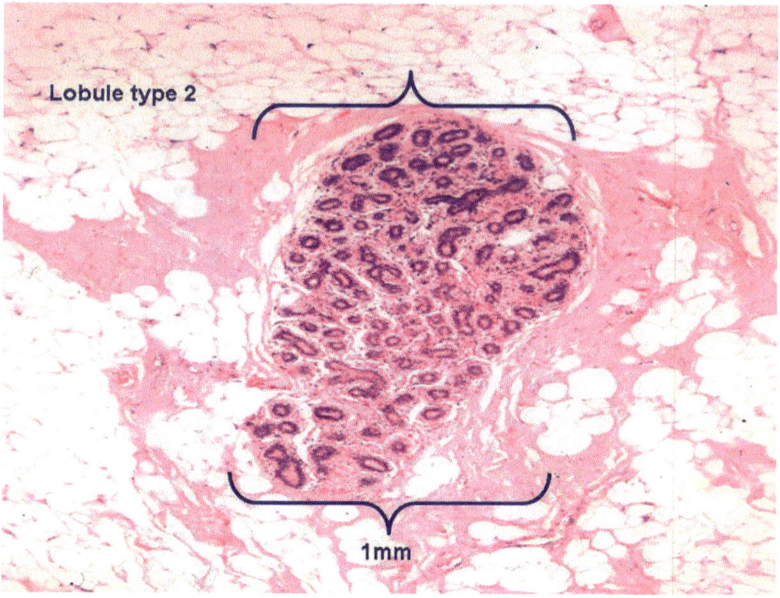

Fig. 4.12 Histological sections of the lobular structure type 2 measuring approximately 1 mm in its larger diameter

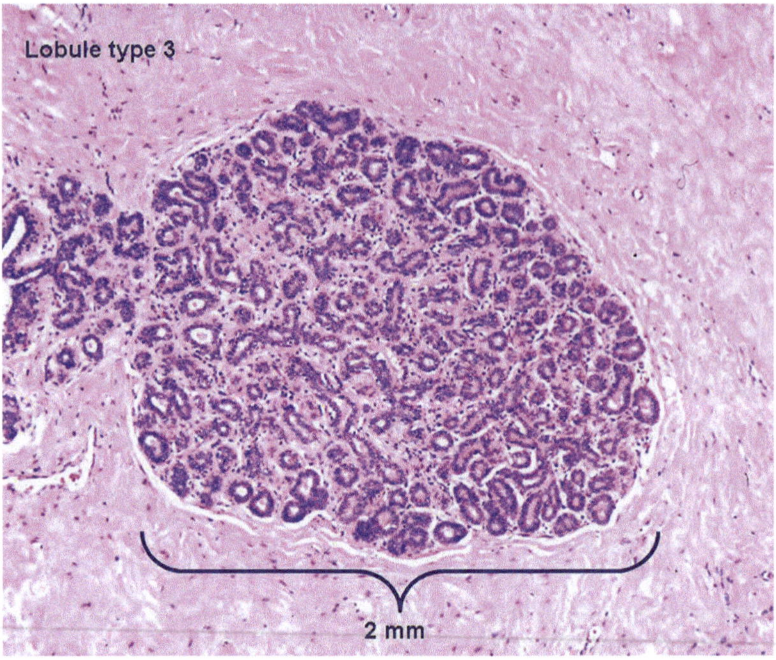

Fig. 4.13 Histological sections of the lobular structure type 3 measuring approximately 2 mm in its larger diameter

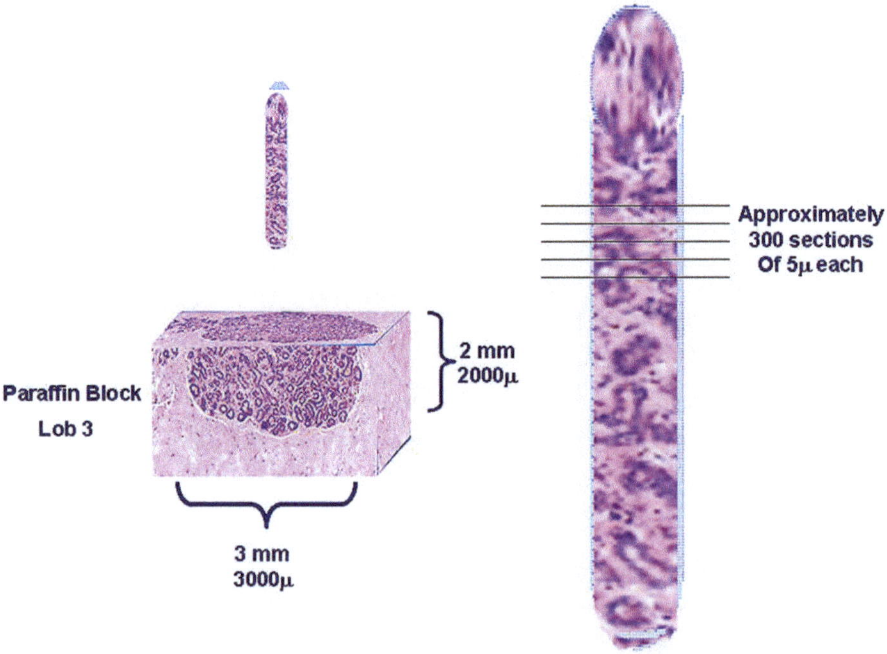

Fig. 4.14 A histologic section of a core containing a lob 3. It is calculated that based on the size of this lobule approximately 300 sections of 5 μm each can be obtained

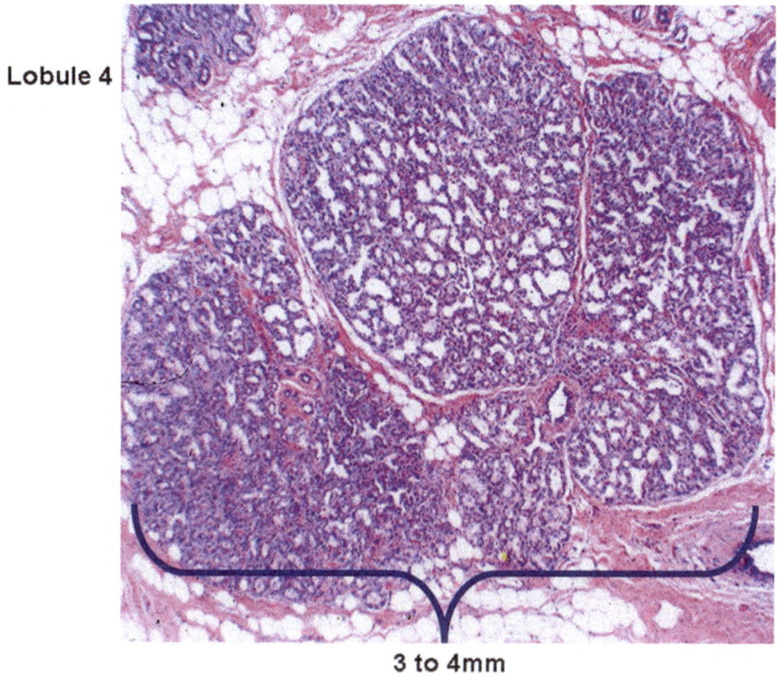

Fig. 4.15 Histological sections of the lobular structure type 4 measuring approximately 3–4 mm in its larger diameter

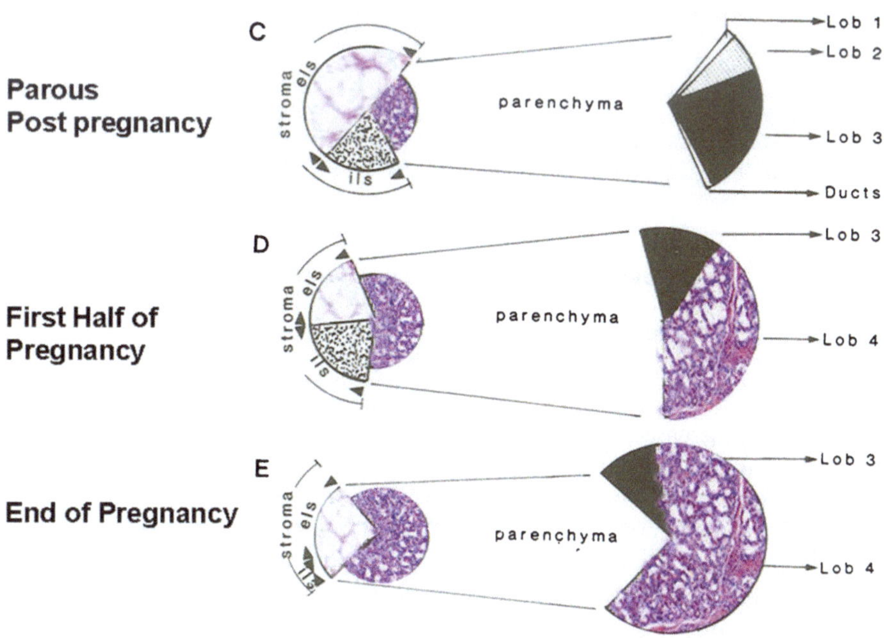

Fig. 4.16 Schematic representation of the percentage of stroma of the breast during pregnancy

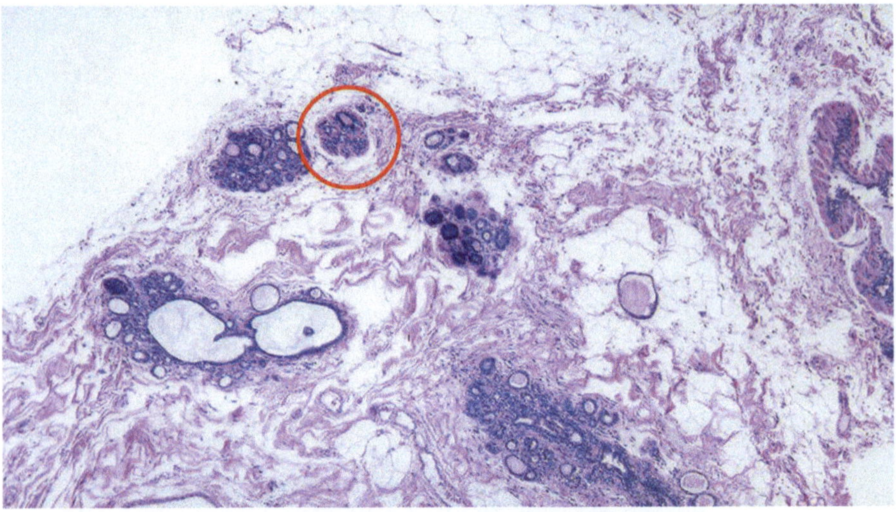

Fig. 4.17 *Circled area* showing a normal Lobule Type 1 that can be used for TMA. The other areas are not suitable for a normal breast TMA

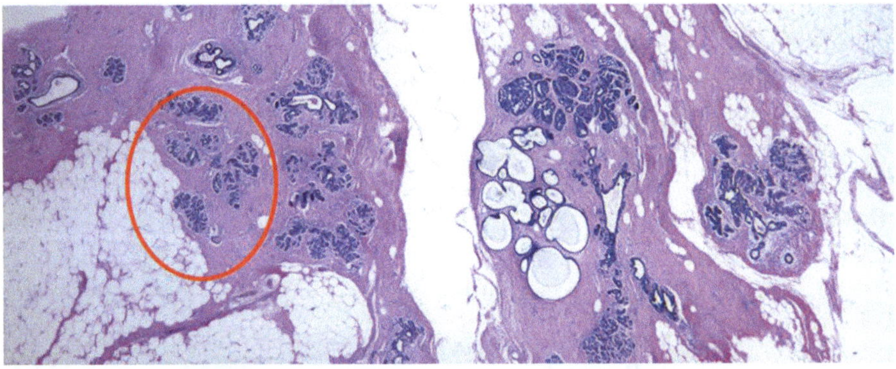

Fig. 4.18 *Circled area* showing a normal lob 1 that can be used for TMA. The other areas are not suitable for a normal breast TMA

In our experience, the limited depth of tissue in donor blocks of normal breast tissue leads to increased rates of individual core loss as sectioning progresses from top to bottom throughout the blocks. In the TMA of normal breast tissue, correct selection during coring (capture rate) is generally good through at least section 20, with only modest improvement realized for 1.0 mm compared to 0.6 mm cores. Correct capture rate for normal tissue cores is also not very good; however, 1.0 mm cores may produce a slight improvement in normal tissue capture rates.

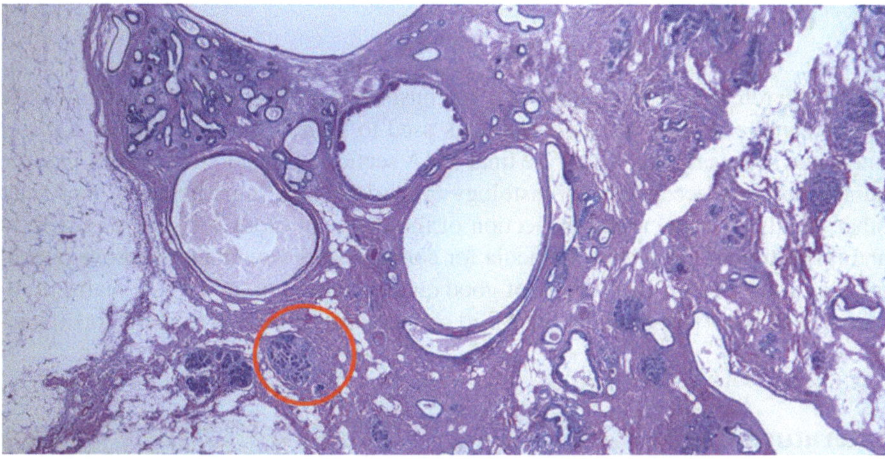

Fig. 4.19 *Circled area* showing a normal lob 1 that can be used for TMA. The other areas are not suitable for a normal breast TMA

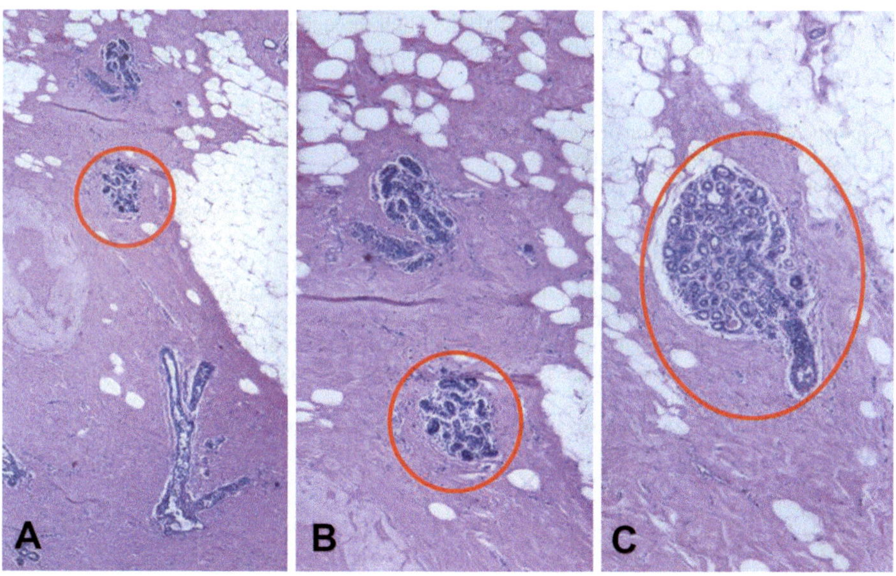

Fig. 4.20 *Encircled areas* shown in (**b**) and (**c**) are adequate for constructing a TMA of normal breast tissue, instead the part (**a**) the lobule type 1 contains fibrous intralobular stroma or mazoplasia that disqualify according to the criteria listed in Table 2.2

4.6 Conclusions

The above referenced techniques used to construct the TMA blocks result in blocks of comparable quality, but the techniques used to section those blocks has a major impact on the overall quality of the final TMA sections obtained from the blocks, the number of cores acceptable for histology or marker study in particular. Attention to other details such as careful inspection of the recipient block for structural defects and inspection of the mounting media for contaminants would also improve overall quality of the TMAs. The capture of good quality normal tissue cores remains problematic and is not substantially improved by increasing core size from 0.6 to 1.0 mm.

Literature Cited

1. Kononen J, Bubendorf L, Kallioniemi A et al (1998) Tissue microarrays for high throughput molecular profiling of tumor specimens. Nat Med 4:844–847
2. Bubendorf L, Kolmer M, Kononen J et al (1999) Hormone therapy failure in human prostate cancer: analysis by complementary DNA and tissue microarrays. J Natl Cancer Inst 91: 1758–1764
3. Moch H, Schraml P, Bubendorf L et al (1999) High-throughput tissue microarray analysis to evaluate genes uncovered by cDNA micrarray screening in renal cell carcinoma. Am J Pathol 154:981–986
4. Mucci NR, Akdas G, Manely S, Rubin MA (1999) Neuroendocrine expression in metastatic prostate cancer: evaluation of high-throughput tissue microarrays to detect heterogeneous protein expression. Hum Pathol 31:406–414
5. Perrone EE, Theoharis C, Mucci NR et al (2000) Tissue microarray assessment of prostate cancer tumor proliferation in African-American and white men. J Natl Cancer Inst 92:937–939
6. Schraml P, Kononen J, Bubendorf L et al (1999) Tissue microarrays for gene amplification surveys in many different tumor types. Clin Cancer Res 5:1966–1975
7. Lee CM, Lee RJ, Hammond E et al (2004) Expression of HER2neu (c-erbB-2) and epidermal growth factor receptor in cervical cancer: prognostic correlation with clinical characteristics, and comparison of manual and automated imaging analysis. Gynecol Oncol 93:209–214
8. Jensen T, Hammond E (2001) The tissue microarray—a technical guide for histologists. J Histotechnol 24(4):283–287
9. Jensen T (2003) Tissue microarray: advanced techniques. J Histotechnol 26(2):101–104

Chapter 5
In Vitro Techniques for Studying the Normal Breast and the Use of Cell Lines in Breast Cancer Research

Jose Russo, Thomas J. Pogash, Janine Gomes, Justin Newton, Ricardo Lopez de Cicco, Yanrong Su, and Irma H. Russo

Keywords Cell culture of cancer cells • Tissue culture collection • MCF10 cells • MCF7 cells • Detection of the IC50 • Colony formation in agar methocel • Ductulogenesis • Invasion assay • Wound healing assay • Tumorigenesis • Metastatic assays • Basal cell lines • Luminal cell lines

5.1 Introduction of History of Cell Culture of Cancer Cells

In the early history of cell culture, researchers such as Alexis Carrel were faced with many obstacles, the most imperative being contamination and an incomplete understanding of the nutrients cells that require to survive. George Gey worked to establish the study of human cells in culture by first perfecting grown medium and sterile technique [1]. The most significant contribution of Gey's lab to the history of cell culture came in February of 1951, when a biopsy was taken from Henrietta Lacks, an American tobacco farmer suffering from cervical cancer, and from which the HeLa cell line was originated. Gey's lab had been trying to establish tissue cultures of both normal and malignant, or otherwise diseased, cells for use in their research [2]. Despite initial concerns that the cells would eventually die, as had the cells of their previous efforts, they found that these HeLa cells grew extremely fast, filling the space they occupied, doubling in number every 24 and accumulating by the millions [1]. The HeLa cell line grew robustly, becoming the first human cancer cell line immortalized in tissue culture [1, 2]. The HeLa, Hela, or hela cell line, is not only the oldest human

J. Russo, M.D. (✉) • T.J. Pogash, B.S. • J. Gomes • J. Newton • R.L. de Cicco, Ph.D.
Y. Su, Ph.D. • I.H. Russo, M.D.
Irma H Russo MD Breast Cancer Research Laboratory, Fox Chase Cancer Center,
Cottman Avenue 333, Philadelphia, PA 19111, USA
e-mail: jose.russo@fccc.edu

J. Russo and I.H. Russo, *Techniques and Methodological Approaches in Breast Cancer Research*, DOI 10.1007/978-1-4939-0718-2_5,

cell line in culture, but also the most common [2]. Apart from the intention of using these cells to eradicate cancer, the cell line's characteristics made it essential in numerous other research efforts. The HeLa cell line was first notably used in the 1950s by Jonas Salk to test the first polio vaccine. Since then, the cell line has been used in a wide variety of research such as, cancer, AIDS, TB, gene mapping, cloning, in vitro fertilization, nanotechnology, the study of toxic agents, and more [1, 2].

HeLa cells were discovered to use an enzyme called telomerase to repair their DNA, allowing them to continue functioning when normal cells would have died [1]. This quality of abnormal cell proliferation is well known in cancer cells. Clinical trials have shown that anticancer drugs work by making the body's immune system fight the cells expressing telomerase. This demonstrates the importance in the discovery of using this cell line in cancer research and continues to allow for the characterization of cancer and the development of treatment options.

The advent of the HeLa cell line raised concerns about an important topic—immortalization. Cells from a regular multicellular organism will typically not have unlimited proliferation; however, some cells break apart from this typical cell characteristic and obtain the capacity to divide indefinitely. These cells are known as immortalized cells. There are various factors that contribute to this characteristic. One factor is the presence of the enzyme telomerase, as previously discussed, which rebuilds the telomeres found at the end of a cell's DNA, whose wearing away causes the aging of cells [1]. In normal cells, this process of telomeres wearing away places a limitation on the proliferation of cells, described by the Hayflick limit [3]. Another factor involves the intricate interplay of calcium within the cell [4]. There is also the consideration that numerous possible factors are capable of damaging DNA and promoting mutations responsible for cell immortalization. Common factors include radiation, chemical carcinogens, or certain oncogenic viruses [4]. These factors induce immortalization via the modification of specific genes, like the p53 gene [5]. There are numerous well-defined mechanisms as well as many that are not yet fully understood.

Immortalized cells, not limited by a defined amount of proliferation, allow for cell culture to be more proficient and research to be more extensive. Thus, techniques have been developed to further explore and promote immortalization of cell lines. However, when it comes to assays studying the effect of compounds on non-invasive, normal-like cells undergoing a transformation process, new concerns arise in that it becomes difficult to determine which compound induces the transformation. For instance, in transforming normal Human Breast Epithelial Cells (HBEC) with the carcinogen benzo[a]pyrene, the cells become immortalized but are described as not being malignantly transformed since they do not form tumors [6]. Nonetheless, they have characteristics of tumor-derived HBECs [6]. Immortalized cells such as these can no longer be studied under the classification of normal epithelial cells, which is why spontaneously immortalized cells, having not been exposed to any carcinogens, are important in research efforts. Efforts in cancer research make use of cell lines characterized as being "normal" as well as cancerous cells, such as the HeLa cell line. Understanding normal epithelial cells is as essential as developing tools to fight transformed cells that lead to the development of cancer. Despite initial controversies, studies rely immensely on the use of immortalized human cells, such as the HeLa cell line and others.

5.2 The Role of Tissue Culture Collection

The main concern with the HeLa cell line is its aggressive growth rendering it able to overtake less robust cell lines, leading it to being a problematic contaminate [1, 2]. This is why private corporations, such as the American Type Culture Collection (ATCC), are so important. In the most basic sense, tissue culture collection focuses on quality control of cell lines. In its mission statement the ATCC describes itself as being a biological resource center that "authenticates microorganisms and cell lines and manages logistics of long-term preservation and distribution of cultures for the scientific community" [7]. The corporation "supports the cultures it acquires and authenticates them with expert technical support, intellectual property management, and characterization data" [7]. ATCC works to acquire, authenticate, preserve, develop, standardize, and distribute biological materials and information, for the advancement and application of scientific knowledge. Corporations such as these provide a standardized system of cells in order to ensure that research within the scientific community can more accurately be developed and interpreted.

Tissue culture collection is done on a national as well as global scale. It is important to remember that successful scientific research lies in the endorsement of collaborations. It is necessary to incorporate multiple facilities into studies requiring a broad range of experimental techniques due to different institutions having certain higher levels of expertise in specific fields. In order to ensure that data is comparable, there needs to be an established base. Tissue culture collection enables established cell lines and protocols to be available to anyone looking to purchase them, something essential for reproducibility of experiments. It is, therefore, crucial for each research institution to have access to the same materials and methods. Since science is defined by its ability to be either proven or disproven, there needs to be a universal standard that allows multiple researchers to conduct the same studies under all the same conditions. Tissue culture collection allows for these necessary aspects of research to be achieved successfully by defining basic standards and distributing the necessary equipment needed.

5.3 The MCF10 Cells as a Prototype of Normal Cells

The MCF-10 cell line was derived from the subcutaneous mastectomy of a 36-year-old woman, who neither had cancer nor a family history of cancer [4, 8, 9]. These cells exhibited a spontaneous immortalized character after extended cultivation in low calcium media [8, 9]. From the MCF-10 cell line, the MCF-10A cell line was sub-derived from the adherent cells in the media, while the MCF-10F cell line was sub-derived from the suspended cells of the same population [8, 9]. These are non-tumorigenic, normal breast epithelial cells characterized by the expression of breast-specific antigens as detected by positive reaction with MFA-Breast and MC-5 monoclonal antibodies, the presence of epithelial sialomucins, cytokeratins, and

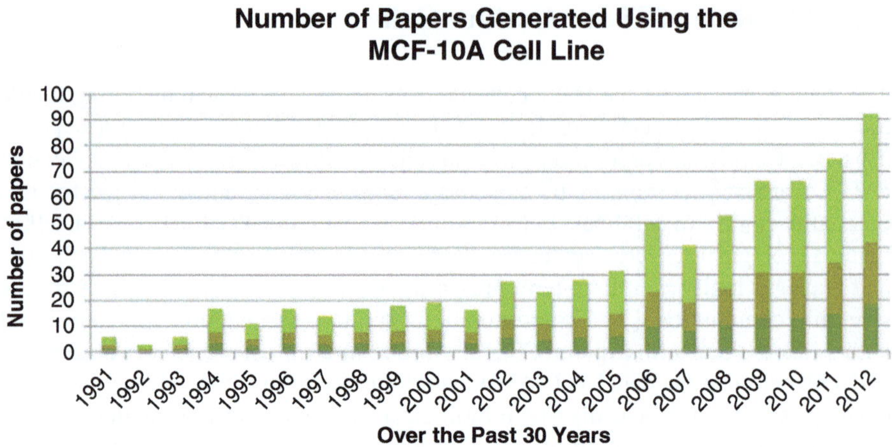

Fig. 5.1 Number of publications generated since 1991 utilizing the MCF-10A cell line. This data was compiled from the information obtained utilizing an analysis of a Web of Knowledge search for the term "MCF-10A" in the topic of published papers. The total number of papers published in this time frame is around 700

milk-fat globule antigen, a response to insulin, glucocorticoids, cholera enterotoxin, and epidermal growth factor (EGF), characteristics of luminal ductal cells but not of myoepithelial cells via electron microscopy analysis, three dimensional growth in collagen, and dome formation in confluent cultures [8, 9].

Characterization of these cell lines can be further understood by comparison to another cell line under the classification of "normal human breast epithelial cells," the 184B5 cell line. The 184B5 cell line was derived from mammary tissue obtained from a normal reduction mammoplasty [6]. The tissue was exposed to benzo(a) pyrene and transformed into an immortal cell line [6, 10]. The 184B5 cell line is described as immortal, but not malignant [6, 10]. Further characterization of the cell line yielded results that demonstrated that it is "mostly normal diploid with only a small percentage of tetraploid cells" [10]. It was further discussed that a combination of deficiencies and duplications characterized the 184B5 cell line. This data is discussed in regard to mutations that took place in the sequence of change from primary cells to immortalized cells [10]. This exemplifies the idea that transformed cells, though they may not be malignant, cannot be entirely characterized as being "normal human like cells." This is why the use of a spontaneously immortalized primary cell line is imperative for the advancement of breast cancer research. MCF-10 cell lines are therefore more advantageous in comparison to the 184B5 cell line in the sense that they have not been transformed with any chemicals, like benzo(a)pyrene, and are both phenotypically and genotypically characterized as normal HBECs.

Both of the MCF-10 cell lines have been used in many novel discoveries in the field of breast cancer. The scope of their use can be graphically represented in Figs. 5.1 and 5.2. In order to get a further grasp for the impact these papers have had, considering the amount of times each paper has been cited allows for a better understanding of how important they have been for other studies (Figs. 5.3, 5.4, and 5.5).

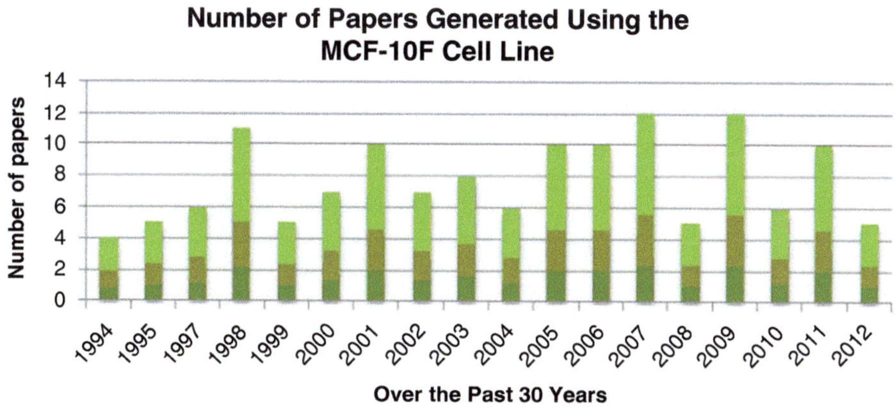

Fig. 5.2 Number of publications generated since 1994 utilizing the MCF-10F cell line. This data was compiled from the information obtained utilizing an analysis of a Web of Knowledge search for the term "MCF-10F" in the topic of published papers. The total number of papers published in this time frame is around 140

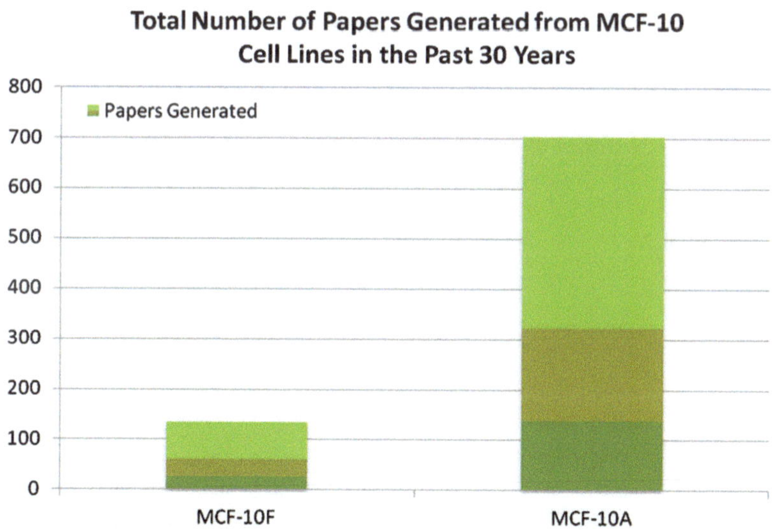

Fig. 5.3 This histogram shows the number of publications generated over the past 30 years utilizing the MCF-10A cell line and the total number of papers published in this time frame is around 700. For MCF10F cells the total number of papers published in this time frame is around 140

Though the MCF-10A cell line is more commonly used, the MCF-10F cell line is classified as being the more normal-like [4, 8, 9] (Figs. 5.6 and 5.7). The MCF-10F cell line has the characteristics of normal HBECs, as shown by its lack of tumorigenicity in Severe Combined Immunodeficiency, SCID mice, its ability to form three-dimensional duct-like structures in collagen, hormone and growth factor

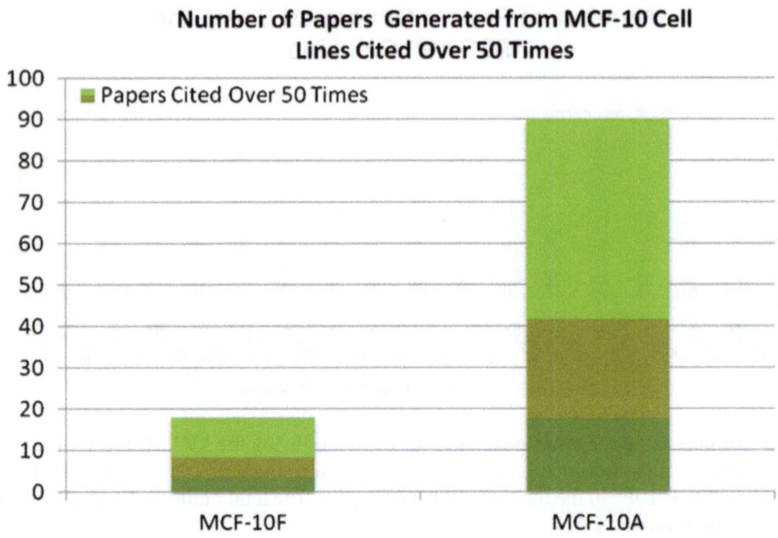

Fig. 5.4 This is a graphical representation of the amount of publications that have been cited over 50 times generated from the use of the MCF-10 cell lines, as reported by a Web of Knowledge search for the term "MCF-10F" and "MCF-10A" in the topic of published papers and noting the times they have been reported to be cited

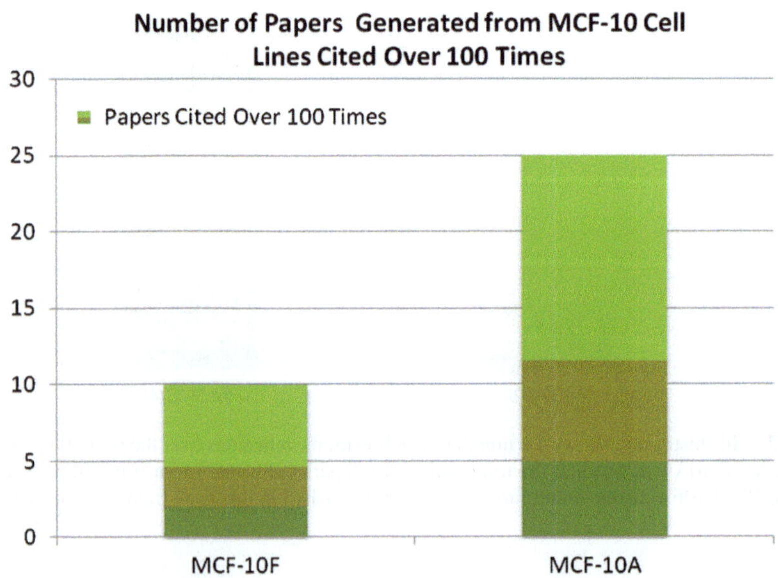

Fig. 5.5 This is a graphical representation of the amount of publications that have been cited over 100 times generated from the use of the MCF-10 cell lines, as reported by a Web of Knowledge search for the term "MCF-10F" and "MCF-10A" in the topic of published papers and noting the times they have been reported to be cited

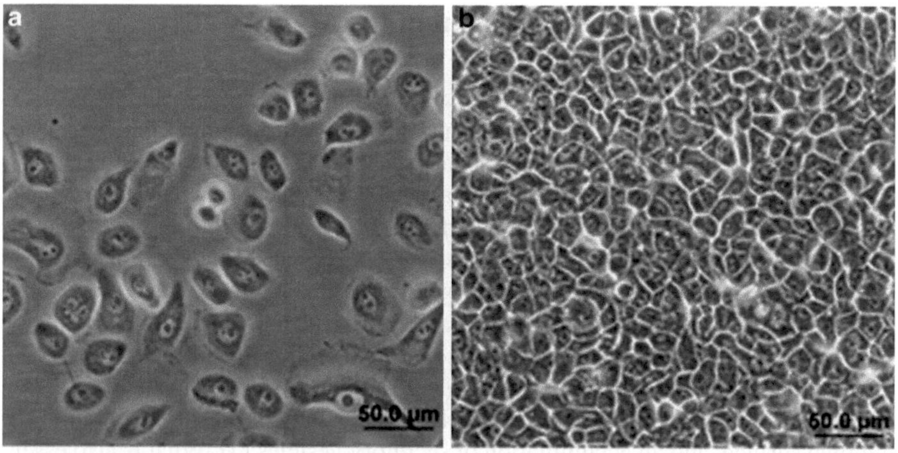

Fig. 5.6 (**a**) Logarithmic growth of MCF-10A cells (**b**) confluent growth of MCF-10A cells. MCF-10A cells were maintained in Low Calcium cell culture medium [8, 9]. Images were acquired with Olympus IMT-2 microscope, Olympus LWDCDPlan 20PL 20X objective, and Olympus DP71 camera

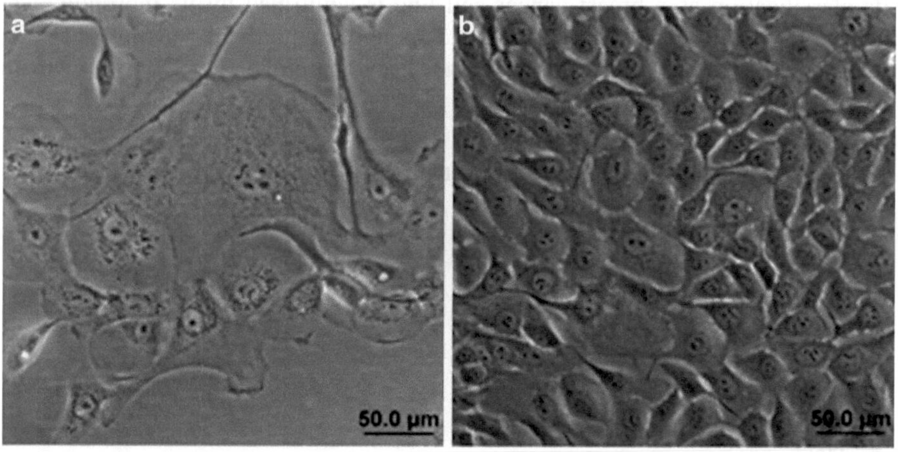

Fig. 5.7 (**a**) Logarithmic growth of MCF-10F cells (**b**) confluent growth of MCF-10F cells. MCF-10F cells were maintained in High Calcium cell culture medium [4]. Images were acquired with Olympus IMT-2 microscope, Olympus LWDCDPlan 20PL 20X objective, and Olympus DP71 camera

dependency for in vitro growth, the lack of anchorage-independent growth, and dome formation in confluent cultures [4, 8, 9]. These cells did not express phenotypes indicative of neoplastic transformation, such that they did not form colonies in agar-methocel [4, 8, 9]. Note that these phenotypic assays, as described in other sections of this chapter, exemplify that these cells express the genetic, cytogenetic, ultrastructural, and phenotypic characteristics of normal human breast epithelium.

5.4 The Historical Development of the MCF7 Cells

An important cell line specific to breast cancer is the MCF-7 cell line, which was derived from a pleural effusion from a patient with metastatic mammary carcinoma, and has been verified as human in both its biochemical characteristics and cytological characteristics [4]. It contains both human alpha-lactalbumin and estrogen receptor protein, two components that support its origin in human breast tissue [4]. It is thus a valuable resource in understanding the biological aspects of malignancy at the cellular level, as can be seen by the abundance of literature generated utilizing this cell line (Fig. 5.8).

In characterizing the MCF-7 cell line as being a true model of HBECs, there are a few important traits that need to be identified. This cell line is comprised of polyhedral epithelial cells that are cuboidal in nature (Fig. 5.9). Their epithelial pattern is also characterized by microvilli, which augment the surface area of the luminal surface, and an accumulation of secretory vesicles [4]. What is also essential to note is that the three markers for breast cells in culture are also commonly found within the cell line: desmosomes, tonofilaments, and intracytoplasmic lumina [4]. These are main points in characterizing the cell line as truly being that of HBECs; originating from a primary effusion of mammary carcinoma, it must also be identified whether these cells can be characterized as being breast cancer cells. Its classification as a breast cancer cell line is confirmed by its morphological behavior when grown in collagen coated sponges and athymic mice [4]. This is representative of the values of malignancy at the cellular level; the growth in

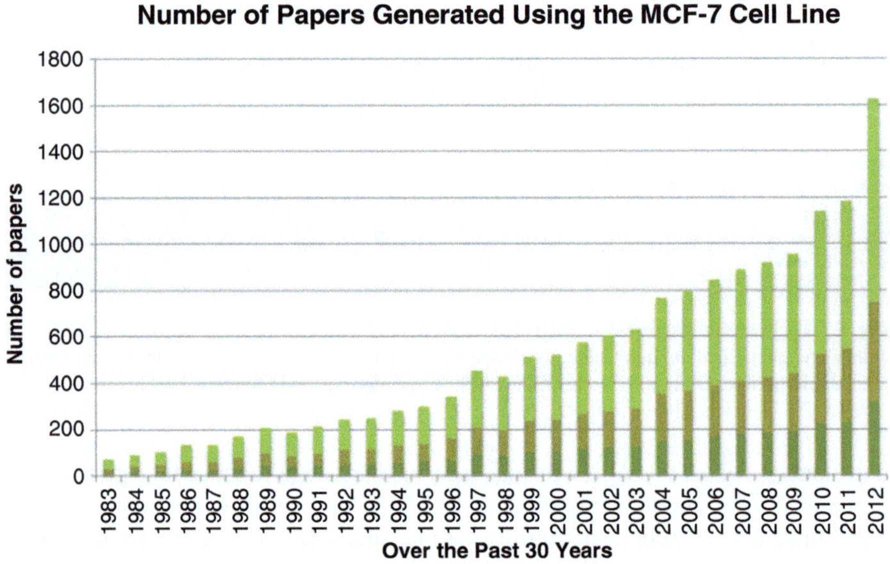

Fig. 5.8 This histogram shows the number of publications generated over the past 30 years utilizing the MCF-7 cell line. This data was compiled from the information provided on PubMed. A search was conducted for the word "MCF-7" in the topic of published papers. The total number of papers published is around 16,000

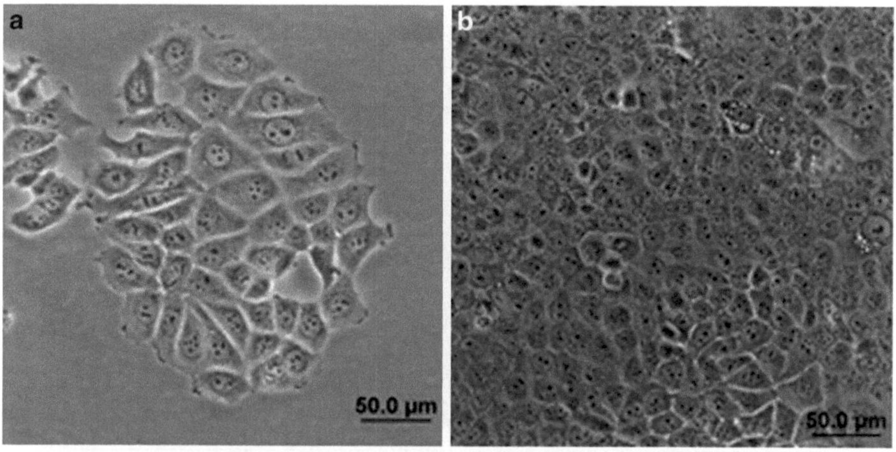

Fig. 5.9 Phase contrast image of the MCF-7 cell line, where panel (**a**) is the log phase and panel (**b**) is the stationary phase. The cells grow together in patches representative of their luminal phenotype. The karyotype is from a human female with modal number = 82; (range = 66–87). The stemline chromosome numbers ranged from hypertriploidy to hypotetraploidy, with the 2S component occurring at 1 %. There were 29–34 marker chromosomes per S metaphase; 24–28 markers occurred in at least 30 % of cells, and generally one large submetacentric (M1) and three large subtelocentric (M2, M3, and M4) markers were recognizable in over 80 % of metaphases. No DM was detected. Chromosome 20 was nullisomic and X was disomic. The base medium for this cell line is ATCC-formulated Eagle's Minimum Essential Medium, Catalog No. 30-2003. To make the complete growth medium, add the following components to the base medium: 0.01 mg/mL bovine insulin; fetal bovine serum to a final concentration of 10 % [8]. Images were acquired with Olympus IMT-2 microscope, Olympus LWDCDPlan 20PL 20X objective, and Olympus DP71 camera

collagen coated sponges is a phenotype previously characterized by Russo and colleagues as being part of the original pattern of a tumor found in the human breast, such as that from which this cell line was derived. It has also been described that a criterion for malignancy of transformed cell lines is their ability to grow in an adequate heterotransplantation system, such as that of the immunologically depressed athymic mice [4]. These characterizations demonstrate that this stable cell line is in fact morphologically similar to breast cancer. Another confirmation of their neoplastic characterization is their ability to form colonies in semisolid like media, agar methocel, which only transformed cells can do [4].

5.5 The Measurement of the Normal and Cancerous Phenotype

5.5.1 Cell Proliferation and Detection of the IC50

The MTT assay is a colorimetric assay for measuring the activity of cellular enzymes that reduce the tetrazolium dye, MTT (3-(4,5-dimethylthiazol-2-yl)-2,5-diphenyltetrazolium bromide, to its insoluble formazan, giving a purple color. The MTT assay

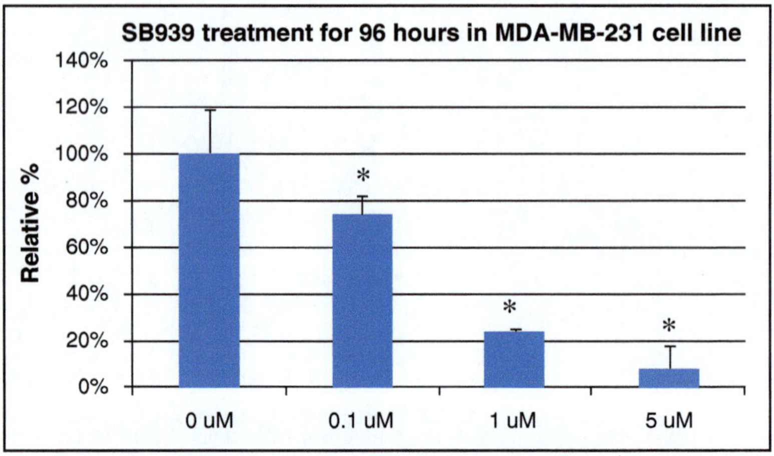

Treatment	0 uM	0.1 uM	1 uM	5 uM	IC50
% Control	100.0%	74.6%	24.1%	7.7%	0.32 uM

Fig. 5.10 Histogram showing the pattern of growth in MDA-MB-231 cells treated with SB939 for 96 h. The table below the figure shows the percentage of growth over the control and the IC50 calculated as described in Sect. 5.5.1

allows for the evaluation of the viability and proliferation of cells via the assessment of metabolic activity. The assay determines the ability of viable cells to convert a soluble tetrazolium salt into an insoluble formazan precipitate, which can be seen as the conversion of a yellow liquid to a purple color. It essentially measures the activity of cellular enzymes involved in this process, which more specifically are the oxidoreductase enzymes found in the cytosol of cells. Basically the procedure consists in seeding cells at an optimal density per cell line, 250–5,000 cells/well, in 100 μL culture medium per well into a 96-well flat bottom tissue culture plate. Cells are treated in adherent conditions on the following day. Media is replaced with 200 μL fresh culture medium. Cell proliferation is assessed every 24 h of treatment using Vybrant MTT Cell Proliferation Assay Kit (Molecular Probes, Eugene, OR). Media is removed and replaced with 100 μL/well phenol-red free media containing 12 mM MTT. The plate is incubated in a 37 °C, 5 % CO_2 incubator for 4 h before 100 μL/well HCl-SDS that is added to solubilize the formazan product. The optical density is read at 570 nm using Epoch Microplate Spectrophotometer (Biotek, Winnoski, VT). Data is plotted and analyzed using Microsoft Excel 2010. Unpaired students *T*-Test is used to calculate a relevant *P*-value. IC50 is calculated using ED50plus v1.0 add-in to Microsoft Excel (http://www.sciencegateway.org/protocols/cellbio/drug/hcic50.htm). Examples are depicted in Figs. 5.10 and 5.11 in which we have tested the effect of two HDAC inhibitors SB939 and Tubastatin A on triple negative breast cancer cell line MDA-MB-231, the cells are sensitive to SB939 but not Tubastatin A.

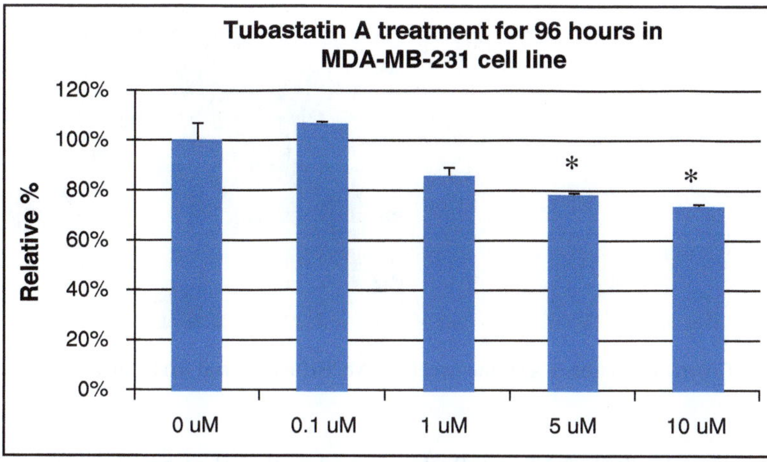

Treatment	0 uM	0.1 uM	1 uM	5 uM	10 uM
% Control	100%	107%	87%	79%	74%

Fig. 5.11 Histogram showing the pattern of growth in MDA-Mb-231 cells treated with Tubastatin A for 96 h. The table below the figure shows the percentage of growth over the control

5.5.2 Colony Formation in Agar Methocel

The concept of clonal growth was reported by Puck et al. by growing mammalian cells at high plating efficiencies in large volumes of cell culture media [11, 12]. Clonal growth required media containing whole serum, dialyzed serum, purified serum components, or undefined dialyzed components from the serum [13, 14]. Ham reported the first successful experiments of clonal growth with a completely synthetic medium, F-12 [12]. MCDB 301 was subsequently developed specifically for clonal growth of CHO cells, and investigations have shown that supplementation of MCDB 301 with insulin or methylcellulose is required for Chinese hamster lung cell growth in low densities [15]. Recent advances in media development technologies have enabled limiting dilution cloning (LDC) and protein production in a serum-free environment. LDC involves diluting a pool of transfected cells down to a single cell per culture well and then expanding that single cell to a larger cell population called a *clonal cell population* or *cell line*. This process has been achieved recently [16] by using medium containing correctly balanced nutrients that promote cell growth at extremely low cell densities [16].

The use of agar methocel is in our hands; it has been a powerful method for separating clones of cells that are transformed. Plating cells in an agar methocel medium allow for the determination of whether or not the cell is transformed because only the clones of transformed cells are capable of forming colonies in

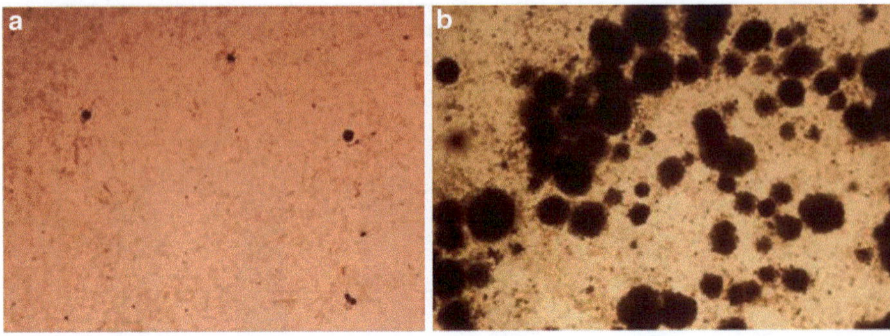

Fig. 5.12 Colony formation in agar methocel, (**a**) MCF10F cells and (**b**) bsMCF10

such a suspension (Fig. 5.12a, b). Both agar and methocel suspensions allow for the determination of anchorage dependence for growth. The formation of colonies is representative of anchorage independent growth, a well-known quality of transformed neoplastic cells [4]. The Irma H. Russo, MD., Breast Cancer Research Laboratory (BCRL) at Fox Chase Cancer Center (FCCC) has characterized the pattern of both transformed and non-transformed MCF-10F cells in agar methocel, showing that DMSO, cholesterol, and progesterone do not alter MCF-10F's inability to form colonies in this medium. However, the treatment of the cells with E_2 does have the effect of transforming the cell line, as demonstrated by the colonies formed in agar. The methodology basically consist in suspending 2×10^4 cells/mL in 2 mL of 0.8 % methocel (Sigma) dissolved in DMEM:F12 (1:1) medium containing 20 % horse serum. Cells are plated into 8 wells of a 24-well tissue culture plate precoated with 0.5 mL 5 % agar base in DMEM:F12 medium. Media is replaced twice a week with 500 L medium. To evaluate colony efficiency, cells are stained with neutral red (1:300) after 21 days in culture. Pictures are taken of the entire well at ×2 magnification with a Nikon TS100 microscope and Olympus DP72 camera. Each colony is measured using MetaMorph software version 7.7.8.0. A normal distribution of area is calculated using Sigma Plot software. The colonies are classified into three different groups regarding the size of the colonies: small (less than 100 m in diameter), medium (between 101 and 200 m), or large (more than 200 m) (Fig. 5.12b). Figure 5.13 depicts an example of the formation of colonies in agar methocel treated with drug of interest. In this case a significant effect in reversion of the phenotype is observed.

5.5.3 Ductulogenesis/Solid Mass Formation Assays

Ductulogenesis is an assay that quantitatively evaluates the ability of a specific cell line to form either tubules or spherical masses in a three-dimensional collagen environment [4, 17–19]. Non-transformed cells form ductile-like structures (Fig. 5.14),

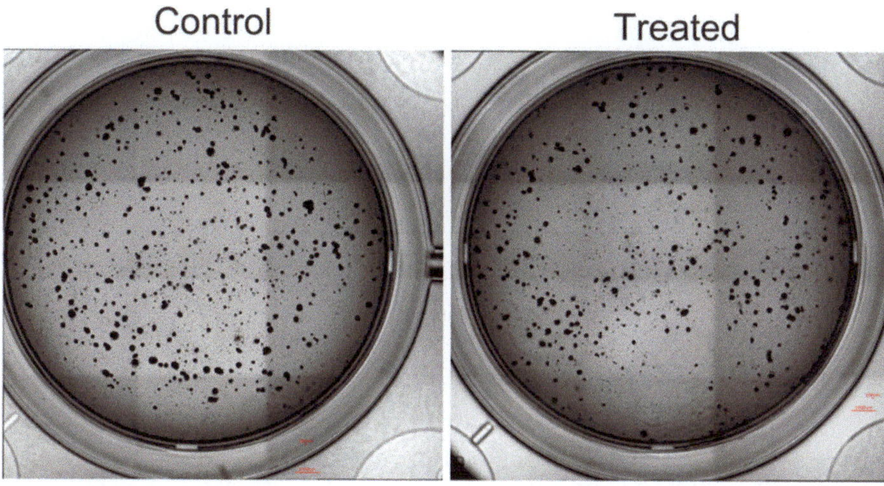

Fig. 5.13 Colony formation in agar methocel of control bsMCF cells and cells treated with drug of interest. The cells were trypsinized and suspended in methyl cellulose solution, then seeded on to agar base in 6-well plate at a density of 1,500 cells per well, and cultured for 14 days. At the end of this period the cells were stained with neutral red solution and images were captured using KEYECE system with Z stack

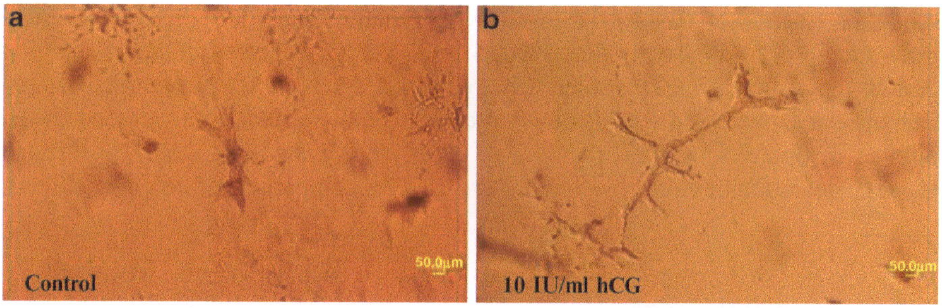

Fig. 5.14 (**a**) MF10 cells growing in collagen gel (control). (**b**) The same cells treated with 10 IU of hCG for 24 h. The photographs were taken after 6 days in culture

Histological sections of paraffin embedded structures growing in collagen show that the normal ducts form ductal like structures made up of cells surrounding a lumen (Fig. 15a), whereas transformed cells form solid masses, visualized as spherical clumps (Fig. 5.15b) or disperse invading cells (Fig. 5.15c) [4, 17–19]. These ductulogenic structures have been shown to be composed of a single layer of cuboidal epithelial cells, which actively proliferate in the collagen medium as detected by immunostaining, with Ki67 [20].

Previous work done by our laboratory has characterized the pattern of both transformed and non-transformed MCF-10F cells. The MCF-10F cell line has a characteristic ductulogenic structure in collagen, and DMSO, cholesterol, and progesterone

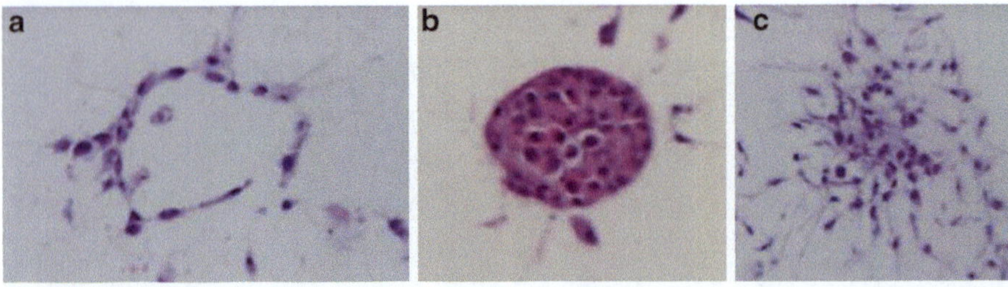

Fig. 5.15 Histological sections of cells growing in a collagen matrix. (**a**) MCF10F; (**b**) tr-MCF10, and (**c**) bcMCF10 cells. Stained with hematoxylin and eosin ×10

have no effect on altering this. However, E_2 does have the effect of transforming the cell line, giving it a new appearance in collagen to that of the solid masses previously described (Fig. 5.15b) [4, 17–19]. In fact, the number of solid masses formed in collagen have been shown to parallel the number of colonies that form in agar methocel. The methodology of growing cells in a collagen matrix basically consists of suspending the cells at a final density of 2×10^4 cells/mL in 89.3 % (Vitrogen^{100}) collagen matrix (Collagen Co., Palo Alto, CA) and plating them into eight 24-well chambers pre-coated with 89.3 % of collagen base. The cells are fed with fresh medium containing 5 % horse serum every 2 days. The cells are examined under an inverted microscope for a period of 10 days for evaluation of the number of duct-like or spherical mass structures as previously described [4, 17, 21, 22]. At the end of the observation period, the structures are counted, photographed, fixed in 10 % neutral buffered formalin, and processed for histological examination. Results are expressed as the number of ductules or solid masses per 3,000 cells plated per well. The mean ± standard deviation of three observations are used for comparison of treatment effects.

5.5.4 Invasion Assay

The invasion assay measures the invasive potential of cells utilizing Boyden-type chambers separated by a porous polycarbonate filter in order to reconstitute basement membrane material [4, 17, 21, 22]. The Boyden chamber was designed over 40 years ago to study chemotaxis, but later modifications resulted in it becoming one of the most popular tools for assessing cell motility and invasion [23]. The classic Boyden chamber consists of two compartments separated by a filter membrane, which, with its microporous surface, provides a barrier so that cells cannot pass through except by active migration. For chemotaxis analysis, cells are seeded onto the upper compartment, and the chemoattractant is placed in the lower compartment. After incubation for a period less than the time taken for cells to divide, the degree of chemotactic response can be microscopically determined by counting the

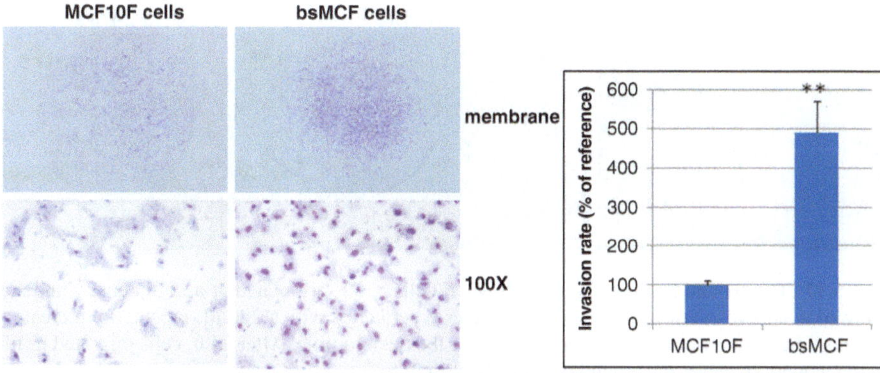

Fig. 5.16 Invasion assay was performed as described in Sect. 5.5.4. The assay was performed in MCF10F and bsMCF cell lines, 25,000 cells were plated and allowed to invade the matrigel trough transwell insert at 8-mm pore size (four replicates for each treatment) for 24 h. The cells on lower surface of chambers were stained with Giemsa stain, photographed under a light microscope, and counted

percentage of transmigrated cells compared to the original number of seeded cells [23]. In order to adapt the Boyden chamber to other applications such as quantitating the invasive potential of tumor cells, various extracellular matrix molecules can be coated onto the membrane. These coating materials can be laminin, collagen, natural basement membrane, or the most widely used substance known as Matrigel™ (BD Biosciences, San Jose, CA, USA; [24–27]). The porous membrane is occluded by Matrigel, therefore mimicking the extracellular environment. Only invasive cells that have digested the gel will reach the underside of the membrane. Cells are placed on filters that have been coated with Matrigel, positioned above the porous membrane and allowed to incubate overnight in order to allow them to migrate toward the chemo-attractant. The cells subsequently found on the underside of the filter, having migrated to the bottom chamber and gone through the porous membrane are what are deemed invasive cells [4, 17, 21, 22]. The cells are quantified by removing the filter, staining the cells, and mounting them on glass slides for counting. The total number of cells that invaded through the Matrigel membrane of the Boyden chamber is counted under a light microscope. The invasion index is expressed as the number of cells that migrated to the lower chamber. The mean ± standard deviations of three observations are used for comparison of treatment effect. Figure 5.16 depicts the appearance of the membranes and high magnification images of normal like breast epithelial cell line MCF10F and high invasive cell line bsMCF.

Currently, there are several commercial and in-house versions of the Boyden chamber in use [28–31]. Most of the Boyden assay systems are prone to common errors such as: (1) the Matrigel coating is often irregular or disrupted, resulting in

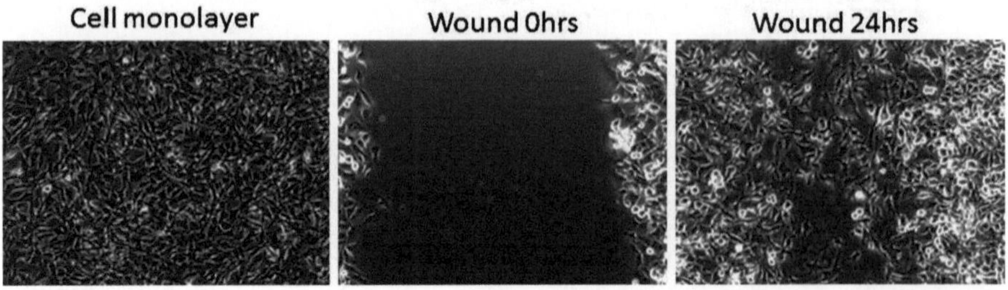

Fig. 5.17 The wound healing assay for bsMCF cells. Cells are plated at a high density and are allowed to adhere overnight. The cell monolayer is scratched with a 200-μL pipette tip to create a wound. Images are acquired over an interval for the desired time. After 24 h, cells migrated to fill in the wound; however the wound is not fully closed

uneven cell invasion; (2) the process of removal of the inserts sometimes leads to cross-contamination and variability in results; (3) the recovery of transmigrated cells is notoriously difficult, especially for inexperienced users; and (4) the manual fixing, cell staining, and cell counting further serve to amplify the inconsistency. As a consequence, many results of the Boyden chamber assays have been shown to vary significantly from one laboratory to another, and even within the same laboratory. To help circumvent some of these problems a new design of the Boyden chamber has been described and tested [32].

5.5.5 Wound Healing Assay

The wound healing assay allows for the determination of the invasive potential of cells because it characterizes cell movement in a unidirectional fashion. Quantifying velocity allows for characterization of the potential of each cell line. This assay is a measure of the ability of the cells to invade, however is less specific than the invasion assay in the matrigel. However, since cell migration is one of the first steps in metastasis it is crucial to study [28], and the wound healing assay is one the most cost effective analytical procedures in which invasive and metastasis potential of cancer cell line can be tested in vitro [33–41]. The wound healing assay mimics cell migration in vivo [39]. The basic procedure consists of first creating a scratch wound on a monolayer of cells, acquiring images at different time periods, and then analyzing the data after the desired time (Fig. 5.17). The introduction of treatment can influence the rate at which the cells migrate to close the wound [40, 41]. Different parameters of cell migration can be quantified using cell tracking software. For example, this assay is suitable for studies on the cell migration, cell-cell interactions, cell-matrix interactions, and monitoring intracellular events [33–41]. Basically the technique consists of seeding (1.0×10^5 cells/

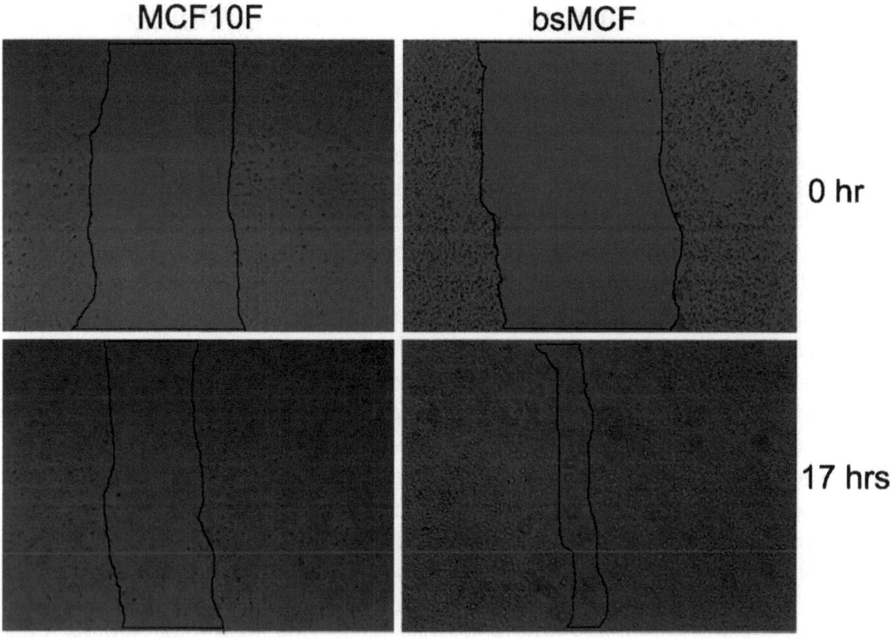

Fig. 5.18 Wound healing assay was performed in MCF10F and bsMCF cells. Cells were plated onto 24-well plate at a density of reaching confluence in 24 h. The wounds were then made and pictures were recorded at the indicated time point. The wound was outlined

well) in 24-well tissue culture plates and incubated overnight to allow cells to adhere. The following day the cell monolayer is scratched with a 200-μL sterile pipette tip and washed twice with normal cell culture medium. Media is replaced with 1-mL medium containing the chosen treatment. Images are acquired in 2-min intervals for 24 h using MetaMorph Advanced 7.7.8.0 software, Olympus IX-71 microscope, prior automated stage, Proscan III controller, LiveCell StageTop incubation unit, and Q-imaging 24-bit camera. To determine cell velocity, cells from the leading edge of the wound are tracked using cell tracking application on MetaMorph Advanced software.

In Fig. 5.18 the images show the wound at 0 hour and 17 hours, the area of wound can be measured by Metamorph and then compared between two cell lines.

5.5.6 Tumorigenesis and Metastatic Assays

These two technologies are described in Chap. 8.

5.6 Cancer Cell Lines Based on Breast Cancer Subtypes

Breast cancer is characterized by a malignancy found in breast epithelial cells, capable of metastasizing to distant areas of the body. Breast cancer is a morphologically heterogeneous disease, encompassing several different entities, each having distinct biological characteristics and thus varying clinical behaviors [42]. Breast cancer can be subdivided into five major subtypes: two of which are hormone receptor positive, luminal A and B; and three of which are hormone receptor negative, normal breast-like, HER-2, and basal-like [43]. Microarray studies have identified that these subtypes have characteristic gene-expression patterns and underlying DNA copy number alterations (CNAs), such as deletions or duplications in regions of their genome [44]. Markers such as estrogen receptor (ER), progesterone receptor (PR), and ERBB2/HER2 are used for prognosis as well as determining the best target therapy [44]. These markers are typically identified using either RT-qPCR or immunohistochemical staining; there is typically agreement between the two methods despite staining not adequately identifying all of the genes present in these markers [45].

The luminal subtypes are ER positive and have the same markers as the luminal epithelial layer of cells lining normal breast ducts [44]. Luminal A and B are distinct in that A is characterized as having a higher expression of estrogen receptor (ER) genes and lower expression of proliferative genes vs. subtype B, thus subtype A has a more favorable prognosis than subtype B [44]. Basal-like tumors are ER negative like the basal cells of normal breast ducts, with frequent chromosome segmental gains/losses [44]. Literature has shown that the basal-like group, whose tumors lack the expression of hormone receptors as well as HER-2, has a more aggressive clinical behavior [46, 47]. Along with having a distinctive metastatic pattern, this subtype, affecting 15 % of all cases, has a poor prognosis despite responding to chemotherapeutic regimes [48, 49]. The luminal A subtype has been characterized as having the most favorable prognosis, with a lower local recurrence rate [37, 50].

5.6.1 The In Vitro Model of Cancer Initiation and Progression

We have developed an in vitro model of HBEC transformation (Fig. 5.19) [17–19, 21, 22] to study the process of initiation and progression of breast cancer. This breast cancer model consists of four cell lines: (a) the spontaneously immortalized cell line MCF-10F, which does not show any characteristics of invasiveness or tumor formation and therefore is considered to be a basal breast epithelial cell line; (b) the transformed trMCF cells; (c) the invasive bsMCF cell line and, (d) the tumor cell lines caMCFs, which show all characteristics of fully malignant breast cancer cell types (Fig. 5.19). The bsMCF and caMCF are highly metastatic cell lines [17–19, 21, 22]. These established cell lines are from the same lineage, all derived from MCF10F, and represent sequential steps of a dynamic process thus creating a

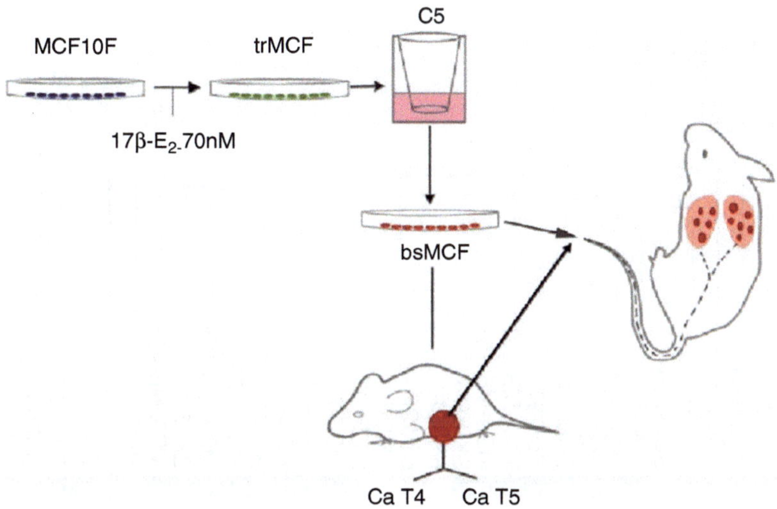

Fig. 5.19 Transformation of MCF-10F cells by 17β-estradiol treatment. Experimental protocol: MCF-10F cells treated with 70 nM 17β-estradiol (E_2) that expressed high colony efficiency (CE) and loss of ductulogenic capacity in collagen-matrix were classified as transformed (trMCF). Transformed cells that were invasive in a Matrigel Boyden type invasion chambers were selected (bsMCF) and plated at low density for cloning (bcMCF). MCF-10F, trMCF, bsMCF, and bcMCF were tested for carcinogenicity by injecting them into the mammary fat pad of 45-day-old female SCID mice. MCF-10F and trMCF cells did not induce tumors; bsMCF, formed solid tumors from which four cell lines, identified as caMCF, were derived and cells proven to be tumorigenic in SCID mice. bsMCF or caMCF when injected in the tail of SCID mice develop metastatic foci in the lung (*from*: *Russo J, Russo IH. Adv Exp Med Biol;720:121–134, 2011*)

new resource for establishing the molecular events from one step to the other and making it easier to pinpoint which genetic and epigenetic events drive each step of breast cancer. In this model (Fig. 5.19), trMCF cells progressively express phenotypes of in vitro cell transformation, including colony formation in agar methocel and loss of ductulogenic capacity that is the earliest phenotype observed, accompanied by increasing cell proliferation and activation of genes related to DNA replication, inhibition of apoptosis, and the expression of genes related to cell polarity, cell positioning, and cellular architecture. Further selection of these trMCF cells for invasiveness in a Matrigel invasion system identified cells (bsMCF) that formed tumors and metastasis to the lung in severe combined immunodeficient (SCID) mice [18]. The cell lines derived from those tumors (caMCF) were poorly differentiated ERα, PR, and Her2-negative adenocarcinomas [18] that when injected into the tail vein of a SCID mouse will induce metastasis to the lung [51] (Fig. 5.19). These characteristics are similar to those reported to be expressed by the human basal cell-like carcinomas or triple negative breast cancer (TNBC) [52–56]. To better understand the molecular events associated with the progressive phenotypic

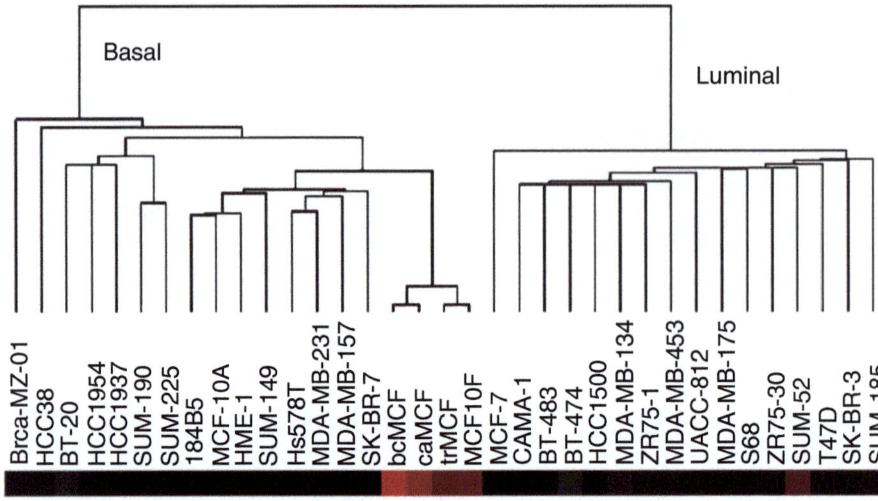

Fig. 5.20 MCF-10F and its derived cell lines trMCF, bsMCF, and caMCF are classified as basal cell lines. The expression data for these cell lines was combined with the published expression values of 364 gene sets and used to classify the breast cell lines by hierarchical clustering. MCF-10F and its derived cell lines occupy a distinct branch of the basal subtype (*left branch*)

changes that were observed during malignant cell transformation, we analyzed chromosomal copy number (CN), loss of heterozygosity (LOH), and gene expression changes that occurred at different stages of cell transformation [12]. By integrating these data we were able to identify associations between CN changes, LOH, transcript expression, and phenotypes of invasion and tumorigenicity, including a strong gene and protein signature of EMT that included the loss of intercellular adhesion (E-cadherin and occludins), down-regulation of epithelial markers (cytokeratins), and up-regulation of mesenchymal markers (vimentin and smooth muscle actin) [18]. CaMCF cells show a complete loss of CDH1, but upregulation of FN1 and vimentin (VIM) [18]; these events are associated with mesenchymal differentiation that enables caMCF cells to invade the extracellular collagen matrix and be metastatic in vivo [51].

The fact that MCF-10F cells are non-transformed, non-tumorigenic, ER-alpha, and PR negative, Her 2 negative, CD44+/CD24+, CRXC1+ and ALDH+ indicates that these cells exhibit all the characteristics of the basal breast stem cell [18, 52–54]. Because malignant transformation of these cells produces poorly differentiated tumors with the characteristics of basal-like carcinomas or TNBC, we chose to classify these cells relative to the breast cell lines described by Charafe-Jauffret et al. [52]. As shown in Fig. 5.20, these MCF-10F and derived cell lines clustered in the branch containing the basal breast cell lines [57]. In our molecular characterization of malignant cell transformation we identified the "intermediate filament" component enriched in Gene Ontology (GO) analysis, separating the non-tumorigenic MCF-10F and trMCF cells from the tumorigenic bsMCF and caMCF cells.

Numerous cytokeratins (KRT) were suppressed or absent, whereas VIM was strongly induced in bsMCF (7.0-fold) and caMCF (8.1-fold). In addition, we found that while the expression of CD44, ALDH, and CXCR1 was maintained, CD24 was lost. EpCAM (tumor-associated calcium signal transducer 1) also known as ESA (epithelial-specific antigen) was completely lost in bsMCF and caMCF, but did not change in trMCF cells [18]. There were 27 genes that were markers of EMT present in our model and gene clustering shows two sample groups and two gene groups. The non-tumorigenic MCF-10F and trMCF cells were grouped into sample cluster κ, while the tumorigenic bsMCF and caMCF cells were grouped into cluster λ. On the other side, the genes were grouped into cluster α and β based on their expression pattern. The epithelial markers *E-cadherin*, *occludin*, *desmoplakin*, *EMA*, *EpCAM*, and *cytokeratins* were decreased, while the mesenchymal markers *FN1*, *VIM,* and *N-cadherin* were increased in bsMCF and caMCF cells [18]. Immunohistochemical (IHC) analyses revealed significant loss of the epithelial markers EMA (also called MUC-1) and E-cadherin and increased expression of the mesenchymal marker VIM in tumorigenic cells. By real-time qRT-PCR, it was confirmed that the expression of *FN1*, *S100A4*, *SNAI2*, *HRAS,* and *TGFβ1* was increased, while *CDH1* (E-cadherin) was decreased in bsMCF and caMCF cells [18]. These findings confirmed the EMT phenotype in our in vitro–in vivo model of cell transformation. We extracted our EMT gene signature from the Genechip expression files of the 30 cell lines that were characterized for invasiveness in a modified Boyden chamber assay [53]. This EMT gene signature classified, with complete concordance to the cell lines into luminal, Basal A, and Basal B subtypes. This result along with the report showing that EMT occurs more frequently in basal-like tumors [54] indicates the relevance of these breast cell lines for molecular analysis of the networks regulating EMT. While the regulation of EMT is not fully understood, a network of several signaling pathways affecting the expression and/or function of a complex hierarchical network of transcription factors (TFs) has been partially elaborated [18]. Known signaling pathways include multiple tyrosine kinase receptors leading to Ras-mediated activation of *MAPK* and *PI3K* pathways, *TGF-β*, *Notch,* and *Wnt.* We have evidence that enhanced *TGF-β*, and *Wnt* signaling pathways are found in the EMT expressing bsMCF and caMCF cells [18]. *TGF-β* acting through *Smad* transcriptional complexes can repress expression of the Id TFs (*Id1*, *Id2*, *Id3*) and activate *HMGA2*, a DNA-binding protein important for chromatin architecture. Expression of *HMGA2* is known to regulate several EMT controlling TFs including *TWIST1*, *SNAI1*, and *SNAI2* (*Slug*) [18, 52, 58]. Analysis of the EMT expressing bsMCF cell line revealed the absence of expression of the secreted frizzled-related protein 1 (*SFRP1*), a repressor of *Wnt* signaling [18, 55]. One allele of SFRP1 was deleted in these cells, with the remaining apparently silenced by methylation, accounting for the 28-fold reduction of this transcript. Loss and epigenetic inactivation of *SFRP1* occurs often in invasive breast cancer and is associated with poor prognosis [55]. We have shown that *EZH2* is also overexposed in the EMT and it is well known that this gene leads to cancer invasion and progression, an indication that this mechanism involves the Polycomb Complex Groups (PCG) *PRC1* and *PRC2* in both normal development and cancer, resulting in H3 methylation modification.

5.6.2 Specific Cell Lines

Breast cancer being a morphologically heterogeneous disease, there are a wide variety of cell lines used to explore its variations. Catalogs of their molecular profile, achieved through whole genome microarray analysis, relate each cell line to a specific breast cancer subtype. Studies have yielded results that demonstrate a retention of expression patterns distinguishing luminal vs. basal subtypes, though the breast cancer cell lines were genetically more complex than tumors (Fig. 5.20) [44]. Such data provides "definitions of cell lines suitable for investigations of subtype-specific pathobiology, cancer stem cell biology, biomarkers, and therapies, and provides a resource for discovery of new breast cancer genes" [44]. Transcriptional profiling identified one luminal cell line subtype displaying an ER signature resembling luminal-A/B tumors. The cell lines associated with the breast cancer basal subtype display mesenchymal and stem/progenitor-cell characteristics [44]. The HER-2 breast cancer subtype is associated with the expression of genes co-amplified with ERBB2, which encodes for HER-2 [44].

5.6.2.1 Basal Cell Lines

Among the typical basal cell lines depicted in Fig. 5.20, some are extremely useful in cancer research. Two of these are the trMCF (Fig. 5.21) and the bsMCF10 derived from trMCF (Fig. 5.22). These cell lines have been described in detail in Sect. 5.6.1 (Table 5.1).

The BT549 cell line (Fig. 5.23) has a human female karyotype with a modal number of 78; (range = 73–80). The cell line is aneuploid with chromosome counts in the hypertriploid range. The X chromosomes are all abnormal. Normal chromosomes N10 and N13 are clearly under-represented, with chromosomes N2, N12, and N17 also tending to be. Chromosome N8 is over-represented with respect to the copy number of other normal chromosomes, and chromosomes N5 and N18 also usually have more copies than most chromosomes. Four marker chromosomes are found: iso(13q), 10q+, del(X)(q22:), del(11)(p11:). The base medium for this cell line is ATCC-formulated RPMI-1640 Medium, Catalog No. 30-2001. To make the complete growth medium the following additives are suggested to the base medium: 0.023 IU/mL insulin; fetal bovine serum to a final concentration of 10 % [7, 8].

The cell line HS578T (Fig. 5.24) is aneuploid with a modal number = 64 (range = 52–68), with chromosome counts in the near-triploid range. Normal chromosomes N8 and N15 are absent. Eleven stable rearranged marker chromosomes

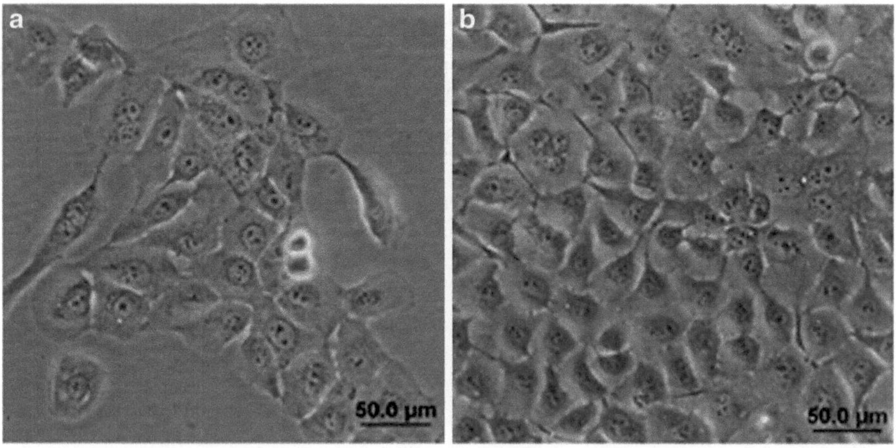

Fig. 5.21 (**a**) Logarithmic growth of trMCF cells; (**b**) confluent growth of trMCF cells. trMCF cells were maintained in High Calcium cell culture medium [4]. Images were acquired with Olympus IMT-2 microscope, Olympus LWDCDPlan 20PL 20X objective, and Olympus DP71 camera

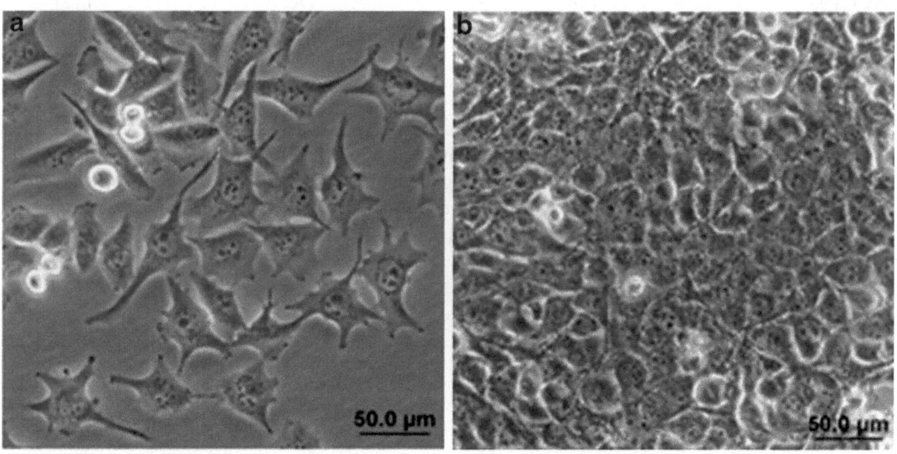

Fig. 5.22 (**a**) Logarithmic growth of bsMCF cells, (**b**) confluent growth of bsMCF cells. bsMCF cells were maintained in High Calcium cell culture medium [4]. Images were acquired with Olympus IMT-2 microscope, Olympus LWDCDPlan 20PL 20X objective, and Olympus DP71 camera

are noted as well as unassignable chromosomes in addition to the majority of autosomes that are trisomic. The base medium for this cell line is ATCC-formulated Leibovitz's L-15 Medium, Catalog No. 30-2008. To make the complete growth medium, the following components to the base medium must be added such as fetal bovine serum to a final concentration of 10 % [7, 43]. Other useful basal cell lines are the MDA-MB-231 (Fig. 5.25) and the SUM 140PT (Fig. 5.26).

Table 5.1 This table characterizes the basal cell lines commonly used in breast cancer research, along with the MCF-10F cell line and its associated derivations, including the trMCF cell line, bsMCF cell line, and caMCF cell line, as described by previous research conducted at the Breast Cancer Research Laboratory of the Fox Chase Cancer Center

Cell line	ER	PR	ERBB2 R2	Tumor type[a]	Reference
MCF-10F	–	–	–	None	[26]
trMCF	–	–	–	None	[26]
bsMCF	–	–	–	Adenocarcinoma	[26]
caMCF	–	–	–	Adenocarcinoma	[26]
MCF-10A	–	–	–	F	[12]
HME-1	/	/	/	Non[a]	[27]
HS578T	–	–	–	C Sar	[12]
BT-549	–	–	–	IDC	[12]
SUM-149	–	–	–	Inf	[12]
SK-BR-7	/	/	/	/	
MDA-MB-157	–	–	–	Med C	[12]
MDA-MB-231	–	–	–	Met AC	[12]
184B5	/	/	/	Non	[27]
SUM-225	–	/	+	/	[27]
SUM-190[a]	–	–	+	Inf	[12]
HCC1937	–	–	–	DC	[12]
HCC1954	–	–	+	DC	[12]
BT-20	–	–	–	AC	[12]
HCC38	–	–	–	DC	[12]
BrCa-Mz-01	+	+	/	Med Ca	[27]

The (+) exemplifies the presence of the respective Estrogen (ER), progesterone (PR), or ERBB2 receptor, whereas the (–) exemplifies it not being present

[a]The tumor type abbreviations correlate with the following: *Met AC* metastatic adenocarcinoma, *Med C* medullary carcinoma, *IDC* invasive ductal carcinoma, *AC* adenocarcinoma, *C Sar* carcinoma sarcoma, *Inf* inflammatory, *F* fibrocystic disease

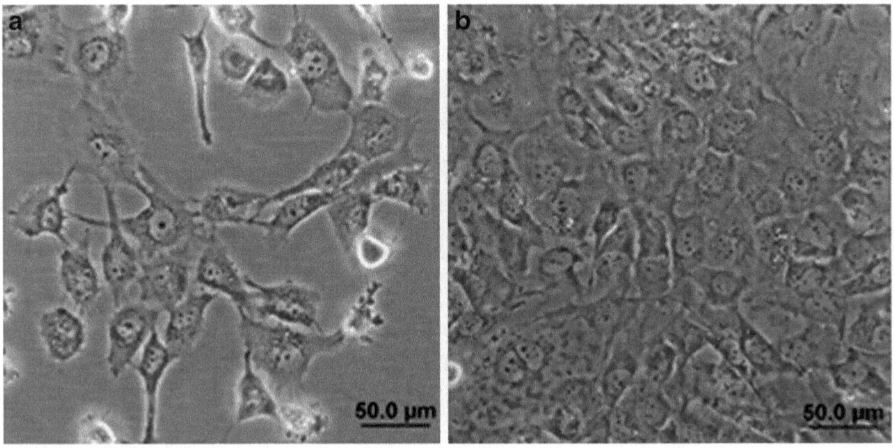

Fig. 5.23 This panel shows the accepted standard morphological appearance of the BT-549 cell line, where panel (**a**) is the log phase and panel (**b**) is the stationary phase. Images were acquired with Olympus IMT-2 microscope, Olympus LWDCDPlan 20PL 20X objective, and Olympus DP71 camera

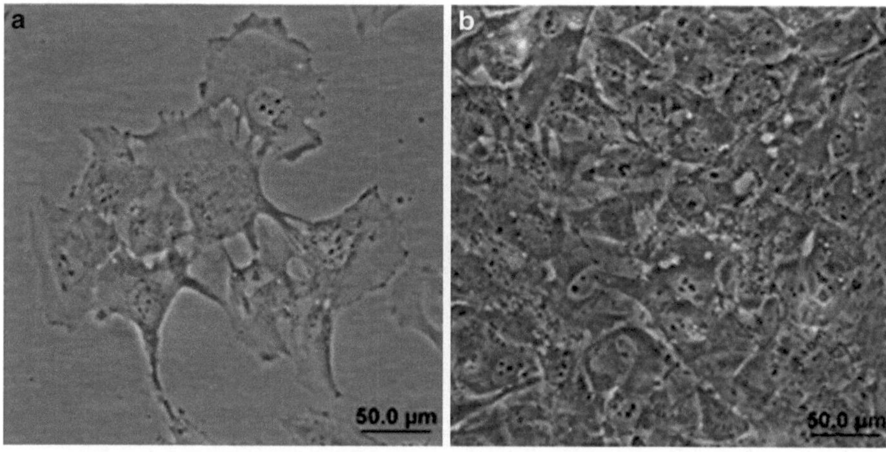

Fig. 5.24 This panel shows the accepted standard morphological appearance of the HS578T cell line, where panel (**a**) is the log phase and panel (**b**) is the stationary phase. Images were acquired with Olympus IMT-2 microscope, Olympus LWDCDPlan 20PL 20X objective, and Olympus DP71 camera

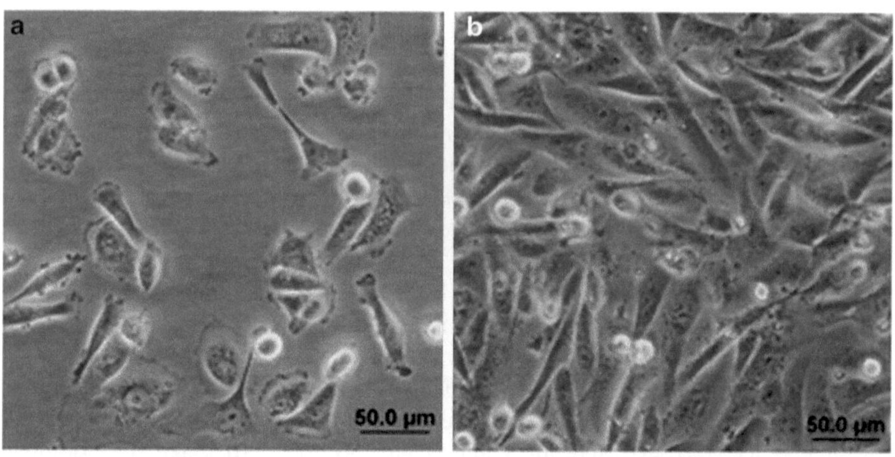

Fig. 5.25 This panel shows the accepted standard morphological appearance of the MDA-MB-231cell line, where panel (**a**) is the log phase and panel (**b**) is the stationary phase. Images were acquired with Olympus IMT-2 microscope, Olympus LWDCDPlan 20PL 20X objective, and Olympus DP71 camera

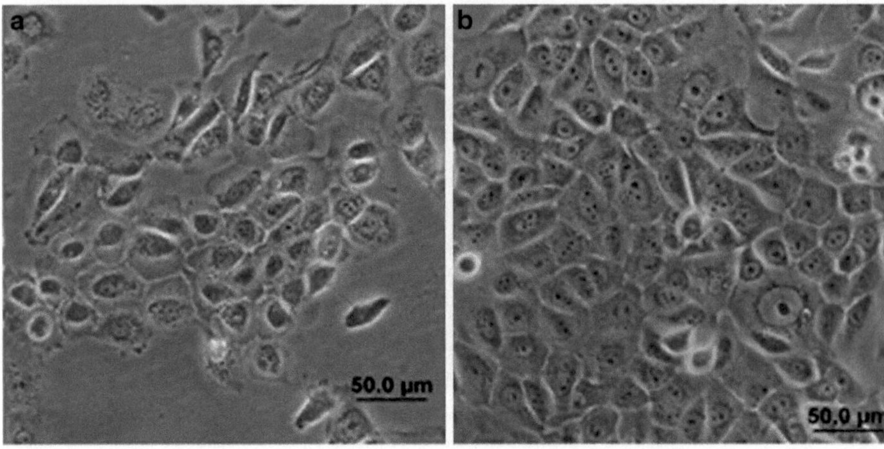

Fig. 5.26 (**a**) Logarithmic growth of SUM-149PT cells (**b**) confluent growth of SUM-149PT cells. SUM-149PT cells were maintained in Ham's F-12 cell culture medium. Images were acquired with Olympus IMT-2 microscope, Olympus LWDCDPlan 20PL 20X objective, and Olympus DP71 camera

5.6.2.2 Luminal Cell Lines

The variety of luminal cell lines is large, as depicted in Fig. 5.19. Among them, one that is highly utilized is T47D (Fig. 5.27) and MCF7 cells (Fig. 5.9). The T-47D cell line is a hypotriploid human cell line. The modal chromosome number is 65 occurring at 50 % and polyploidy at 0.8 %. Eighteen marker chromosomes are common to most cells, of which 7 are paired and 11 are single-copied. The t(8q14q), t(9q17q), and t(10q17p) are among the seven paired markers common to most cells. N7, N9, and N10 are absent and N11 is generally present in four copies. DM's occurred, but infrequently. Q-band examination did not show the presence of a Y chromosome. T47D contains the following markers: calcitonin, expressed, positive/androgen receptor expressed positive/glucocorticoid receptor, expressed positive/prolactin expressed positive. The base medium for this cell line is ATCC-formulated RPMI-1640 Medium, Catalog No. 30-2001. To make the complete growth medium, the following components are added to the base medium: 0.2 Units/mL bovine insulin; fetal bovine serum to a final concentration of 10 % [7, 45] (Tables 5.2 and 5.3).

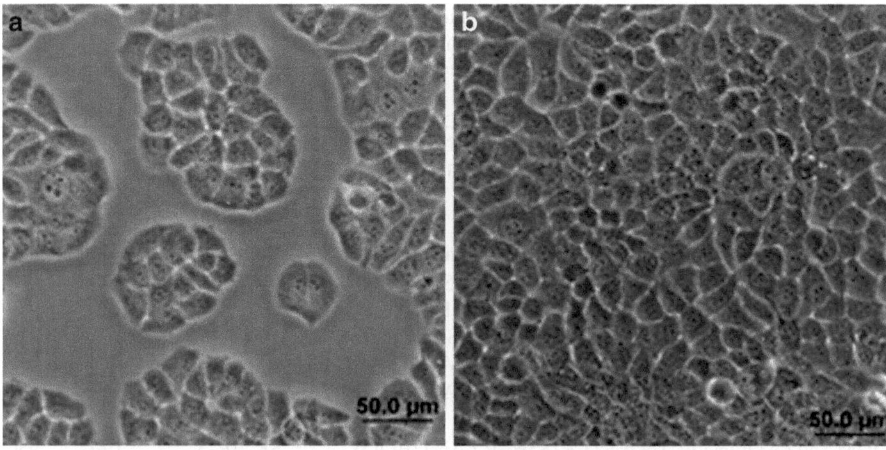

Fig. 5.27 This panel shows the accepted standard morphological appearance of the T-47D cell line, where panel (**a**) is the log phase and panel (**b**) is the stationary phase. Images were acquired with Olympus IMT-2 microscope, Olympus LWDCDPlan 20PL 20X objective, and Olympus DP71 camera

Table 5.2 This table characterizes the luminal cell lines commonly used in breast cancer research

Cell line	ER	PR	ERBB2/HER2	Tumor type[a]	Reference
MCF-7	+	+	–	Met AC	[12]
T47D	+	+	–	IDC	[12]
SK-BR-3	–	–	+	AC	[12]
SUM-185	–	–	+	AP	[27]
SUM-52	+	–	+	Met C	[12]
ZR75-30	+	–	+	IDC	[12]
S68	/	/	/	/	[27]
MDA-MB-175	+	–	–	IDC	[12]
UACC-812	+	–	+	IDC	[12]
MDA-MB-453	–	–	+	Met C	[12]
ZR75-1	+	–	–	IDC	[12]
MDA-MB-134	+	–	–	IDC	[12]
HCC1500	+	+	–	DC	[12]
BT-474	+	+	+	IDC	[12]
BT-483	+	+	–	IDC	[12]
CAMA-1	+	/	/	AC	[27]

The (+) exemplifies the presence of the respective Estrogen (ER), progesterone (PR), or ERBB2 receptor, whereas the (–) exemplifies it not being present

[a]The tumor type abbreviations correlate with the following: *Met AC* metastatic adenocarcinoma, *IDC* invasive ductal carcinoma, *DC* ductal carcinoma, *AC* adenocarcinoma, *C Sar* carcinoma sarcoma, *Inf* inflammatory, *AP* anaplastic carcinoma

Table 5.3 This table is representative of the main articles generated from the cell lines found in the list of both basal and luminal breast epithelial cell lines

Paper title	Cell line(s) used	Date	Times cited
THE E-SCREEN ASSAY AS A TOOL TO IDENTIFY ESTROGENS—AN UPDATE ON ESTROGENIC ENVIRONMENTAL-POLLUTANTS	MCF-7	1995	1,123
Caspase-3 is required for DNA fragmentation and morphological changes associated with apoptosis	MCF-7	1998	1,070
APOPTOTIC DEATH IN EPITHELIAL-CELLS—CLEAVAGE OF DNA TO 300 AND OR 50 KB FRAGMENTS PRIOR TO OR IN THE ABSENCE OF INTERNUCLEOSOMAL FRAGMENTATION	MCF-7	1993	1,061
A multidrug resistance transporter from human MCF-7 breast cancer cells	MCF-7	1998	1,051
HUMANIZATION OF AN ANTI-P185HER2 ANTIBODY FOR HUMAN CANCER-THERAPY	SK-BR-3	1992	956
ACCUMULATION OF P53 TUMOR SUPPRESSOR GENE PROTEIN—AN INDEPENDENT MARKER OF PROGNOSIS IN BREAST CANCERS	MDA-MB-231	1992	773
Recombinant humanized anti-HER2 antibody (Herceptin (TM)) enhances the antitumor activity of paclitaxel and doxorubicin against HER2/neu overexpressing human breast cancer xenografts	BT-474	1998	648
Membrane type 1 matrix metalloproteinase digests interstitial collagens and other extracellular matrix macromolecules	MDA-MB-237	1997	610
Resveratrol, a polyphenolic compound found in grapes and wine, is an agonist for the estrogen receptor	T47D	1997	578
Surface modification of superparamagnetic magnetite nanoparticles and their intracellular uptake	BT20	2002	541
NF-kappa B function in growth control: Regulation of cyclin D1 expression and G(0)G(1)-to-S-Phase transition	T47D	1999	466
INSULIN-LIKE GROWTH-FACTOR (IGF)-INDEPENDENT ACTION OF IGF-BINDING PROTEIN-3 IN HS578T HUMAN BREAST-CANCER CELLS—CELL-SURFACE BINDING AND GROWTH-INHIBITION	Hs578T	1993	453
Identification of a human mitotic checkpoint gene: hsMAD2	T47D	1996	436
TAT peptide on the surface of liposomes affords their efficient intracellular delivery even at low temperature and in the presence of metabolic inhibitors	BT20	2001	422
Chemopreventive agent resveratrol, a natural product derived from grapes, triggers CD95 signaling-dependent apoptosis in human tumor cells	T47D	1998	418
ASSOCIATION OF INCREASED BASEMENT-MEMBRANE INVASIVENESS WITH ABSENCE OF ESTROGEN-RECEPTOR AND EXPRESSION OF VIMENTIN IN HUMAN BREAST-CANCER CELL-LINES	Hs578T	1992	408

(continued)

Table 5.3 (continued)

Paper title	Cell line(s) used	Date	Times cited
ASSOCIATION OF INCREASED BASEMENT-MEMBRANE INVASIVENESS WITH ABSENCE OF ESTROGEN-RECEPTOR AND EXPRESSION OF VIMENTIN IN HUMAN BREAST-CANCER CELL-LINES	SK-BR-3	1992	408
hCds1-mediated phosphorylation of BRCA1 regulates the DNA damage response	HCC1937	2000	362
Rational combinations of trastuzumab with chemotherapeutic drugs used in the treatment of breast cancer	MDA-MB-453	2004	267
P53 MUTATIONS IN HUMAN IMMORTALIZED EPITHELIAL-CELL LINES	184B5	1993	261
Activation of HER4 by heparin-binding EGF-like growth factor stimulates chemotaxis but not proliferation	MDA-MB-453	1997	245
BRCA2 is required for ionizing radiation-induced assembly of rad51 complex in vivo	HCC1937	1999	227
Cyclooxygenase-2 is overexpressed in HER-2/neu-positive breast cancer—Evidence for involvement of AP-1 and PEA3	184B5	2002	212
Differential expression of CYP1A1 and CYP1B1 in human breast epithelial cells and breast tumor cells	MDA-MB-157	1998	193
Induction of cyclo-oxygenase-2 mRNA by prostaglandin E-2 in human prostatic carcinoma cells	MDA-MB-134	1997	163
Enhanced invasiveness of breast cancer cell lines upon co-cultivation with macrophages is due to TNF-alpha dependent up-regulation of matrix metalloproteases	HME-1	2004	131
COMPLEMENTARY-DNA CLONING OF A NOVEL EPITHELIAL-CELL MARKER PROTEIN, HME1, THAT MAY BE DOWN-REGULATED IN NEOPLASTIC MAMMARY CELLS	HME-1	1992	129
A central role for HER3 in HER2-amplified breast cancer: implications for targeted therapy	MDA-MB-175	2008	124
Cytotoxic effects of adenovirus-mediated wild-type p53 protein expression in normal and tumor mammary epithelial cells	MDA-MB-157	1995	110
INFRARED OBSERVATIONS OF DUST CLOUD STRUCTURE IN YOUNG R-ASSOCIATIONS—NGC-1333, S68, AND NGC-7129	S68	1984	110
RNA interference-mediated knockdown of DNA methyltransferase 1 leads to promoter demethylation and gene re-expression in human lung and breast cancer cells	HCC1954	2004	99
Estrogen promotes growth of human thyroid tumor cells by different molecular mechanisms	CAMA-1	2001	88
Chromosomal alterations in 15 breast cancer cell lines by comparative genomic hybridization and spectral karyotyping	CAMA-1	2000	85
BRCA1 and BRCA2 mRNA levels are coordinately elevated in human breast cancer cells in response to estrogen	BT-483	1996	82

This table was made utilizing a Web of Knowledge search using the name of each cell line to search for topic and the number of times cited as a measure of their degree of significance. The top three papers (cited at least 100 times) generated utilizing each cell line was used to narrow the selection. This table shows the main articles cited, listed in decreasing order

Literature Cited

1. Skloot R (2010) The immortal life of Henrietta lacks. Crown Publishers, New York
2. Rahbari R, Sheahan T, Modes V et al (2009) A novel L1 retrotransposon marker for HeLa cell line identification. Biotechniques 46:277–284
3. Shay JW, Wright WE (2000) Hayflick, his limit, and cellular ageing. Nat Rev Mol Cell Biol 1:72–76
4. Russo J, Russo IH (2004) In vitro models for human breast cancer. In: Molecular basis of breast cancer: prevention and treatment. Springer, Berlin, pp 227–280
5. Lehman TA, Modali R, Boukamp P et al (1993) p53 Mutations in human immortalized epithelial cell lines. Carcinogenesis 14:833–839
6. Stampfer MR, Bartley JC (1985) Induction of transformation and continuous cell lines from normal human mammary epithelial cells after exposure to benzo[a]pyrene. Proc Natl Acad Sci U S A 82:2394–2398
7. ATCC The Global Bioresource Center. http://www.atcc.org/
8. Soule HD, Maloney TM, Wolman SR et al (1990) Isolation and characterization of a spontaneously immortalized human breast epithelial cell line, MCF-10. Cancer Res 50:6075–6086
9. Tait L, Soule HD, Russo J (1990) Ultrastructural and immunocytochemical characterization of an immortalized human breast epithelial cell line, MCF-10. Cancer Res 50:6087–6094
10. Walen KH, Stampfer MR (1989) Chromosome analyses of human mammary epithelial cells at stages of chemical-induced transformation progression to immortality. Cancer Genet Cytogenet 37:249–261
11. Puck TT, Markus PI, Cieciura SJ (1956) Clonal growth of mammalian cells in vitro: growth characteristics of colonies from single HeLa cells with and without a "feeder" layer. J Exp Med 103(2):273–283
12. Ham RG (1965) Clonal growth of mammalian cells in a chemically defined, synthetic medium. Proc Natl Acad Sci U S A 53:288–293
13. Fisher HW, Puck TT, Sato G (1959) Molecular growth requirements of single mammalian cells: quantitative colonial growth of single S3 cells in a medium containing synthetic small molecular constituents and two purified protein fractions. J Exp Med 109:649–660
14. Gwatkin RB (1960) Are macromolecules required for growth of single isolated mammalian cells? Nature 186:984–985
15. Hamilton GW, Ham RG (1977) Clonal growth of Chinese hamster cell lines in protein-free media. In Vitro 13:537–547
16. Dhulipala P, Liu M, Beatty S, Sabourin M, Piras G, Barrett S, Hassett R, Gorfien S (2011) Differential cell culture media for single-cell cloning. Bioprocess Int 9(11):44–51
17. Russo J, Fernandez SV, Russo PA et al (2006) 17-Beta-estradiol induces transformation and tumorigenesis in human breast epithelial cells. FASEB J 20:1622–1634
18. Huang Y, Fernandez SV, Goodwin S et al (2007) Epithelial to mesenchymal transition in human breast epithelial cells transformed by 17beta-estradiol. Cancer Res 67:11147–11157
19. Tiezzi DG, Fernandez SV, Russo J (2007) Epithelial mesenchymal transition during the neoplastic transformation of human breast epithelial cells by estrogen. Int J Oncol 31:823–827
20. Kocdor H, Kocdor MA, Russo J et al (2009) Human chorionic gonadotropin (hCG) prevents the transformed phenotypes induced by 17 beta-estradiol in human breast epithelial cells. Cell Biol Int 33:1135–1143
21. Russo J, Hasan Lareef M, Balogh G et al (2003) Estrogen and its metabolites are carcinogenic agents in human breast epithelial cells. J Steroid Biochem Mol Biol 87:1–25
22. Lareef MH, Garber J, Russo PA et al (2005) The estrogen antagonist ICI-182-780 does not inhibit the transformation phenotypes induced by 17-beta-estradiol and 4-OH estradiol in human breast epithelial cells. Int J Oncol 26:423–429
23. Boyden S (1962) The chemotactic effect of mixtures of antibody and antigen on polymorphonuclear leucocytes. J Exp Med 115:453–466

24. Kleinman HK, McGarvey ML, Liotta LA, Robey PG, Tryggvason K, Martin GR (1982) Isolation and characterization of type IV procollagen, laminin, and heparin sulfate proteoglycan from the EHS sarcoma. Biochemistry 21:6188–6193
25. Terranova VP, Hujanen ES, Loeb DM, Martin GR, Thornburg L, Glushko V (1986) Use of a reconstituted basement membrane to measure cell invasiveness and select for highly invasive tumor cells. Proc Natl Acad Sci U S A 83:465–469
26. Terranova VP, Hujanen ES, Martin GR (1986) Basement membrane and the invasive activity of metastatic tumor cells. J Natl Cancer Inst 77:311–316
27. Livant DL, Linn S, Markwart S, Shuster J (1995) Invasion of selectively permeable sea urchin embryo basement membranes by metastatic tumor cells, but not by their normal counterparts. Cancer Res 55:5085–5093
28. Palmer TD, Ashby WJ, Lewis JD et al (2011) Targeting tumor cell motility to prevent metastasis. Adv Drug Deliv Rev 63:568–581
29. Albini A, Iwamoto Y, Kleinman HK, Martin GR, Aaronson SA, Kozlowski JM, McEwan RN (1987) A rapid in vitro assay for quantitating the invasive potential of tumor cells. Cancer Res 47:3239–3245
30. Gehlsen KR, Wagner HN, Hendrix MJC (1984) Membrane invasion culture system (MICS). Med Instrum 18:268–271
31. Hendrix MJC, Seftor EA, Seftor REB, Fidler IJ (1987) A simple quantitative assay for studying the invasive potential of high and low human metastatic variants. Cancer Lett 38:137–147
32. Chou R-H, Lin K-C, Lin S-C, Cheng J-Y, Wu C-W, Chang W-SW (2004) Cost-effective trapezoidal modified Boyden chamber with comparable accuracy to a commercial apparatus. Biotechniques 37(5):724–726
33. Bise R, Kanade T, Yin Z et al (2011) Automatic cell tracking applied to analysis of cell migration in wound healing assay. Conf Proc IEEE Eng Med Biol Soc 2011:6174–6179
34. Cory G (2011) Scratch-wound assay. Methods Mol Biol 769:25–30
35. Fotos JS, Patel VP, Karin NJ et al (2006) Automated time-lapse microscopy and high-resolution tracking of cell migration. Cytotechnology 51:7–19
36. Goetsch KP, Niesler CU (2011) Optimization of the scratch assay for in vitro skeletal muscle wound healing analysis. Anal Biochem 411:158–160
37. Liang CC, Park AY, Guan JL (2007) In vitro scratch assay: a convenient and inexpensive method for analysis of cell migration in vitro. Nat Protoc 2:329–333
38. Niinaka Y, Haga A, Raz A (2001) Quantification of cell motility: gold colloidal phagokinetic track assay and wound healing assay. Methods Mol Med 58:55–60
39. Rodriguez LG, Wu X, Guan JL (2005) Wound-healing assay. Methods Mol Biol 294:23–29
40. Rose MT (2012) Effect of growth factors on the migration of equine oral and limb fibroblasts using an in vitro scratch assay. Vet J 193:539–544
41. Vultur A, Buettner R, Kowolik C et al (2008) SKI-606 (bosutinib), a novel Src kinase inhibitor, suppresses migration and invasion of human breast cancer cells. Mol Cancer Ther 7:1185–1194
42. Perou CM, Sorlie T, Eisen MB et al (2000) Molecular portraits of human breast tumours. Nature 406:747–752
43. Ibrahim E, Al-Gahmi AM, Zeenelin AA et al (2009) Basal vs. luminal A breast cancer subtypes: a matched case-control study using estrogen receptor, progesterone receptor, and HER-2 as surrogate markers. Med Oncol 26:372–378
44. Kao J, Salari K, Bocanegra M et al (2009) Molecular profiling of breast cancer cell lines defines relevant tumor models and provides a resource for cancer gene discovery. PLoS One 4:e6146
45. Bastien RR, Rodriguez-Lescure A, Ebbert MT et al (2012) PAM50 breast cancer subtyping by RT-qPCR and concordance with standard clinical molecular markers. BMC Med Genomics 5:44
46. Brenton JD, Carey LA, Ahmed AA et al (2005) Molecular classification and molecular forecasting of breast cancer: ready for clinical application? J Clin Oncol 23:7350–7360
47. Sorlie T, Perou CM, Tibshirani R et al (2001) Gene expression patterns of breast carcinomas distinguish tumor subclasses with clinical implications. Proc Natl Acad Sci U S A 98:10869–10874
48. Carey LA, Dees EC, Sawyer L et al (2007) The triple negative paradox: primary tumor chemosensitivity of breast cancer subtypes. Clin Cancer Res 13:2329–2334

49. Rouzier R, Perou CM, Symmans WF et al (2005) Breast cancer molecular subtypes respond differently to preoperative chemotherapy. Clin Cancer Res 11:5678–5685
50. Nguyen PL, Taghian AG, Katz MS et al (2008) Breast cancer subtype approximated by estrogen receptor, progesterone receptor, and HER-2 is associated with local and distant recurrence after breast-conserving therapy. J Clin Oncol 26:2373–2378
51. Russo J, Russo IH (2011) The role of the basal stem cell of the human breast in normal development and cancer. Adv Exp Med Biol 720:121–134
52. Charafe-Jauffret E, Ginestier C, Monville F et al (2006) Gene expression profiling of breast cell lines identifies potential new basal markers. Oncogene 25:2273–2284
53. Neve RM, Chin K, Fridlyand J et al (2006) A collection of breast cancer cell lines for the study of functionally distinct cancer subtypes. Cancer Cell 10:515–527
54. Sarrio D, Rodriguez-Pinilla SM, Hardisson D et al (2008) Epithelial-mesenchymal transition in breast cancer relates to the basal-like phenotype. Cancer Res 68:989–997
55. Klopocki E, Kristiansen G, Wild PJ et al (2004) Loss of SFRP1 is associated with breast cancer progression and poor prognosis in early stage tumors. Int J Oncol 25:641–649
56. Thiery JP (2003) Epithelial-mesenchymal transitions in development and pathologies. Curr Opin Cell Biol 15:740–746
57. Althuis MD, Brogan DD, Coates RJ et al (2003) Breast cancers among very young premenopausal women. Cancer Causes Control 14:151–160
58. Zhai Y, Santucci-Pereira J, Lopez De Cicco R et al (2013) An in vitro–in vivo model of epithelial mesenchymal transition in triple negative breast cancer. Drug Discov Today Dis Mech 9:e35–e40

Chapter 6
In Situ Methods for Identifying the Stem Cell of the Normal and Cancerous Breast

Yanrong Su, Pedro J. Gutiérrez Diez, Julia Santucci-Pereira, Irma H. Russo, and Jose Russo

Keywords Cancer stem cells • ALDH1 • CD24 • CD44 • Cytokeratin 5 • EpCAM • CD29/β1 integrin • CD61 • CD49f/α6 integrin • CD133/2

6.1 Introduction

Stem cells are unspecialized cells with the ability of self-renewal, a high potential for proliferation, and the ability to become a variety of cell types in the body. There are basically four types of stem cells: Embryonic stem cells, which are isolated from the inner cell mass of blastocysts; adult stem cells, which are found in various tissues including umbilical cord blood, bone marrow, mammary, brain, endothelium, etc.; amniotic stem cells, which are found in amniotic fluid; and inducible pluripotent stem cells—reprogrammed cells (e.g. epithelial cells) given pluripotent capabilities. Mammary stem cells belong to adult stem cells; these cells provide the source of cells for the growth of the mammary gland during puberty and gestation. Single such cells can give rise to both luminal and myoepithelial cell types within the gland, and have the ability to regenerate the entire organ in mice. The practical definition of a stem cell is the functional definition—a cell that has the potential of self-renewal and to regenerate tissue over a lifetime. The stem cell markers used are genes or products used to isolate and identify stem cells.

Y. Su, Ph.D. • J. Santucci-Pereira, Ph.D. • I.H. Russo, M.D. • J. Russo, M.D. (✉)
Irma H Russo MD Breast Cancer Research Laboratory, Fox Chase Cancer Center, Cottman Avenue 333, Philadelphia, PA 19111, USA
e-mail: jose.russo@fccc.edu

P.J.G. Diez, Ph.D.
School of Economics, University of Valladolid, Valladolid, Spain

J. Russo and I.H. Russo, *Techniques and Methodological Approaches in Breast Cancer Research*, DOI 10.1007/978-1-4939-0718-2_6,

6.2 Cancer Stem Cells

Cancer stem cells (CSCs) are cancer cells that can be found within tumors or hematological cancers. These cells differ from other non-tumorigenic cells (tumor-forming) by possessing characteristics of self-renewal and, specifically, the ability to give rise to all cell types found in a particular cancer sample. Such cells are proposed to persist in tumors that are usually resistant to normal chemotherapy and radiotherapy as a distinct population and cause relapse and metastasis by initiating new tumors. The presence of stem cells in the breast has been postulated [1] and several methods have been developed in order to study breast stem cells. Among theses methods are flow cytometry (FCM) or FACS, in vitro culture to generate mammospheres, colony formation assay, 3D morphogenesis assay (3D culture in matrigel), and in vivo repopulation assay [1].

6.3 Accepted Markers for Identifying Mammary Stem Cells or Breast Cancer Stem Cells

Our objective in this chapter is to discuss the use of immunocytochemical markers for the evaluation of the localization of these putative stem cells in primary breast cancer and the markers' possible relation to stages of the disease, tumor type, lymph node involvement, and distant metastasis. The markers suitable for paraffin-embedded tissue for in situ localization of stem cells are listed in Table 6.1.

6.4 Calibration of the Antibodies for ICC Using a Test Tissue Microarray

Table 6.1 depicts the histological appearance of primary breast cancer immunostained with several antibodies suitable for detecting stem cells in paraffin-embedded tissue. We have identified stem cell/s in primary breast cancer tissue embedded in paraffin utilizing immunocytochemical markers of stemness (Table 6.1). We have also aimed to establish a correlation between clinical stage and the presence or absence of metastasis at the time of surgery. For this purpose we have analyzed around 16 samples identified as triple negative breast cancer and around 32 samples classified as ER negative. The cells were quantified based on the presence of staining. The stain indicated the presence of the protein, and therefore occurred in the cell membrane and cytoplasm, or nuclei. Each tumor cell was counted as either positive or negative by using Image J to label the cells as either Type 1 for positive, or Type 2 for negative. This process was repeated for all of the samples. After quantification, the results were transferred to an Excel spreadsheet. Through the use of Excel, a percentage breakdown was calculated for both positive and negative cells. This method was chosen in order to clearly show the amount of stem cell activity within each tumor. Additionally, if a tumor had multiple image cores to be

Table 6.1 Calibration of immunocytochemical reagents for stemness

Antibody name	Source	Species reactivity	Concentration	Incubation time (min)	Example
ALDH1	Mouse monoclonal; BD Biosciences	Human	0.31 μg/mL	30	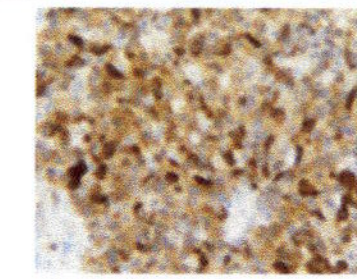
Anti-integrin beta 1	Rabbit monoclonal; abcam	Mouse, rat, human	1:400 (original concentration not determined)	30	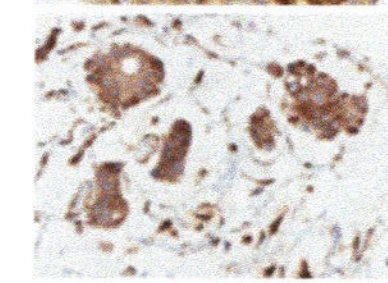
Anti-ITGA6/ CD49f	Rabbit polyclonal; Lifespan Biosciences	Human	5 μg/mL	30	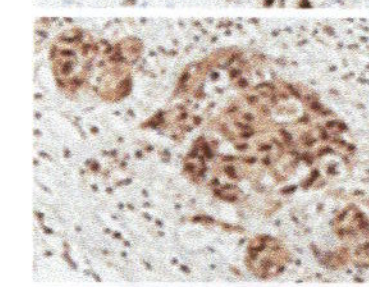
CD24	Rabbit polyclonal Ig; ABBIOTEC	Human, mouse, rat	6.7 μg/mL	30	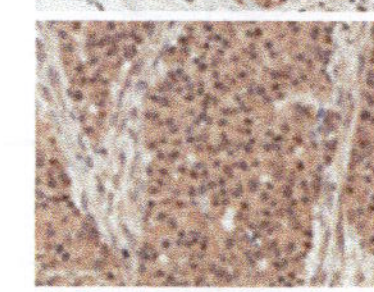
CD44	Mouse monoclonal; BioGenex	Human	Prediluted Ab (concentration not determined)	30	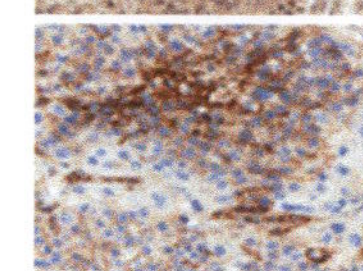

(continued)

Table 6.1 (continued)

Antibody name	Source	Species reactivity	Concentration	Incubation time (min)	Example
CD61	Rabbit IgG; BioGenex	Human	Prediluted Ab (concentration not determined)	30	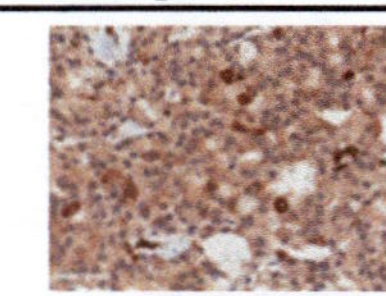
Cytokeratin 5	Rabbit IgG; BioGenex	Human	Prediluted Ab (concentration not determined)	30	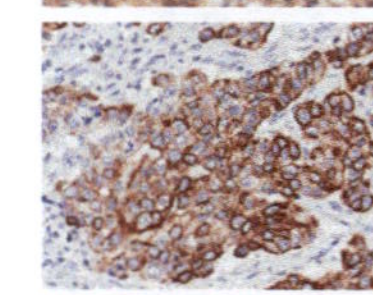
EpCAM	Rabbit IgG; BioGenex	Human	Prediluted Ab (concentration not determined)	30	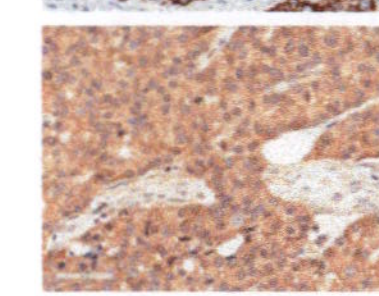

quantified then the results from all of the tumor's images were added together and the final percentages were calculated. For some of the genes, the stain was evaluated based on the staining intensity on each cancer cell and scored from 0 to 3+, where 0 was for negative staining, 1+ for weak staining, 2+ for moderate staining, and 3+ for strong staining. Upon the completion of all the images and once the final counts and percentages were obtained, statistical analysis was performed on the data by a statistician. We used regression analysis to study how the intensity of staining was related to the different stages. To do so, we considered the selected staining degree or intensity in percentage as regressand and the dummy variables defined by the occurrence of the different stages as regressors, assigning the value 1 if a particular stage happens and 0 otherwise. The coefficients associated with the different stages therefore measure the sensibility of the regressand (the selected staining degree) to the presence of the specific associated stage [2].

6.4.1 *The ALDH1 as a Marker of Stem Cell in Primary Breast Cancer*

ALDH1 is a detoxifying enzyme responsible for the oxidation of intracellular aldehydes, important for normal development, and homoeostasis in several organs and crucial during embryogenesis. It is an important detoxifying enzyme in the liver, also expressed in kidney as well as hematopoetic progenitor cells. Isolated ALDH1 positive cells from reduction mammoplasties (as detected by the Aldefluor assay) showed containing the subpopulation with stem cell properties, such as it can form mammospheres, differentiate into luminal and myoepithelial lineages, and has in vivo outgrowth potential. Immunostaining of Aldefluor sorted cells showed only 15 % cells were ALDH1 positive. The in vivo outgrowth potential of normal human breast epithelial cells sorted by the ALDEFLUOR can be assayed in NOD/scid mice fat pads. In the normal breast tissue, ALDH1-positive cells appeared to form a bridge in the lumen that could be located at the bifurcation point of side branches in the TDLUs breast. This marker can also be used in human breast tumors and only 2 of the 577 tumors analyzed were fully positive for ALDH1 [3]. Patients with tumors positive for ALDH1 staining had a poor prognosis compared to patients with tumors negative for ALDH1 staining, indicating ALDH1 is a predictor of poor clinical outcome. According to Park [4] in normal breast, ALDH1 was expressed heterogeneously in stromal, luminal, and basal cells, and in some lobules, all cells were ALDH1+. ALDH1 expression varied depending on the area of the tissue. ER negative tumors were more likely to be ALDH1 positive than ER positive tumors. The reactivity of human breast cancer cell lines also varies as demonstrated by Wendy [5].

In our study there was no relationship between the expression of ALDH1 and clinical stage in both triple and ER negative tumors. When samples negative for lymph node involvement were compared to those with lymph node involvement and distant metastasis, no significant difference was observed in this sample size. The samples with a higher number of ALDH1 positive cells were usually positive for

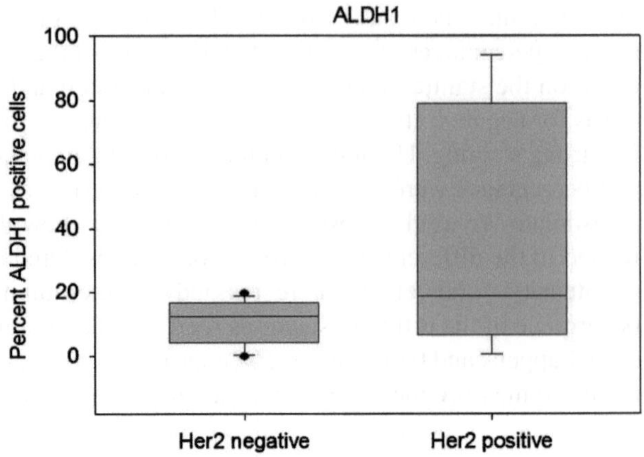

Fig. 6.1 Box plot shows in ER negative tumors, the expression of ALDH1 tends to relate to HER2 status (Mann–Whitney Rank Sum Test, $P=0.108$)

Table 6.2 Expression of ALDH1 in triple negative and ER negative tumors

Group	Sample size	Median (%)	Min (%)	Max (%)	25 %	75 %
%ALDH1+ cells						
Triple negative tumors	14	12.6	0.0	19.5	4.3	16.1
ER negative tumors	29	13.0	0.0	94.1	4.8	18.4
%ALDH1+ cells in ER negative tumors						
Lymph node negative	8	14.4	0.0	73.9	8.6	26.4
Lymph node positive or distant metastasis	16	12.5	0.7	94.1	6.6	18.1
P value: 0.646						

HER2 expression (Fig. 6.1). In triple negative tumors (14 samples), the percentage of ALDH1 positive cells ranged from 0 to 19.5 %, with the median of 12.6 %. In ER negative tumors (29 samples), the range was between 0 and 94.1 % positive, with the median of 13.0 % (Table 6.2 and Figs. 6.2 and 6.3).

6.4.2 *CD24*

CD24 is known as heat stable antigen (HSA) and it is the only known ligand for P-selectin. It is expressed at the surface of most B lymphocytes and differentiating neuroblasts, and in hematological malignancies as well as in a large variety of solid tumors, including breast cancer. CD44+/CD24– has been used as breast CSC markers in human whereas lin-CD29high/CD24high is more suitable to identify mammary stem cells/CSCs in mouse [6]. Shackleton found CD29hiCD24+ cells showed the highest rate of outgrowth formation; transplantation of a single Lin-CD29hiCD24+ cell was capable of generating a functional mammary gland [6]. Zhang et al. [7]

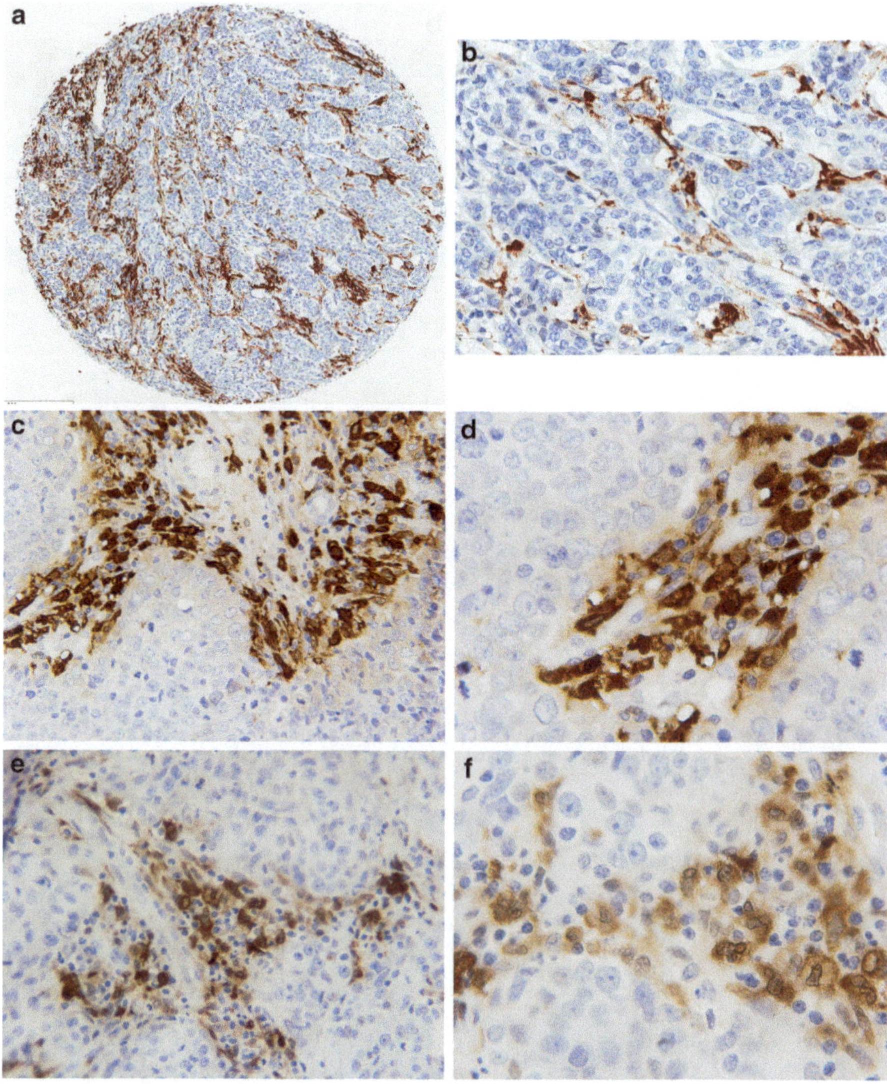

Fig. 6.2 ALDH1 expression in triple negative cancer. (**a**) (×8) and (**b**) (×40), stage 3, 1-11-01090. Reactivity to ALDH1: 17.9 %. (**c**) (×20) and (**d**) (×40), stage 2a, 1-02-02836. Reactivity to ALDH1: 15.2 %. (**e**) (×20) and (**f**) (×40), stage 4, 1-11-00572. Reactivity to ALDH1: 0.5 %

obtained tumor-initiating cells enriched in Lin-CD29^{H}CD24^{H} subpopulation in the P53 null mouse model of breast cancer. This population generated tumors that mimic the parent tumors in vivo and showed increased number of secondary mammospheres in vitro. One discrepancy is that Sleeman [8] reported transplantation of FVB mouse CD24Low populations into female syngeneic mice has the best chance

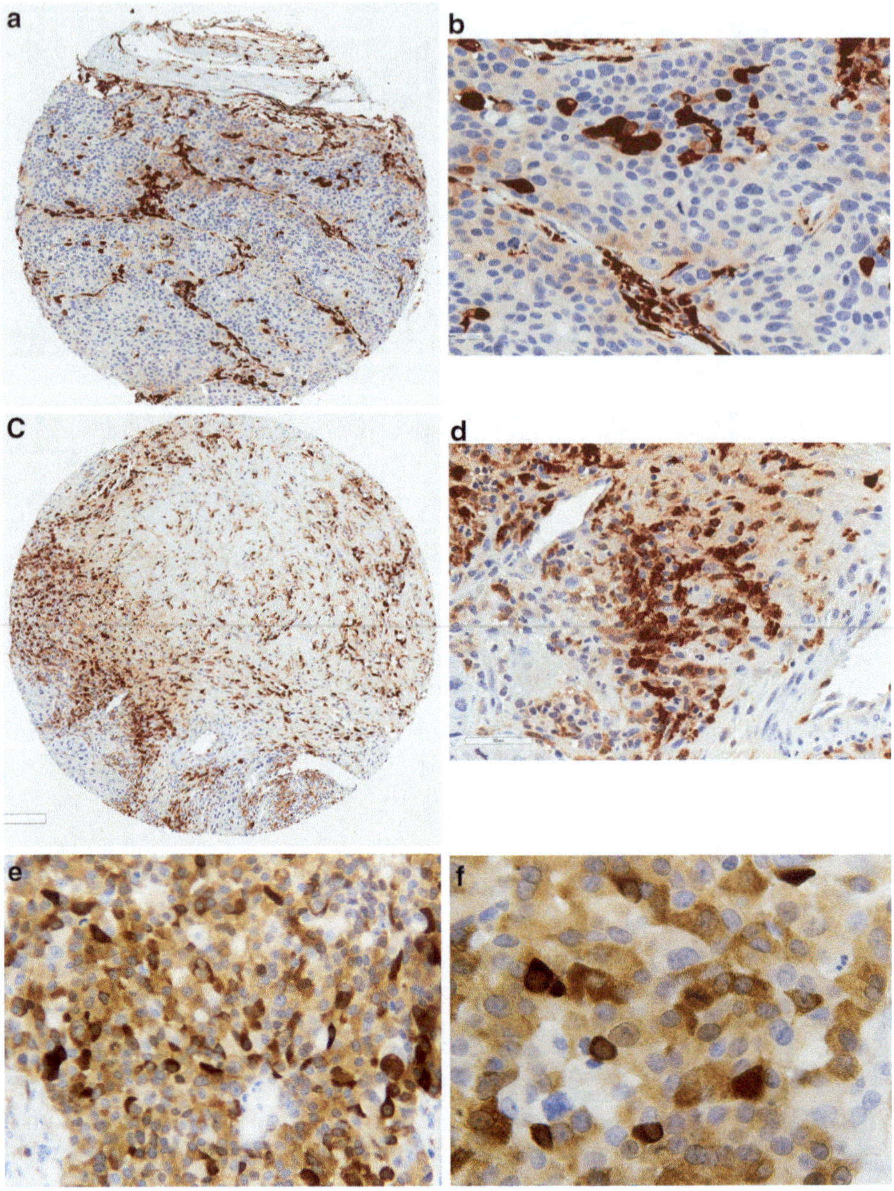

Fig. 6.3 ALDH1 expression in ER negative breast cancer. (**a**) (×8) and (**b**) (×40), stage 3, 1-08-05137. Reactivity to ALDH1: 15 %. (**c**) (×8) and (**d**) (×40), stage 3, 1-06-01378. Reactivity to ALDH1: 73.9 %. (**e**) (×20) and (**f**) (×40), stage 4, 1-11-00281. Reactivity to ALDH1: 84.4 %

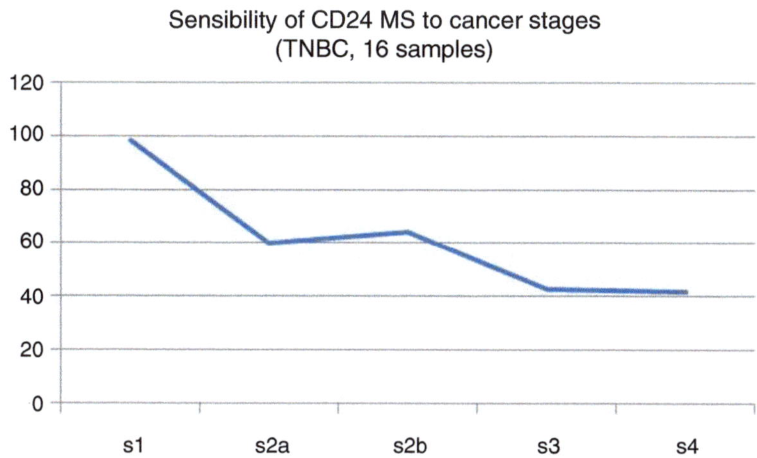

Fig. 6.4 The expression of CD24 (protein level) was reduced as tumors progressed in triple negative breast cancer (regression analysis, $R^2=0.63$). Graph shows the sensibility of moderate to strong staining of CD24 to stages in triple negative cancer by regression analysis. *Y* axis represents sensibility

to repopulate the normal mammary gland compared to that of CD24 High and CD24 Negative populations. We have found the HCG treatment of rats resulting in down regulation of CD24 expression in mammary gland. In the normal human breast, CD24 is located in the luminal cells [9]. In breast cancer 84.6 % (170/201) of tumors are CD24 positive. Cytoplasmic CD24 positivity was associated with shorter disease-free survival ($P=0.0008$) and overall survival ($P=0.031$) [10]. CD24− cells have a high rate of forming tumors after transplantation of human breast cancer cells. CD44+CD24− cancer cells are highly tumorigenic and could be passaged in culture [11].

In our study, CD24 expression level (quantified by intensity scoring) was significantly associated with clinical stage in triple negative breast cancer, the protein level was dramatically reduced in stage 3–4 tumors compared to the stage 1–2 tumors (Figs. 6.4, 6.5, and 6.6). However, the percentage of CD24 positive cells didn't change too much; more than 90 % of tumor cells still had CD24 expression even in stage 4 but the staining was much weaker compared to that at stage 1–2 which showed moderate staining in the majority of samples, indicating that the reduction of CD24 is important for the progression of the disease in triple negative breast cancer. In ER negative tumors, both CD24 expression level and percentage of positive cells were not related to the clinical stage. Noteworthy, the expression of CD24 in ER negative tumors was associated with positive HER2 status where 7 out of 9 samples positive for HER2 showed more than 50 % of tumor cells were moderately stained, whereas only 4 out of 14 samples negative for HER2 showed the same staining for CD24 (Chi-square, $P=0.033$).

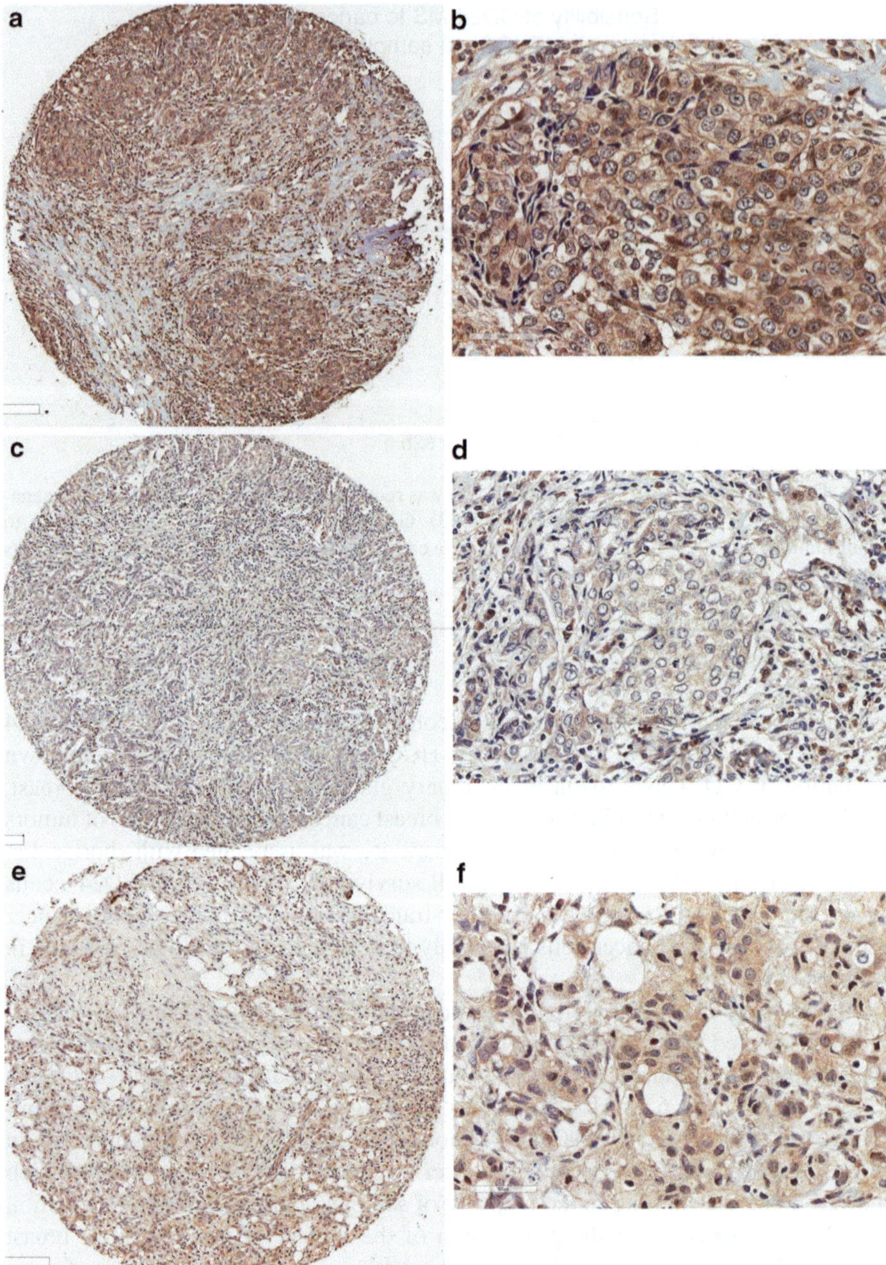

Fig. 6.5 CD24 expression in triple negative cancer. (**a**) (×8) and (**b**) (×40), stage 1, 1-02-02641. CD24 reactivity: 3.2 % weak, 95.4 % moderate, 1.4 % strong. (**c**) (×8) and (**d**) (×40), stage 3b, 1-11-01090. CD 24 reactivity: 2.9 % negative, 97.1 % weak. (**e**) (×8) and (**f**) (×40), stage 4, 1-10-02789. CD24 reactivity: 2.3 % negative, 73.9 % weak, 23.8 % moderate

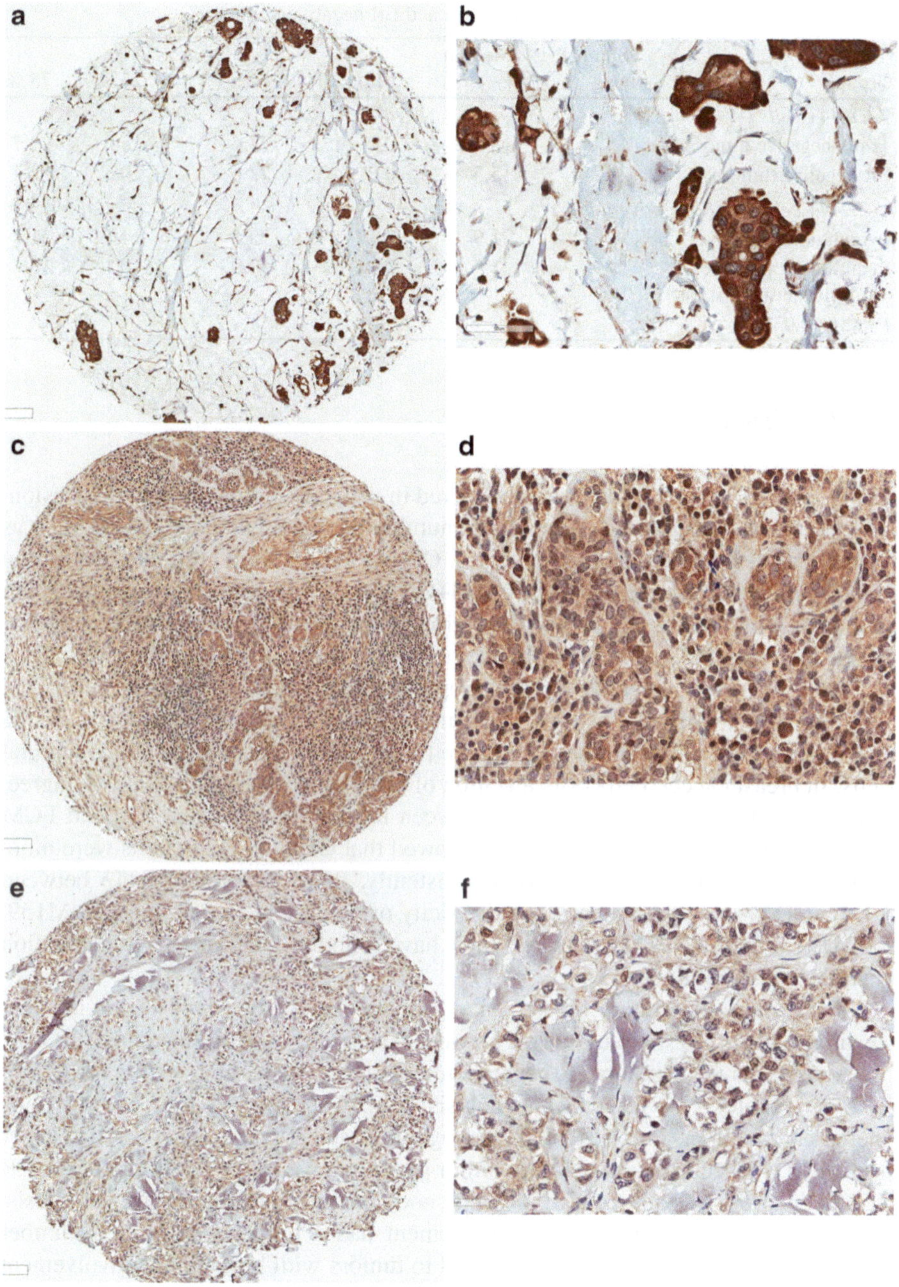

Fig. 6.6 CD24 expression in ER negative cancer. (**a**) (×8) and (**b**) (×40), stage 2b, 1-02-02484. CD 24 reactivity: 100 % strong. (**c**) (×8) and (**d**) (×40), stage 3, 1-06-01378. CD24 reactivity: 1.2 % weak, 98.8 % moderate, 0 % strong. (**e**) (×8) and (**f**) (×40), stage 4, 1-02-02566. CD24 reactivity: 0.7 % negative, 99.3 % weak, 0 % moderate, 0 % strong

Table 6.3 Expression of CD44 in triple negative and ER negative tumors

Group	Sample size	Median (%)	Min (%)	Max (%)	25 %	75 %
%CD44+ cells						
Triple negative tumors	15	18.4	0.0	73.6	5.0	22.7
ER negative tumors	29	15.5	0.0	73.6	0.3	22.4
%CD44+ cells in ER negative tumors						
Lymph node negative	8	21.9	2.4	73.6	14.1	31.7
Lymph node positive or distant metastasis	21	3.5	0.0	57.9	0.1	18.6
P value: 0.033						

6.4.3 CD44

CD44 is a cell-surface glycoprotein involved in cell–cell interactions, cell adhesion, and migration. It is expressed in a large number of mammalian cell types, such as epithelium, leukocytes, and erythrocytes. CD44+ cells have a high rate of forming tumors after transplantation [11]. CD44+CD24– cancer cells are highly tumorigenic and could be passaged, this population is capable of generating tumors resembling the phenotypic complexity of the tumors from which they are derived. Thus CD44+CD24– cells seem to exhibit properties of CSCs [11]. Perrone et al. [12] described the distribution of CD44 as being mainly in the membrane of basal cells in the normal breast. In our work we have found staining in both basal and luminal cells. In breast cancer, cell co-localization of CD44 and CD24 do not always agree. Perrone et al. found a disagreement between immunofluorescence (IF) and FCM [12]. Immunostaining of breast cancer showed that CD44+CD24– cells were more common in basal-like tumors [4]. Consistently, there is a relationship between CD44+CD24– population and tumorigenicity of breast cancer cell lines, SUM159, SUM1315, and MDA-MB-231 cell lines have a higher CD44+CD24– population than the other cell lines [13].

In our study of primary breast cancer, the percentage of CD44+ cells in triple negative tumors (15 samples) varied between 0 and 73.6 % with a median of 18.4 %. The range was the same in ER negative tumors (29 samples) but with a median of 15.5 % (Table 6.3). Regression analysis (Fig. 6.7) showed that CD44– cells increased as tumors progressed from stage 1 to stage 2 and stage 3–4 (in other words, CD44+ cells decreased with tumor progression). CD44+ cell number in ER negative tumors was also related to lymph node involvement and distant metastasis, tumors negative for lymph node involvement had a higher CD44+ cell number (mean 26.9 %, median 21.9 %) compared to tumors with lymph node involvement and distance metastasis (mean 13.0 %, median 3.5 %) (Table 6.3, and Figs. 6.8, 6.9, and 6.10). As CD24$^{-/low}$ and CD44+ are often used in combination to isolate mammary or CSCs in human, we did regression analysis of CD24$^{-/low}$CD44+ combined, which showed that CD24$^{-/low}$CD44+ staining was increased from stage 1 to 3, then reduced from stage 3 to 4 in triple negative tumors (Fig. 6.11).

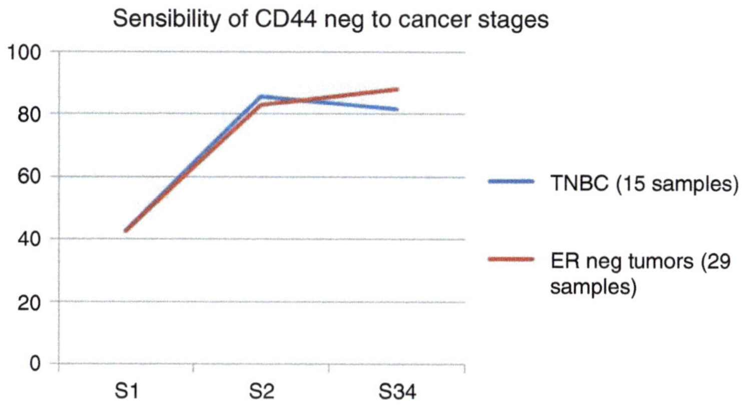

Fig. 6.7 The proportion of CD44 negative cells was increased as tumors progressed (regression analysis, R^2 was 0.47 for TNBC and 0.35 for ER negative tumors)

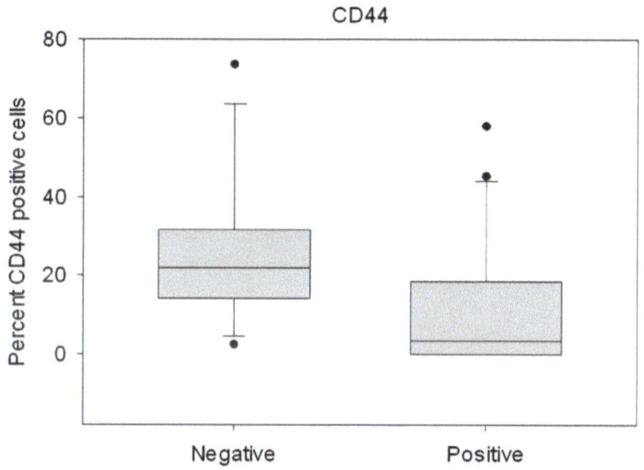

Fig. 6.8 The proportion of CD44 positive cells was reduced in ER negative tumors positive for lymph node and distant metastasis. Mann–Whitney Rank Sum Test, $P=0.033$

6.4.4 *Cytokeratin 5 (CK5)*

Immunostaining of human normal breast cells showed CK5 expressed in both luminal and basal cells (see Chap. 1). hTERT immortal hMEC exhibits stem/progenitor cell properties and express markers for well-known stem cell signature pathway. These cells fall into two distinct types—a K5+/K19− type and a K5+/K19+ type. Both types of progenitor cells have self-renewal and differentiation ability [14]. Lim et al. also showed that CD49f+EpCAM+ luminal progenitor cells and $CD49f^{hi}$EpCAM− basal and MaSC cells are positive for CK5 staining [15]. The CK5+ cells in the ducts of normal postmenopausal women were 75.7 ± 13.8 %

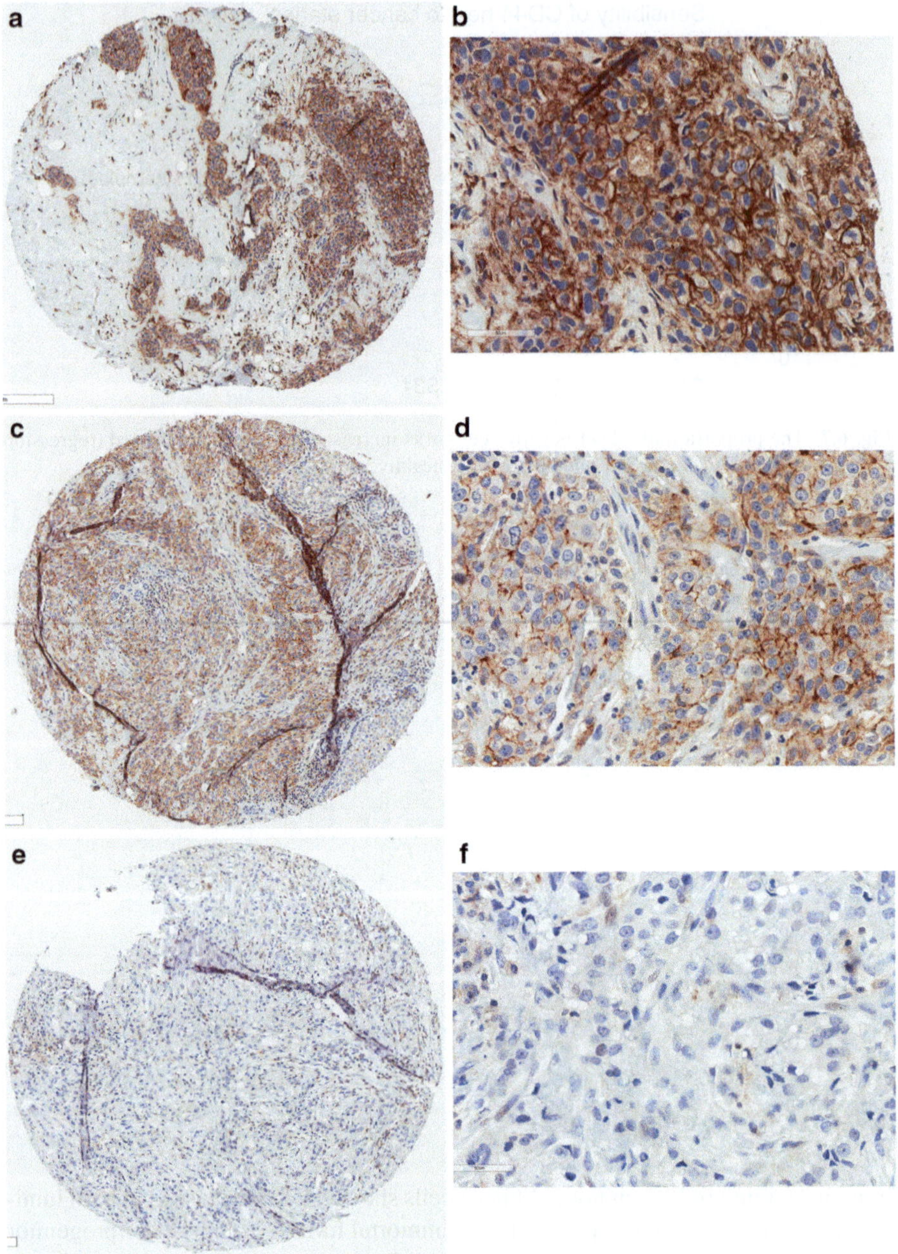

Fig. 6.9 CD44 expression in triple negative cancer. (**a**) (×8) and (**b**) (×40), stage 1, 1-06-01388. CD44 reactivity: 73.6 % positive. (**c**) (×8) and (**d**) (×40), stage 4, 1-11-00572. CD44 reactivity: 58 % positive. (**e**) (×8) and (**f**) (×40), stage 4, 1-10-02789. CD44 reactivity: 0 % positive

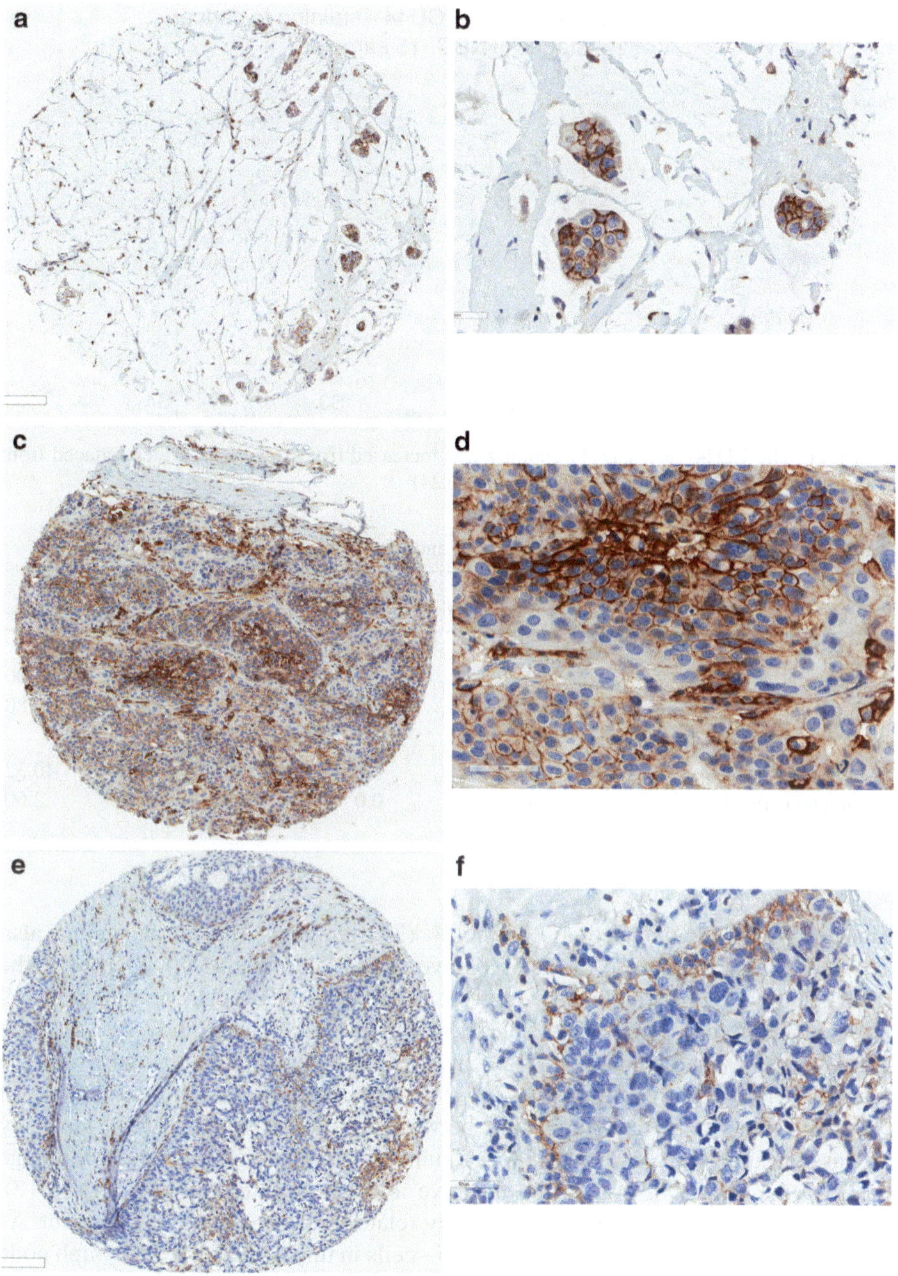

Fig. 6.10 CD44 expression in ER negative cancer. (**a**) (×8) and (**b**) (×40), stage 1, 1-06-02484. CD44 reactivity: 45.2 % positive. (**c**) (×8) and (**d**) (×40), stage 3a, 1-08-01537. CD44 reactivity: 43.4 % positive. (**e**) (×8) and (**f**) (×40), stage 3b, 1-09-01742. CD44 reactivity: 8.9 % positive

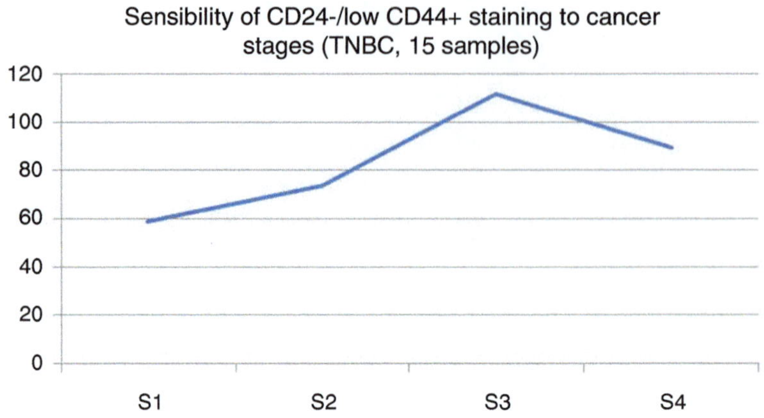

Fig. 6.11 The CD24−/low CD44+ staining was increased from stage 1 to 3, then reduced from stage 3 to 4 in TNBC (regression analysis, $R^2=0.24$)

Table 6.4 Expression of CK5 in triple negative and ER negative tumors

Group	Sample size	Median (%)	Min (%)	Max (%)	25 %	75 %
%CK5+ cells						
Triple negative tumors	19	3.22	0.0	76.1	1.35	26.00
ER negative tumors	40	2.02	0.0	76.1	0.77	11.80
%CK5+ cells in ER negative tumors						
Lymph node negative	16	9.7	0.0	76.1	0.67	40.32
Lymph node positive	20	1.5	0.0	43.1	0.65	2.60
P value: 0.085						

(mean ± SD), with the median of 78.6 % (24 samples). The fibroadenoma also showed a high percentage of CK5-positive cells (data not shown). Markedly, the CK5+ cells reduced significantly in triple negative and ER negative tumors. In 19 triple negative tumors we analyzed, CK5+ cells were between 0 and 76.1 %, with mean of 17.4 % and median of 3.2 %. In 40 ER negative tumors, the range was the same as triple negative tumors (0–76.1 %), with mean of 11.7 % and median of 2.0 % (Table 6.4). The relationship between CK5+ cells and clinical stage by regression analyses showed CK5+ cell number dropped significantly from stage 1 to further stages in both triple negative and ER negative tumors (Fig. 6.12). In addition, CK5+ cells might be inversely related to lymph node involvement. As shown in the Table 6.4, the median of CK5+ cells in tumors negative for lymph node was 9.7 % whereas it dropped to 1.5 % in tumors with lymph node involvement. Taken together, the loss of CK5 expression might be an early event in the development of breast cancer and the lower expression of CK5 was related to the lymph node involvement (Figs. 6.13 and 6.14). One explanation is that CK5 might be a marker of more differentiated stem cells, such as normal breast stem cells, upon the

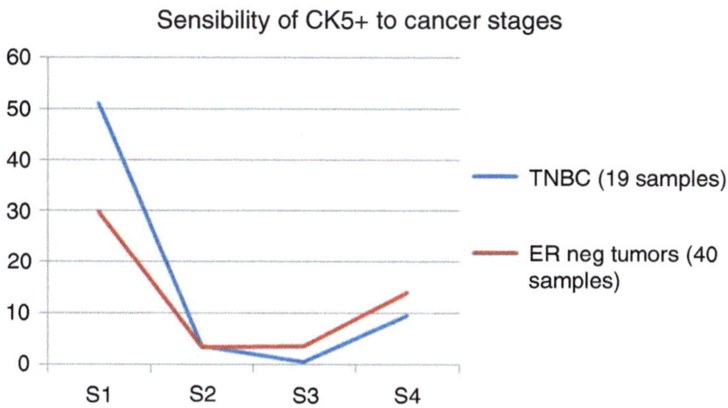

Fig. 6.12 The expression of CK5 was significantly reduced from stage 1 to further stages (regression analysis, R^2 was 0.87 for TNBC and 0.35 for ER negative cancer)

development of the breast cancer, this population was decreased whereas the CSCs (for example, ALDH1+ cells) were increased. This idea was supported by the finding that duct carcinoma in situ (DCIS) exhibited more cases positive for CK5 than DCIS accompanied with invasive breast cancers [16]. Functionally, CK5 is usually expressed in the basal layer of breast ducts or lobules, the loss of CK5 cells resulted in the disruption of basal layer and facilitated the invasion of tumor cells.

6.4.5 EpCAM

EpCAM/ESA is a pan-epithelial differentiation antigen expressed on most normal epithelial cells, almost all carcinomas, and progenitor and stem cells. According to Lim's study, CD49f+EpCAM+ population was considered as luminal progenitor, and CD49f^{hi}EpCAM− population was considered as stem cells and/or bipotent progenitor cells [15]. The luminal progenitor signature genes are more associated with basal cancers than other types [15]. Breast cancer cell lines contain CD44+/CD24−/ESA+ breast cancer stem-like cells [13]. CD44+/CD24−/low/ESA+ cells are enriched for tumorigenic potential as compared with the bulk of the cell line [13]. CD44+/CD24−/low/ESA+ cells transplantation form the tumors and recapitulate tumors derived from parental tumor [13]. ESA/EpCAM is highly expressed in metastatic breast cancers compared to matched primary breast cancers [17].

In our study, EpCAM was expressed in all triple negative and ER negative samples and no correlation was observed between the expression of EpCAM and clinical stage in triple negative and ER negative breast cancer from the sample size that we have evaluated (Figs. 6.15 and 6.16). The strong expression of EpCAM might be related to the lymph node involvement and distant metastasis in triple negative tumors ($P = 0.077$) (Table 6.5).

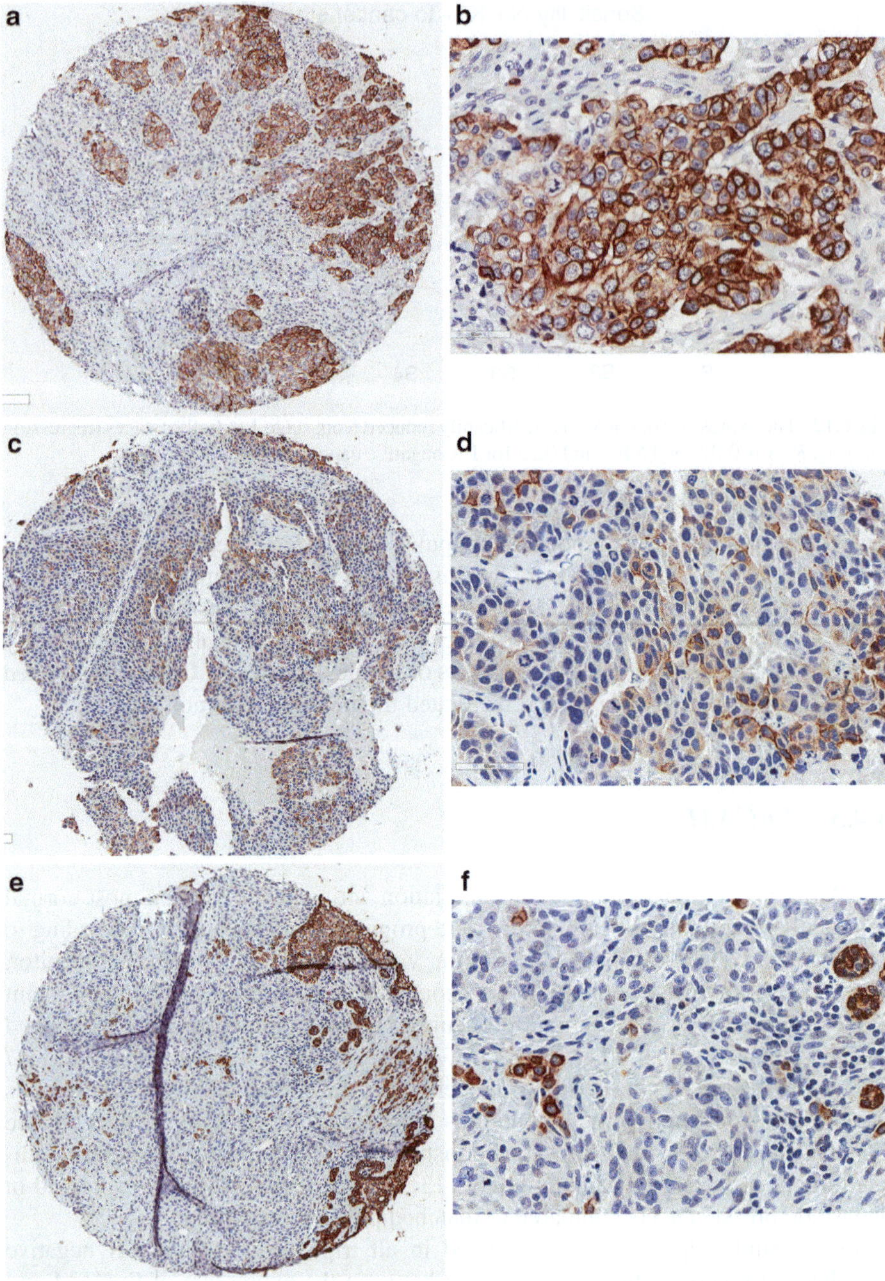

Fig. 6.13 CK5 expression in triple negative cancer. (**a**) (×8) and (**b**) (×40), stage 1, 1-02-02641. CK5 reactivity: 61.6 % positive. (**c**) (×8) and (**d**) (×40), stage 2a, 1-02-01660. CK5 reactivity: 11.9 % positive. (**e**) (×8) and (**f**) (×40), stage 4, 1-11-00572. CK5 reactivity: 17.2 % positive

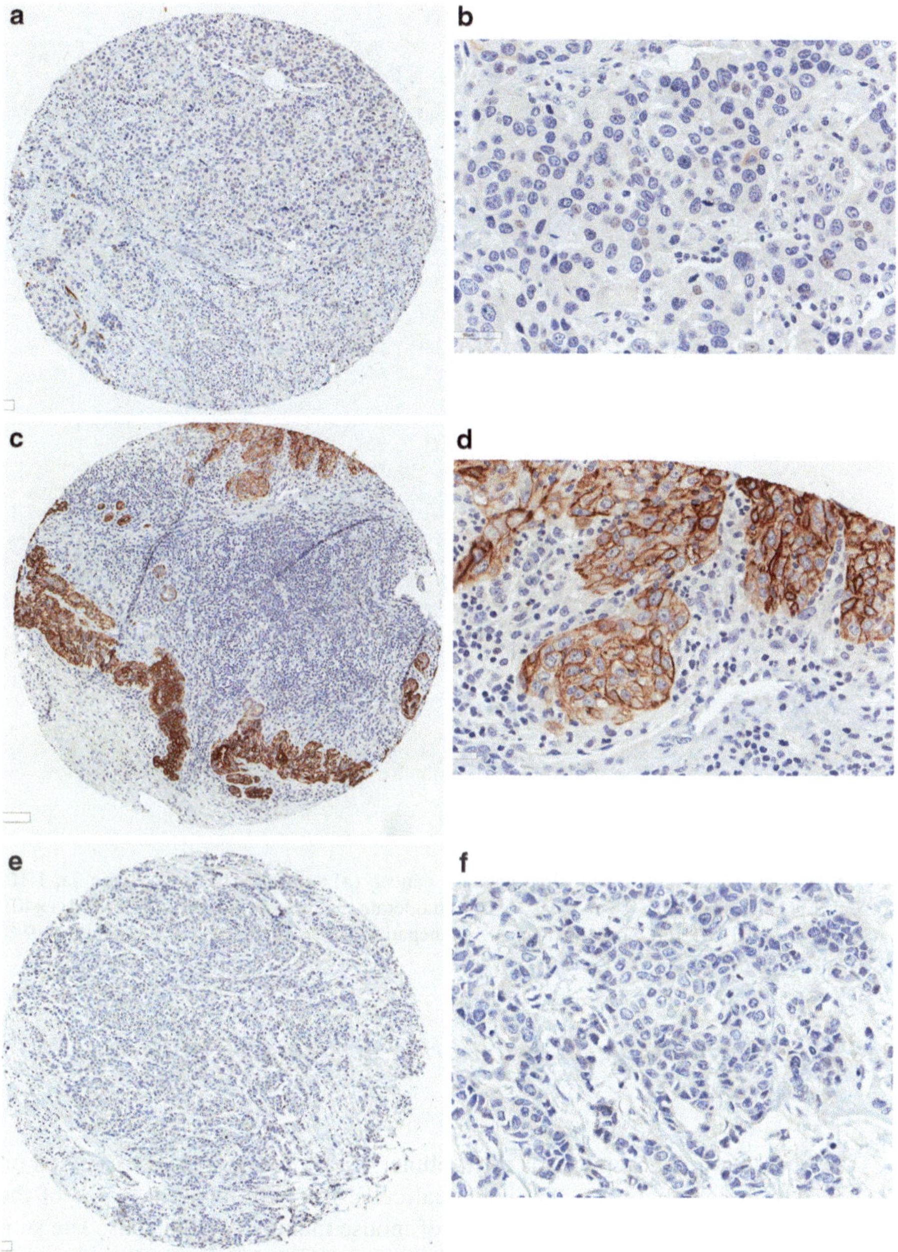

Fig. 6.14 CK5 expression in ER negative cancer. (**a**) (×8) and (**b**) (×40), stage 2a, 1-02-02341. CK5 reactivity: 1 % positive. (**c**) (×8) and (**d**) (×40), stage 3b, 1-06-01378. CK5 reactivity: 11.4 % positive. (**e**) (×8) and (**f**) (×40), stage 3b, 1-06-01429. CK5 reactivity: 0 % positive

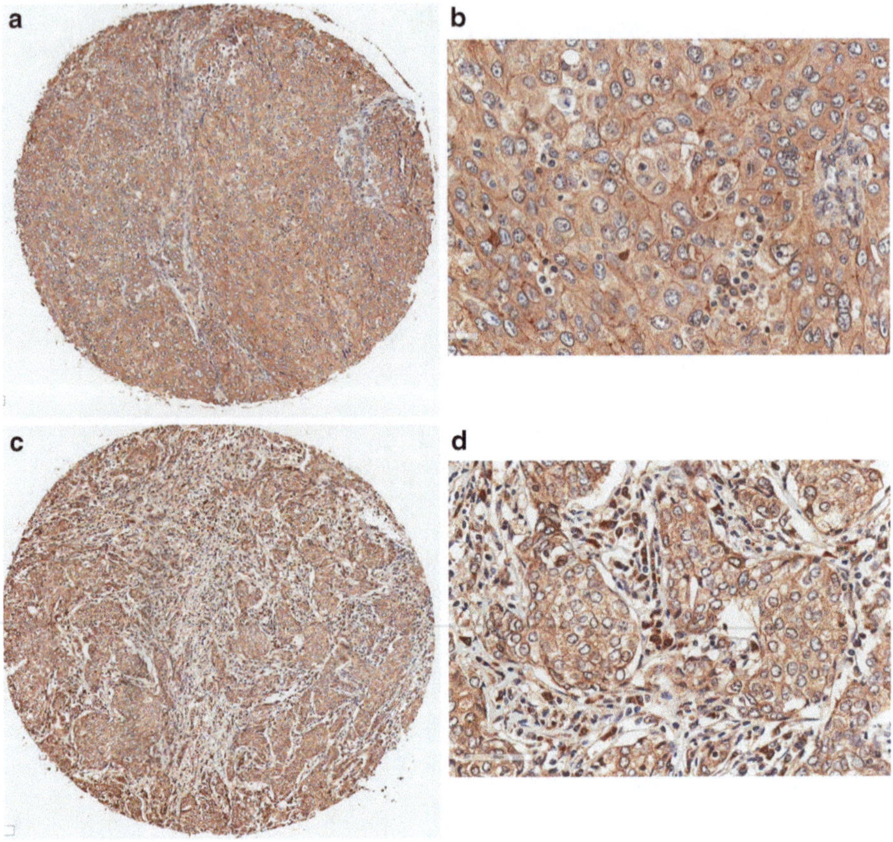

Fig. 6.15 EpCAM expression in triple negative cancer. (**a**) (×8) and (**b**) (×40), stage 2a, 1-10-02772. EpCAM reactivity: 0.4 % weak, 68.1 % moderate, 31.5 % strong. (**c**) (×8) and (**d**) (×40), stage 3b, 1-11-01090. EpCAM reactivity: 0.1 % negative, 0.3 % weak, 53.8 % moderate, 45.9 % strong

6.4.6 CD29/β1 Integrin as a Marker of Stem Cell in Breast Cancer

CD29/β1 integrin is an important extracellular matrix receptor. It is expressed on leukocytes, endothelial cells, and epithelial cells. CD29/β1 integrin is one of the most important markers for the isolation of mouse mammary stem cells. The subpopulation of CD29low/CD24+ was shown enriched for luminal progenitors whereas CD29high/CD24+ enriched for mammary stem cells which have the highest rate of outgrowth formation in mouse [6]. β1 Integrin (CD29) is required for mouse mammary epithelial morphogenesis. Conditional deletion of β1 integrin showed that it is essential for alveolar development and controls mammary epithelial differentiation. Cultured β1-integrin null cells displayed abnormal focal

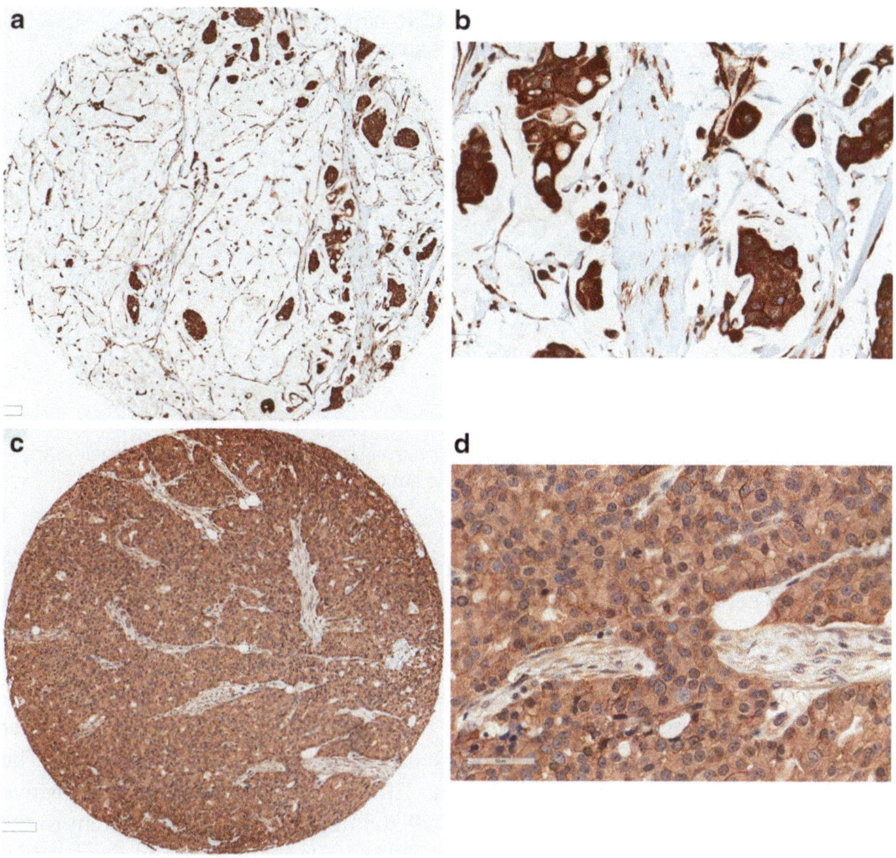

Fig. 6.16 EpCAM expression in ER negative cancer. (a) (×8) and (b) (×40), stage 2b, 1-02-02484, EpCAM reactivity: 0.7 % moderate, 99.3 % strong. (c) (×8) and (d) (×40), stage 4, 1-11-00281. EpCAM reactivity: 56.7 % moderate, 43.2 % strong

Table 6.5 Expression of CD61 in triple negative and ER negative tumors

Group	Sample size	Median (%)	Min (%)	Max (%)	25 %	75 %
%CD61 strong expression cells in triple negative tumors						
Lymph node negative	6	3.1	0.9	5.6	2.2	4.0
Lymph node positive or distant metastasis	8	8.3	2.7	10.3	3.1	9.8
$P = 0.027$						
%CD61+ cells in ER negative tumors						
Stage 1–2	12	83.8	17.6	97.1	72.9	95.3
Stage 3–4	13	97.3	78.3	99.6	94.0	99.4
P value: 0.005						

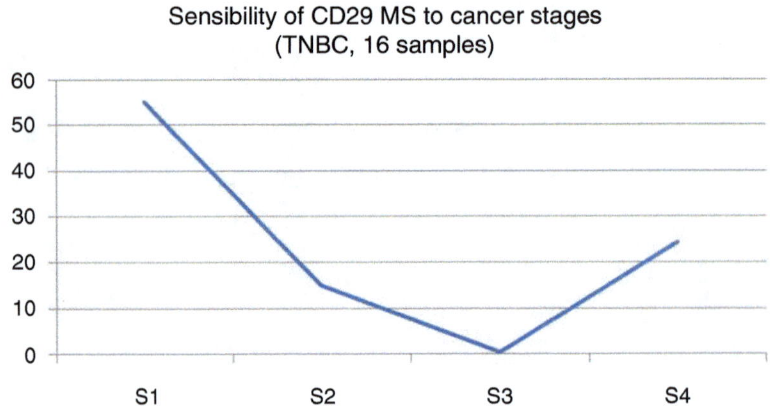

Fig. 6.17 Expression of CD29/β1Integrin beta1 is stronger in stage 1 than other stages in triple negative cancer (regression analysis of moderate to strong expression of CD29, $R^2=0.37$)

adhesion function and signal transduction and could not form or maintain polarized acini [18]. There are conflicting data regarding the relationship of the expression of CD29 with the overall survival and metastasis of breast cancer in published literatures [19–21]. Zutter reported that β1 integrin is reduced in invasive breast cancer compared to normal breast and fibroadenoma [22]. Our data generated by the intensity scoring showed that the expression of CD29 was stronger in stage 1 than other stages in triple negative (16 samples) breast cancer (Figs. 6.17, 6.18, and 6.19). The expression of CD29 was also related to the ER status in our staining, ER positive tumors showed negative or very weak CD29 staining, which was consistent to the observation in the mouse that CD29low/CD24+ cells were enriched for luminal progenitor cells [6].

6.4.7 CD61/β3 Integrin

CD61/β3 integrin is a member of the beta integrin family and it is expressed by a wide variety of cells, including leukocytes, platelets, endothelial and smooth muscle cells. In the mouse model, the CD29low/CD24+ subpopulation can be further subdivided into CD61+ luminal progenitors and CD61– mature luminal cells [23]. In the mouse mammary tumor virus (MMTV)-Her2/neu transgenic mouse model, the CD49fhighCD61high subpopulation exhibits TIC's characteristics: (a) Enhanced tumor sphere formation; (b) Increased clonogenic ability; and (c) the early onset of in vivo tumor formation [24]. Wnt-1 confers stem-like properties on luminal progenitor cells in preneoplastic mammary tissue. Generation of hyperplastic outgrowths from the stem cell-enriched (CD29high/CD24+) and luminal fractions (CD29low/CD24+/CD61+) isolated from MMTV-wnt-1 preneoplastic mammary tissue has been reported [25]. In humans, CD61 was broadly expressed in all tissues

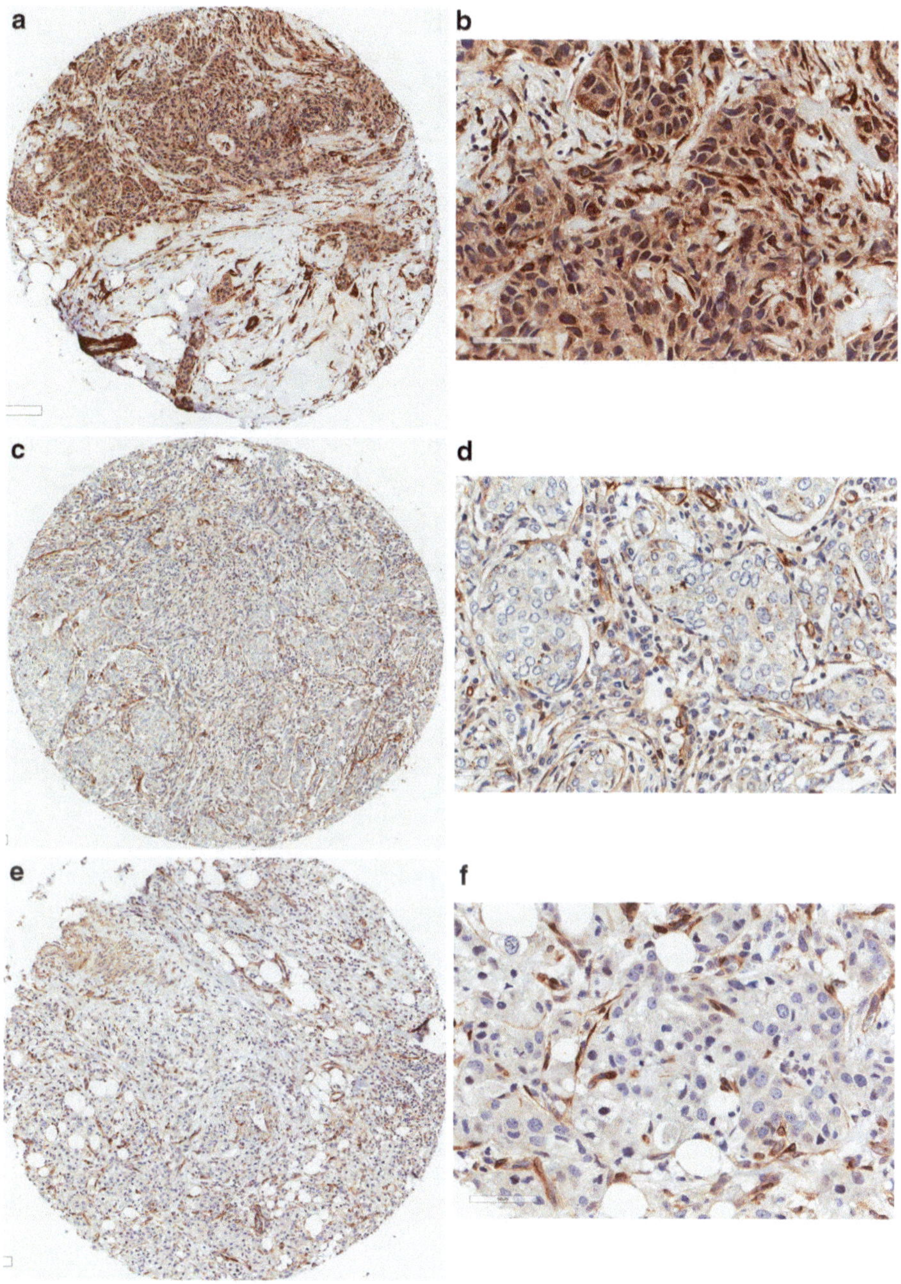

Fig. 6.18 CD29 expression in triple negative cancer. (**a**) (×8) and (**b**) (×40), stage 1, 1-06-01388. CD29 reactivity: 0.0 % negative, 15.9 % weak, 81.1 % moderate, 3.0 % strong. (**c**) (×8) and (**d**) (×40), stage 3b, 1-11-01090. CD29 reactivity: 99.7 % negative, 0.3 % weak, 0.0 % moderate, 0.0 % strong. (**e**) (×8) and (**f**) (×40), stage 4, 1-10-02789. CD29 reactivity: 92.2 % negative, 6.0 % weak, 1.3 % moderate, 0.5 % strong

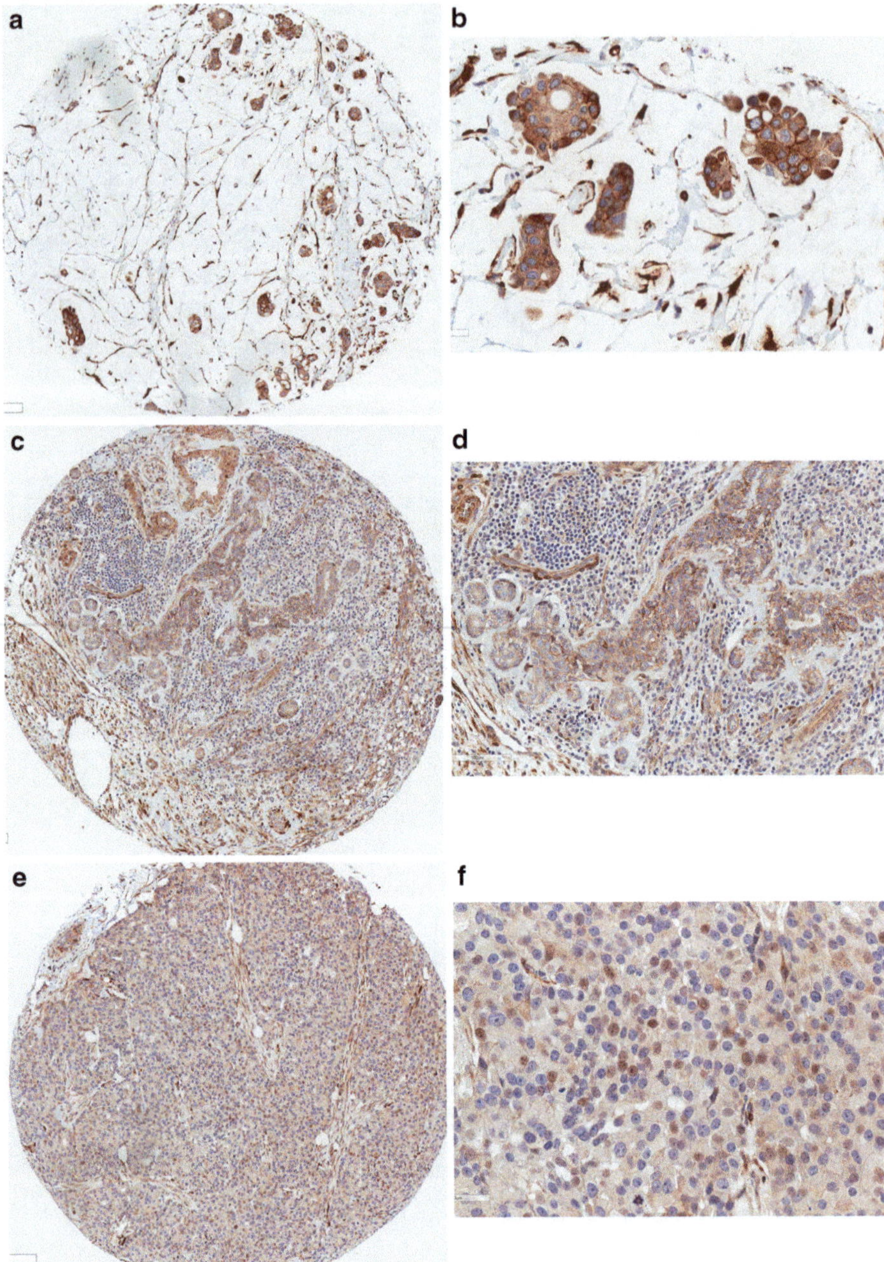

Fig. 6.19 CD29 expression in ER negative cancer. (**a**) (×8) and (**b**) (×40), stage 2b, 1-02-02484. CD29 reactivity: 15.6 % negative, 7.1 % weak, 33 % moderate, 44.3 % strong. (**c**) (×8) and (**d**) (×40), stage 3b, 1-06-01378. CD29 reactivity: 74.6 % negative, 21.7 % weak, 3.3 % moderate, 0.4 % strong. (**e**) (×8) and (**f**) (×40), stage 4, 1-11-00281. CD29 reactivity: 84.6 % negative, 0.3 % weak, 0.0 % moderate, 15.0 % strong

Table 6.6 Expression of EpCAM in triple negative tumors

Group	Samples with more than 40 % cells showing strong EpCAM expression	Samples with less than 40 % cells showing strong EpCAM expression
Lymph node negative	2	3
Lymph node positive or distant metastasis	9	1
Fisher exact test, $P=0.077$		

examined. When looking at CD61 staining in the human breast tumors by IHC, it was expressed in almost all the tumors we studied, but the majority of cells are only weakly or moderately stained. In the contrast, the staining in a small percentage of cells was very strong. The quantification of these CD61 strongly expressed cells showed a positive correlation to the lymph node involvement and distant metastasis in triple negative samples ($P=0.027$), which was 3.1 % (median) in the lymph node negative samples and 8.3 % (median) in node positive and distant metastasis samples. This correlation wasn't seen in ER negative tumors. However, quantification of CD61 positive cells showed that stage 3–4 tumors had a higher CD61+ cells number than stage 1–2 ER negative tumors, although both of them were at a high percentage (median 97.3 % vs. 83.8 %, respectively; $P=0.005$) (Table 6.6 and Fig. 6.20).

6.4.8 CD49f/α6 Integrin

CD49f/α6 integrin is expressed on T cells, monocytes, epithelial cells, and endothelial cells. Similar to CD29, CD49f is another important marker used to isolate mouse mammary stem cells. Sorted mouse CD49f+ fraction is able to form mammospheres at a frequency of 4.2 ± 1.5 spheres/10,000 cells, compared to CD49f– fraction at 0.7 ± 0.2 spheres/10,000 cells [26]. Lin-/CD24med/CD49fhigh population was designated as mammary repopulating units (MRUs) (mouse). A single Lin-/CD24med/CD49fhigh cell injected into fat pad is able to regenerate a mouse mammary gland [27]. Lin-CD49f+ CD24– subpopulation was also obtained from the fresh human breast surgical specimens and the frequencies were not significantly different between benign and tumor samples [28]. CD49f was required for the tumorigenicity of a stem cell-like subpopulation within the MCF7 cells [29]. In mouse mammary gland, CD49f expression is higher in basal cells than in luminal cells whereas in human is more basal located. Although no relationship between the expression of CD49f and clinical stage was observed in the triple negative tumors (15 samples), in ER negative tumors the CD49f+ cells were increased at stages 3–4 compared to stages 1–2. CD49f+ was not associated with lymph node involvement and distant metastasis in the samples we analyzed (Table 6.7 and Figs. 6.21 and 6.22). Regression analysis of combined $CD24^{-/low}$CD49f+ showed this staining pattern

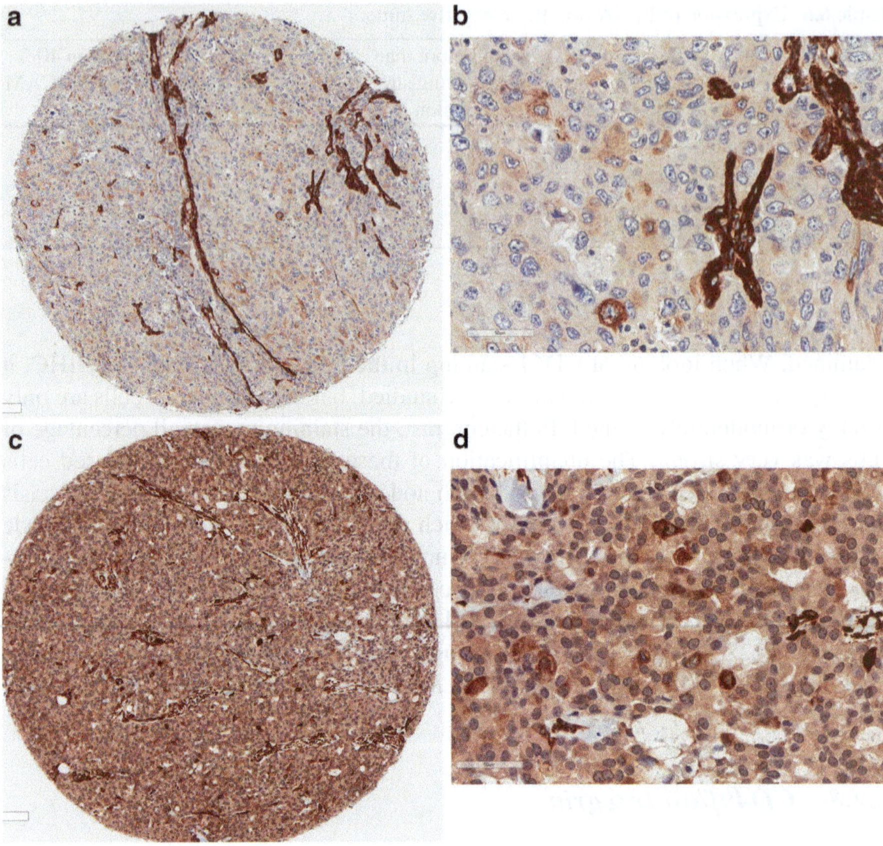

Fig. 6.20 CD61 expression. (**a**) (×8) and (**b**) (×40), triple negative cancer, stage 2a, 1-10-02772. CD61 reactivity: 82.4 % negative, 4.3 % weak, 3.2 % moderate, 10.1 % strong. (**c**) (×8) and (**d**) (×40), ER negative cancer, stage 4, 1-11-00281. CD61 reactivity: 0.4 % negative, 80.4 % weak, 13.4 % moderate, 5.7 % strong

Table 6.7 Expression of CD49f in triple negative and ER negative tumors

Group	Sample size	Median (%)	Min (%)	Max (%)	25 %	75 %
%CD49f+ cells						
Triple negative tumors	15	47.6	0.0	88.7	19.2	78.4
ER negative tumors	28	61.2	0.0	100.0	31.5	85.8
Group	*Sample size*	*Mean*	*SD (%)*	*Min (%)*	*Max (%)*	
%CD49f+ cells in ER negative tumors						
Stage 1–2	12	40.9	25.6	3.5	88.7	
Stage 3–4	16	67.1	32.6	0	100.0	
P value = 0.030						

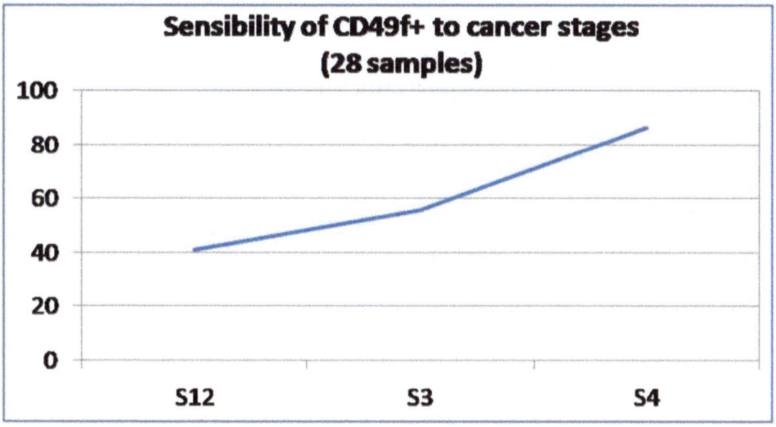

Fig. 6.21 The number of CD49f positive cells is increased in stage 3–4 ER negative tumors compared to stage 1–2 tumors (28 samples, regression analysis, $R^2 = 0.30$)

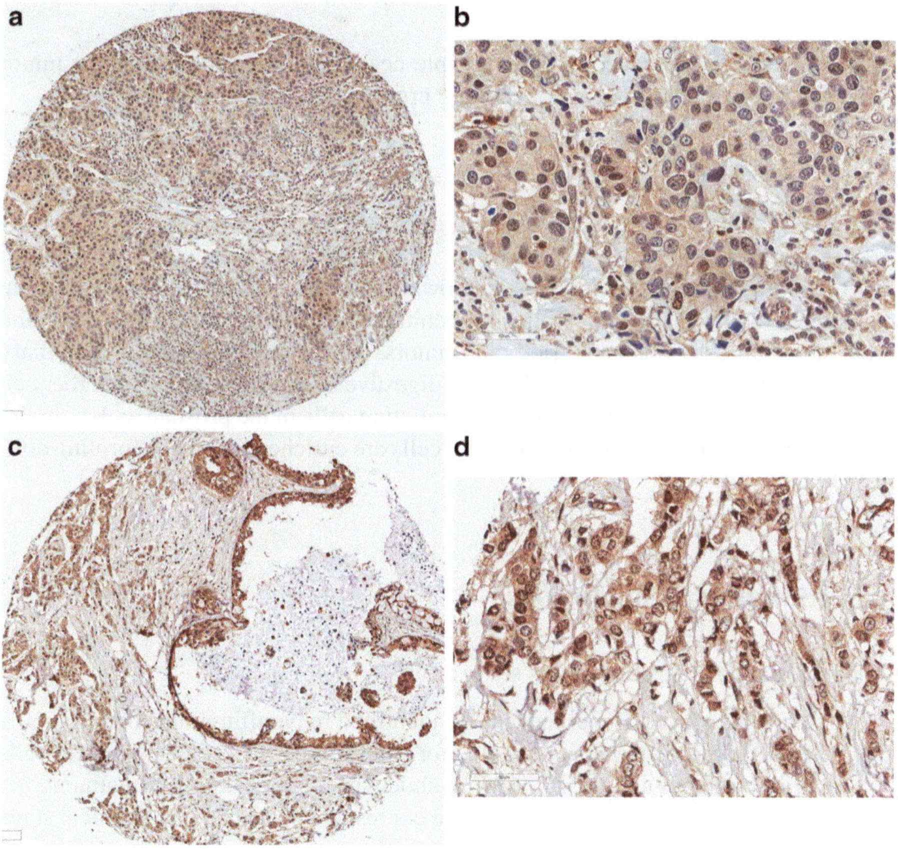

Fig. 6.22 CD49f expression. (**a**) (×8) and (**b**) (×40), ER negative cancer, stage 2a, 1-02-02341. CD49f reactivity: 31.5 % positive. (**c**) (×8) and (**d**) (×40), ER negative cancer, stage 3a, 1-02-00008. CD49f reactivity: 67.7 % positive

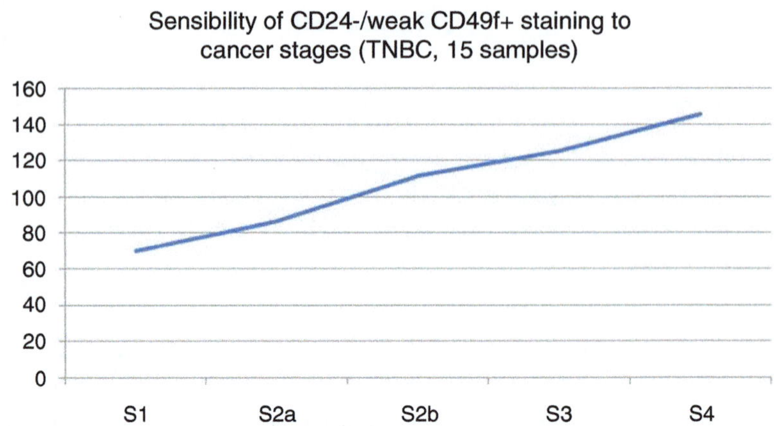

Fig. 6.23 The CD24–/low CD49f+ staining is positively correlated to stages in triple negative cancer by regression analysis ($R^2=0.58$)

was correlated to the cancer stages in triple negative cancer; the higher the tumor stage, the more likely this kind of staining presented in tumors (Fig. 6.23).

6.4.9 CD133/2

Currently the function of CD133 is unknown. It is expressed in hematopoietic stem cells, endothelial cells, endothelial progenitor cells, glioblastoma, neuronal, and glial stem cells, various pediatric brain tumors, as well as adult kidney, mammary glands, trachea, salivary glands, placenta, digestive tract, testes, and some other cell types. CD133 is expressed by xenograft-initiating cells in the prostate, colon, brain, and kidney. CD44posCD49fhiCD133/2hi cells are enriched for tumor-forming ability [30].

6.5 Summary Conclusions

The important conclusions of this study are:

1. Established stem cell markers can be identified in paraffin-embedded tissue.
2. The use of these markers is of value for determining the presence of stem cells in primary tumor and their follow up under specific neoadjuvant treatment.
3. *There* was no relationship between the expression of ALDH1 and clinical stage in both triple and ER negative tumors. The samples with higher number of ALDH1 positive cells usually were positive for HER2 expression (Table 6.8).

Table 6.8 Summary of ALDH1 reactivity in primary breast cancer

ALDH1 expression is not related to the clinical stage in the triple negative and ER negative breast cancer (14 triple negative and 29 ER negative samples)
The range of ALDH1 expression in triple negative breast cancer is from 0 to 19.5 % positive, with the median of 12.6 % positive
The range of ALDH1 expression in ER negative breast cancer is from 0 to 94.1 % positive, with the median of 13.0 % positive
The expression of ALDH1 might relate to the Her2 status in ER negative breast cancer, Her2 positive ER negative tumors tend to have more ALDH1 positive cells ($P=0.108$)

Table 6.9 Summary of CD24 reactivity in primary breast cancer

CD24 expression level is reduced as tumor progressed in triple negative breast cancer, indicating that the reduction of CD24 is important for the progression of the disease
CD24 level is positively correlated to the expression of HER2 in ER negative cancer
There is no difference in the number of CD24 positive cells between stage 3–4 and stage 1–2 in triple negative and ER negative breast cancer

Table 6.10 Summary of CD44 reactivity in primary breast cancer

The proportion of CD44 positive cells is reduced in stage 2–4 tumors compared to stage 1 tumors in both triple negative (15 samples) and ER negative (29 samples) breast cancer
The range of CD44 expression in triple negative breast cancers is from 0 to 73.6 % positive, with the median of 18.4 % positive
The range of CD44 expression in ER negative breast cancers is from 0 to 73.6 % positive, with the median of 15.5 % positive
The proportion of CD44 positive cells is reduced in tumors with lymph nodes or distant metastasis in ER negative breast cancer
The CD24–/low CD44+ staining was increased from stage 1 to stage 3, and then reduced from stage 3 to stage 4 in triple negative cancer

4. *The* number of CD24 positive cells didn't change too much, more than 90 % of tumor cells still had CD24 expression even in stage 4 but the staining was gradually reduced as tumor progressed in triple negative cancer, indicating the reduction of CD24 is important for the progression of this disease (Table 6.9).
5. *The* CD44+ cells were significantly reduced in stage 2–4 tumors compared to stage 1 in both triple negative and ER negative cancer. However, combined analysis showed CD24–/low CD44+ staining was increased from stage 1–3, and then reduced from stage 3 to stage 4 (Table 6.10).
6. *The* CK5+ cells were significantly reduced in stage 2–4 tumors compared to stage 1 tumors in both triple negative and ER negative cancer. In addition, CK5+ cells tend to be reduced in the tumors with lymph node involvement (Table 6.11).
7. *EpCAM* was expressed in all triple negative and ER negative tumors (Table 6.12).
8. *CD61* strongly expressed cells showed a positive correlation to the lymph node involvement and distant metastasis in triple negative cancer ($P=0.027$) (Table 6.13).

Table 6.11 Summary of CK5 reactivity in primary breast cancer

The percentage of CK5 positive cells in ducts is 77.95 % ± 15.61 % in parous and 72.52 % ± 13.83 % in nulliparous postmenopausal women. In ductules, it is 53.5 % ± 15.13 % in parous and 54.99 % ± 15.3 % in nulliparous women. The intensity of CK5 is reduced in nulliparous ducts compared to parous ducts
The percentage of CK5 positive cells is reduced in triple negative and ER negative invasive breast cancer compared to normal breast tissues and fibroadenoma. In 40 ER negative samples that we analyzed, the mean CK5 positive cells is 11.7 %, median is 2.02 % and maximal expression is 76.1 %. Twenty-five percent samples have the CK5 expression in less than 0.77 % of tumor cells
CK5 positive cells tend to be reduced in the tumors with lymph node involvement
CK5 positive cells are significantly reduced in stage 2–4 tumors compared to stage 1 tumors both in triple and ER negative breast cancer
The loss of CK5 expression might be an early event in the development of breast cancer and the lower expression of CK5 is related to the lymph node involvement

Table 6.12 Summary of EpCAM reactivity in primary breast cancer

EpCAM is highly expressed in triple negative and ER negative breast cancer. We didn't observe the correlation between the expression of EpCAM and clinical stage from the sample size we have studied
The overexpression of EpCAM may positively relate to the clinical stage and lymph node involvement as well as distant metastasis if larger sample sizes are studied

Table 6.13 Summary of CD61 reactivity in primary breast cancer

There are more CD61 positive cells in stage 3–4 ER negative tumors than stage 1–2 tumors
The strong expression of CD61 is correlated to the lymph node involvement and distant metastasis in triple negative breast cancer

Table 6.14 Summary of CD29/β1 integrin reactivity in primary breast cancer

There are conflicting data regarding the relationship of the expression of β1 integrin with the overall survival and metastasis of breast cancer
Our data showed that the expression of β1 integrin is reduced in the stage 2–4 triple negative tumors compared to stage 1 tumors
Our data also suggested that the expression of β1 integrin may relate to the ER status, the ER positive cancers tend to have lower level of β1 integrin expression
Fibroadenoma showed high frequency and intensity of β1 integrin positive cells compared to invasive carcinoma

9. *CD29/β1 integrin* expression was reduced in the stage 2–4 triple negative (16 samples) tumors compared to stage 1 tumors (Table 6.14).
10. *CD49f/α6 integrin* showed no relationship between its expression and clinical stage in the triple negative cancer (15 samples), but it showed that in ER

Table 6.15 Summary of CD49f reactivity in primary breast cancer

Literatures showed that the expression of CD49f in tumor cells is important for maintaining tumorigenicity of tumor cells, but in normal breast it is abundantly expressed in polarized lobular/alveolar cells
We observed a strong expression of CD49f in fibroadenoma, whereas the expression level reduced in both ER negative (including triple negative tumors) and ER positive tumors
In ER negative tumors, CD49f positive cells were increased in stage 3–4 tumors compared to stage 1–2 tumors
CD24–/weak CD49f+ staining is positively correlated to cancer stages in triple negative cancer

negative cancer the CD49f+ cells were increased in the stage 3–4 compared to stage 1–2. CD24–/weak CD49f+ staining was positively correlated to cancer stages in triple negative cancer (Table 6.15).

Acknowledgements The authors would like to acknowledge Theresa Nguyen for TMA preparation and IHC staining; Nathan Hopfinger for extracting images and analyses; Fritzi Thomas, Joseph Silverberg, Viviana Serrato, Riddhi Amin, Edward Wadell, Victoria Hall, Courtney Michner and Cameron Jeffers for the quantification of the staining. We also acknowledge the histopathology facility at FCCC for the scanning of the slides.

Literature Cited

1. Shipitsin M, Polyak K (2008) The cancer stem cell hypothesis: in search of definitions, markers, and relevance. Lab Invest 88(5):459–463
2. Gutiérrez PJ, Russo IH, and Russo J (2012) The evolution of the use of mathematics in cancer research. Springer New York, pp 66-102, pp 130–141
3. Ginestier C, Hur MH et al (2007) ALDH1 is a marker of normal and malignant human mammary stem cells and a predictor of poor clinical outcome. Cell Stem Cell 1(5):555–567
4. Park SY, Lee HE et al (2010) Heterogeneity for stem cell-related markers according to tumor subtype and histologic stage in breast cancer. Clin Cancer Res 16(3):876–887
5. Hwang-Verslues WW, Kuo WH et al (2009) Multiple lineages of human breast cancer stem/ progenitor cells identified by profiling with stem cell markers. PLoS One 4(12):e8377
6. Shackleton M, Vaillant F et al (2006) Generation of a functional mammary gland from a single stem cell. Nature 439(7072):84–88
7. Zhang M, Behbod F et al (2008) Identification of tumor-initiating cells in a p53-null mouse model of breast cancer. Cancer Res 68(12):4674–4682
8. Sleeman KE, Kendrick H et al (2006) CD24 staining of mouse mammary gland cells defines luminal epithelial, myoepithelial/basal and non-epithelial cells. Breast Cancer Res 8(1):R7
9. Jones C, Mackay A et al (2004) Expression profiling of purified normal human luminal and myoepithelial breast cells: identification of novel prognostic markers for breast cancer. Cancer Res 64(9):3037–3045
10. Kristiansen G, Winzer KJ et al (2003) CD24 expression is a new prognostic marker in breast cancer. Clin Cancer Res 9(13):4906–4913
11. Al-Hajj M, Wicha MS et al (2003) Prospective identification of tumorigenic breast cancer cells. Proc Natl Acad Sci USA 100(7):3983–3988

12. Perrone G, Gaeta LM et al (2012) In situ identification of CD44+/CD24- cancer cells in primary human breast carcinomas. PLoS One 7(9):e43110
13. Fillmore CM, Kuperwasser C (2008) Human breast cancer cell lines contain stem-like cells that self-renew, give rise to phenotypically diverse progeny and survive chemotherapy. Breast Cancer Res 10(2):R25
14. Zhao X, Malhotra GK et al (2010) Telomerast—immortalized human mammary stem/progenitor cells with ability to self-renew and differentiate. Proc Natl Acad Sci USA 107(32): 14146–14151
15. Lim E, Vaillant F et al (2009) Aberrant luminal progenitors as the candidate target population for basal tumor development in BRCA1 mutation carriers. Nat Med 15(8):907–913
16. Aguiar FN, Mendes HN et al (2013) Basal cytokeratin as a potential marker of low risk of invasion in ductal carcinoma in situ. Clinics (Sao Paulo) 68(5):638–643
17. Cimino A, Halushka M et al (2010) Epithelial cell adhesion molecule (EpCAM) is overexpressed in breast cancer metastases. Breast Cancer Res Treat 123(3):701–708
18. Naylor MJ, Li N et al (2005) Ablation of beta1 integrin in mammary epithelium reveals a key role for integrin in glandular morphogenesis and differentiation. J Cell Biol 171(4):717–728
19. Dos Santos PB, Zanetti JS et al (2012) Beta 1 integrin predicts survival in breast cancer: a clinicopathological and immunohistochemical study. Diagn Pathol 7:104
20. Yao ES, Zhang H et al (2007) Increased beta1 integrin is associated with decreased survival in invasive breast cancer. Cancer Res 67(2):659–664
21. Gonzalez MA, Pinder SE et al (1999) An immunohistochemical examination of the expression of E-cadherin, alpha- and beta/gamma-catenins, and alpha2- and beta1-integrins in invasive breast cancer. J Pathol 187(5):523–529
22. Zutter MM, Mazoujian G et al (1990) Decreased expression of integrin adhesive protein receptors in adenocarcinoma of the breast. Am J Pathol 137(4):863–870
23. Asselin-Labat ML, Sutherland KD et al (2007) Gata-3 is an essential regulator of mammary-gland morphogenesis and luminal-cell differentiation. Nat Cell Biol 9(2):201–209
24. Lo PK, Kanojia D et al (2012) CD49f and CD61 identify Her2/neu-induced mammary tumor-initiating cells that are potentially derived from luminal progenitors and maintained by the integrin-TGFbeta signaling. Oncogene 31(21):2614–2626
25. Vaillant F, Asselin-Labat ML et al (2008) The mammary progenitor marker CD61/beta3 integrin identifies cancer stem cells in mouse models of mammary tumorigenesis. Cancer Res 68(19):7711–7717
26. Liao MJ, Zhang CC et al (2007) Enrichment of a population of mammary gland cells that form mammospheres and have in vivo repopulating activity. Cancer Res 67(17):8131–8138
27. Stingl J, Eirew P et al (2006) Purification and unique properties of mammary epithelial stem cells. Nature 439(7079):993–997
28. Pommier SJ, Hernandez A et al (2012) Fresh surgical specimens yield breast stem/progenitor cells and reveal their oncogenic abnormalities. Ann Surg Oncol 19(2):527–535
29. Cariati M, Naderi A et al (2008) Alpha-6 integrin is necessary for the tumourigenicity of a stem cell-like subpopulation within the MCF7 breast cancer cell line. Int J Cancer 122(2):298–304
30. Meyer MJ, Fleming JM et al (2010) CD44posCD49fhiCD133/2hi defines xenograft-initiating cells in estrogen receptor-negative breast cancer. Cancer Res 70(11):4624–4633

Chapter 7
Imaging Techniques for Evaluation In Vitro Behavior of Normal and Cancerous Breast Tissue

Jose Russo, Ricardo Lopez de Cicco, Yanrong Su, Thomas J. Pogash, Nathan Hopfinger, and Irma H. Russo

Keywords Use of FCM • Separation of stem cells from cell lines • Advantages and disadvantages of FCM sorting • Immunocytochemistry • Immunofluorescence • Double immunohistochemistry and fluorescence • In situ hybridization • Comparative genomic hybridization (CGH) • Fluorescence in situ hybridization (FISH)

7.1 Introduction

Among the in vitro imaging techniques of greatest use in cancer research are flow cytometry (FCM), immunofluorescence microscopy, and in situ hybridization (ISH).

7.2 Use of FCM in Normal and Cancerous Breast

FCM is a technique that can be used in a wide range of applications to quantify and sort cells based on size, shape, composition, membrane permeability, as well as a variety of selected biomarkers [1]. FCM can be used for the quantification of cells, sorting of cell populations, analysis of DNA and RNA, and the detection of biomarkers based on fluorescent activity [1, 2]. With the ability to analyze cells without inflicting damage, it is an accessory method in cancer research for in vitro and in vivo models [2]. In order to use FCM, it is first necessary to select a fluorescent-labeled

J. Russo, M.D. (✉) • R.L. de Cicco, Ph.D. • Y. Su, Ph.D. • T.J. Pogash, B.S.
N. Hopfinger, B.S. • I.H. Russo, M.D.
Irma H Russo MD Breast Cancer Research Laboratory, Fox Chase Cancer Center, Cottman Avenue 333, Philadelphia, PA 19111, USA
e-mail: jose.russo@fccc.edu; Irma.Russo@fccc.edu

J. Russo and I.H. Russo, *Techniques and Methodological Approaches in Breast Cancer Research*, DOI 10.1007/978-1-4939-0718-2_7,

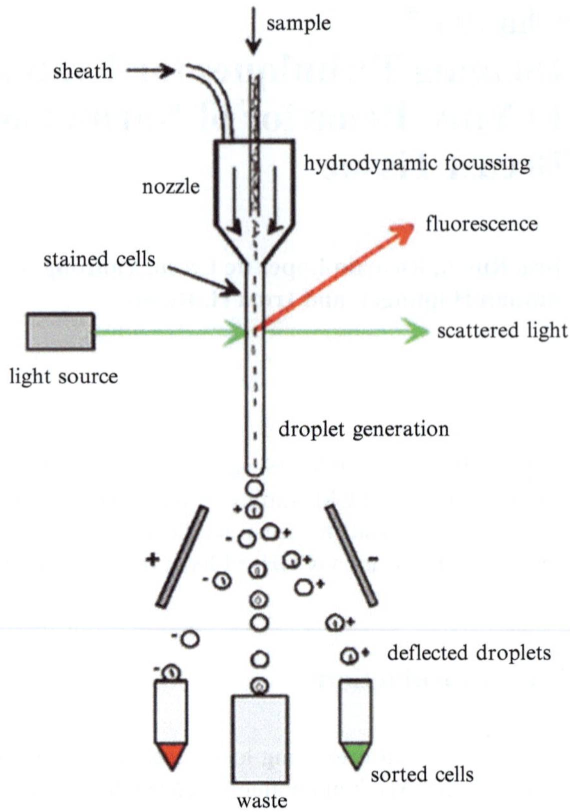

Fig. 7.1 A diagram of FACS analysis. The sample is loaded into the machine where it is combined with a sheath fluid. The population is then run in a single stream through a projected laser. Fluorescence and scatter data are collected by photodetectors to be analyzed by computer software. The stream is then broken into droplets and sorted by population using electric charges (*Source*: http://missinglink.ucsf.edu/lm/molecularmethods/flow.htm)

antibody, or fluorescent dye, which is specific to the biomarker under study [1, 2]. FCM requires a single cell suspension for proper analysis, so tissue must be processed before analyzing [2]. Such processing is dependent on the source organ and cell type to be isolated. After tissue processing, cells can be suspended and stained with the selected fluorescent antibody or dye according to the respective protocol. After staining, cells are suspended in a buffer (usually saline) [3, 4] at a relatively high concentration (~10^5 to 10^6 cells/mL) [5], then injected into the flow cytometer where they combine with a pressurized sheath liquid [1–6]. Cells are run in a single stream though the machine, and then passed though a projected laser beam (Fig. 7.1), most commonly an argon laser [1–5]. Commonly used fluorochromes with an argon laser include fluorescein isothiocyanate (FITC), phycoerythrin, PE-Cy5, Perc-P, propidium iodide, DAPI, Hoechst 33342, chromomycin, indo-1, nile red, and thiazol orange [4, 6]. When the fluorochrome passes through the laser, it becomes excited and subsequently releases light at a particular wavelength which is then collected by photodetectors [2–5]. The purpose of these photodetectors is to collect and convert the corresponding light into electrical impulses that can be read by the computer software [4]. Because different antibodies and dyes can emit at different wavelengths, it is also possible to sort for multiple markers in a single sorting [2, 5].

In addition to sorting for antibodies and dyes, additional data in the form of forward scatter, and side scatter is collected as determined by how the individual cell scatters the light from the laser beam. Forward scatter (5° from plane of laser) is proportional to the diameter or size of the cells while the side scatter (90° angle from plane of laser) is proportional to the granularity of cells [1–5]. One method of visualizing the data gathered from FCM is in the form of a scatter plot [4, 5]. Data that shows high variation in signals such as fluorescence is commonly graphed as a histogram using log scale, while data with little variation in signals is generally graphed linearly, or as a dot plot comparing two parameters [4, 5]. In a histogram display, cell number is generally plotted on the *y*-axis while the *x*-axis represents the intensity of the collected fluorescence [4]. In a dot plot, each point represents the features of a single cell. The graph can then be separated into regions based on a control population to determine which cells are positive for the markers of interest [2, 4]. From one or more of these regions, a gate or low limit is then selected; this gate identifies all of the cells that will be used for the final analysis [4].

It has been well described in a recent article of Smith [1], that the number of parameters that can be measured simultaneously in one assay will depend largely on how many lasers and detectors the flow cytometer is equipped with. The wavelengths/colors of the lasers it employs will dictate the excitation frequencies of the fluorophores that can be used. The possible number of fluorophore colors will also be determined by the number of detectors at the respective emission wavelengths. In addition, detectors for scattered light will provide estimates of the dimensions of the cells (or particles). For example, the Gallios™ flow cytometer from Beckman Coulter (Indianapolis, IN; www.beckmancoulter.com) is equipped with up to four lasers and ten detectors, allowing multicolor assays of up to ten colors. Two of the lasers are the standard red and blue, with optional violet (405 nm excitation wavelength) and yellow (561 nm excitation wavelength) lasers also possible. The yellow laser can improve detection of red fluorescent proteins. Six fluorescence detectors are standard, with the option of adding four more, making it possible to assay ten fluorescent signals simultaneously [1].

According to Smith [1], the number of lasers used will also affect how many ways are needed to sort the cells, for example, into two, four, or six sorted populations. The more sorted groups, the greater the need to think about where they should be placed in the sample vessels. Typically, flow cytometers sort cells into sample tubes or microplates of up to 384 wells. While it is possible to run single samples on most flow cytometers, the ability to use microplates makes higher throughput possible. Automation can augment throughput further still. Many variations and degrees of automation are available from automated liquid handling systems that carry out particular sampling and dispensing jobs to fully automated robotic systems that execute all the functions of the work flow, including moving plates and samples around the flow cytometer. Other considerations include sample volume and cell counting [1]. Most flow cytometers use sample sizes ranging from several

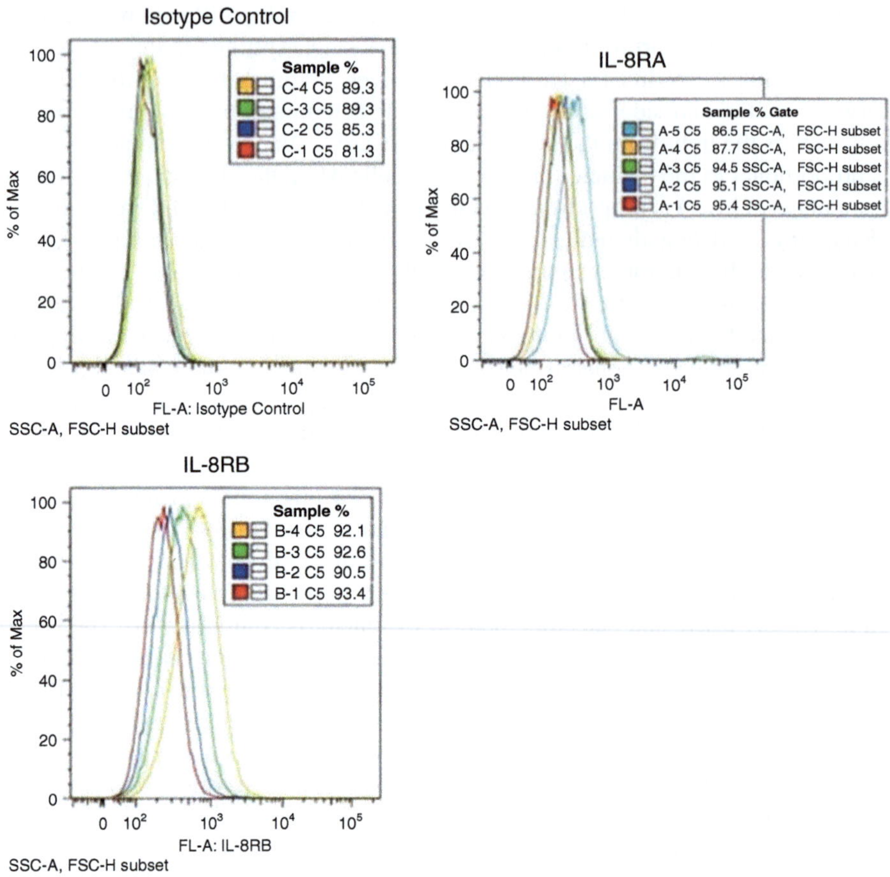

Fig. 7.2 Titration of IL-8RA and IL-8RB antibodies to be used in FACS analysis. One million C5 or SUM149 cells are incubated with different concentrations of the corresponding antibody (5, 10, 20, 40 μL) in 100 μL of PBS buffer. Cells are incubated at 4 °C for 1 h. At the end of incubation, cells are washed and resuspended in PBS and then subjected to FACS analysis

microliters to hundreds of microliters. Also, some offer absolute cell counting features, while others reach cell counts using reference bead standards.

Fluorescence-activated cell sorting (FACS) analysis works as an isolating mechanism by processing cells using FCM and, then breaking the stream into electrostatically charged droplets based on their fluorescent intensity, which are then directed to collecting chambers [2, 7]. FCM can be used to sort tumor cells and identify cells with markers of high invasiveness or increased migratory properties, such as those that express IL-8RA or IL-8RB (Figs. 7.2, 7.3, and 7.4). This allows for a determination of the risk of tumor metastasis based on the number of these cells present in a sample.

In vitro, FCM can be used to isolate cells with particular properties and this allows researchers to test compounds and therapies on selected populations of cells.

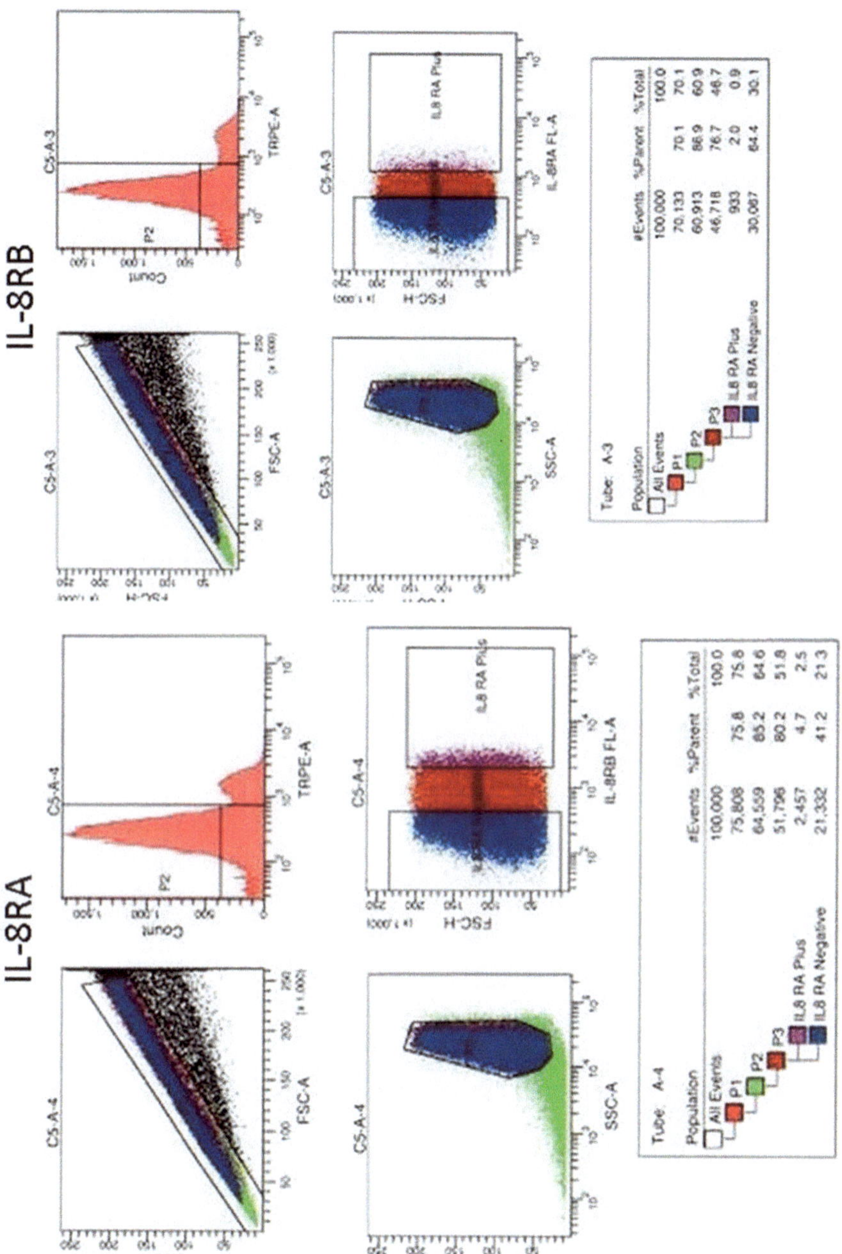

Fig. 7.3 Sorting of IL-RA (*left panel*) and IL-8RB (*right panel*) positive C5 cells. Using the optimal concentration of antibodies, IL-8RA or IL-8RB expressing-C5 cells were sorted using FACS

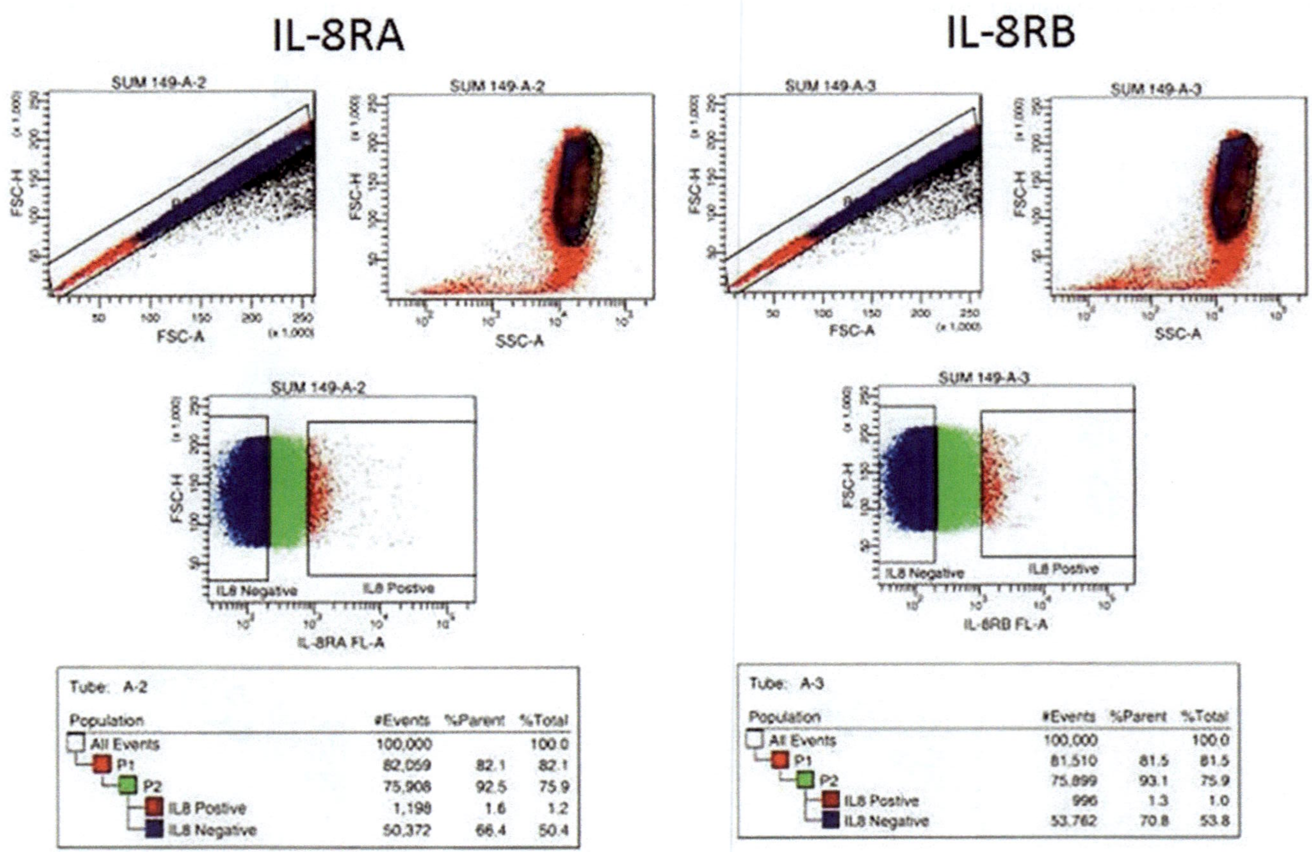

Fig. 7.4 Sorting of IL-RA (*left panel*) and IL-8RB (*right panel*) positive SUM149 cells. Using the optimal concentration of antibodies, IL-8RA or IL-8RB expressing-SUM149 cells were sorted using FACS

In addition, data acquired from FCM can be used to aid researchers in the search for pathways for further investigation. For example, FCM has been used to discover pathways, DNA fragments, and markers that differ between normal tissue cells and tumor cells, the results of which can then be used to study a particular target.

Several considerations must be taken when rare or precious samples, or trying to isolate a cell population of low abundance from a large mixture, it is necessary in this case to look for particular FCM solutions to these challenges. A preparatory purification or enrichment step prior to FCM can make it easier to find the target cells [www.miltenyibiotech.com]. Samples that are particularly vulnerable to environmental influences, or that require sterile conditions, might benefit from an entirely closed FCM system (www.partec.com). If the experiments require fluorescence imaging during cell sorting, a flow cytometer with integrated imaging capabilities may be useful [www.emdmillipore.com]. This instrument can take multiple images of each cell so that the intracellular location of the fluorescence can be detected. Another way to conserve precious resources is to use a much smaller volume with a microfluidics-based FCM chip. This is like a flow cytometer in miniature. The microfabricated chip contains microfluidic channels into which a tiny sample is injected. Pneumatically controlled pumps and valves move the sample along the channels and single-file through laser(s) in a sorting area, where the scattered and fluorescent light is collected. Pumps accomplish the tailor-made movement instructions for individual cells-reroute the "red" cell to its group, the "green" cell to its group, and a "red and green" cell to its group. Microfluidic chips boast lower background noise in general, but they have very low recovery rates.

7.2.1 Markers Used for Separating Breast Cancer Stem Cells

A recent topic of interest in the breast cancer research field has been the presence of cancer stem cells (CSCs) in tumors and their importance in the development and perpetuation of different forms of human cancer [7, 8]. CSCs are cancer cells, which exhibit similar characteristics to stem cells including the ability of self-renewal, the ability to develop into multiple lineages, increased invasiveness, and the potential to proliferate extensively [7]. It is believed that CSCs arise from mutated existing normal stem cells, or they arise from mutations in progenitor cells [8, 9]. Studies over the last 40 years have begun to shed light on the fact that cells containing stem cell capabilities are present in some forms of human cancers including cancers of the hematopoietic system, brain and breast [8, 9]. Research provides support for the premise that the properties of these CSCs may account for what gives many types of cancer their malignant phenotype, allow invasiveness and metastasis, and resistance to anticancer therapy. Literature varies on the importance of CSCs in tumor growth and malignancy with some reports supporting that CSCs are essential for the growth of the tumor and others reporting that CSCs only play a role in invasiveness, resistance to therapy and relapse. Because of these stem-like characteristics and their resistance to current treatments, CSCs are a recent target of cancer treatment strategies. Although CSCs are believed to play an important role in cancer biology, literature reports that they can comprise a very small population of the mass of tumor cells. Because of the way that the CSCs divide (similar to stem cells, with one daughter cell retaining the same properties of the parent), as generations pass, the concentration of CSCs is reduced, diluted by the mass of more rapidly proliferating cells. This low level of proliferation creates difficulty in maintaining a culture of complete CSCs, a technique that requires much further study.

FCM uses biomarkers to sort stem cells. For the purpose of CSCs, there are several markers that are useful in order for the cells to be isolated from the mass population. The most common cell surface markers used in FCM to isolate the CSC population are CD19, CD34, CD44, CD117, and CD133 [9]. CD19 is a protein generally found on the surface of B lymphocytes [10]. CD34 is a cell surface molecular cluster that plays an important role in cell–cell adhesion and mediates the attachment of stem cells to bone marrow and extracellular matrix [11]. CD44 is a cell surface molecule which is involved in the processes of cell proliferation, cell differentiation, cell migration, and angiogenesis and important in signaling for cell survival [12]. CD117 is a transmembrane tyrosine kinase receptor which, when bound to its specific ligand, stem cell growth factor, plays a role in cell survival, proliferation, and differentiation [13]. CD133 is a glycoprotein that currently has an unknown function but is found on the surface of hematopoietic stem cells, endothelial progenitor cells, glioblastoma, as well as neuronal and glial stem cells [14]. While these five markers are used for sorting by detection of their presence, CD24, when in combination with active CD44 is commonly used as a negative selection marker and CSCs are selected through its absence or inactivation. In addition to surface makers, functional properties are also commonly used to sort CSCs (Fig. 7.5). Common functional properties include Hoechst 33342 efflux and high aldehyde dehydrogenase (ALDH) activity [10] (Figs. 7.6 and 7.7). All of the listed cell markers, as well as functional properties, are present in normal stem cell populations in addition to CSC populations, indicating a correlation between the two cell types. It is however, important to note that not all CSCs express every marker and/or functional property; there are also many additional markers which may also be used to sort for CSCs (Figs. 7.5, 7.6, and 7.7).

As mentioned above, although CSCs are hypothesized to play a highly important role in breast cancer metastasis, growth, and resistance, data has shown that CSCs can only account for a very small fraction of the bulk tumor or tumor cell line population. A marker commonly used in CSC studies and sorting is enzymatic ALDH activity, with the most common assessment of CSCs done using the ALDEFLUOR assay [15–18] (Figs. 7.6 and 7.7). Breast cancer cell populations sorted for CSCs using only ALDEFLUOR+ FCM showed populations of anywhere from 0 to 100 % [18]. When looking at cell populations possessing more than one marker, such as both ALDEFLUOR+ and CD44+/CD24-populations dropped to <0.5 % [19, 20]. Studies have shown that the most common phenotype used for CSC sorting in breast cancer is CDC44+/CD24− [16, 17, 21, 22]. The above data is only applicable to basal type cell lines. Much work in CSC studies are done using basal type breast cancer cell lines, which have been found through sorting to contain

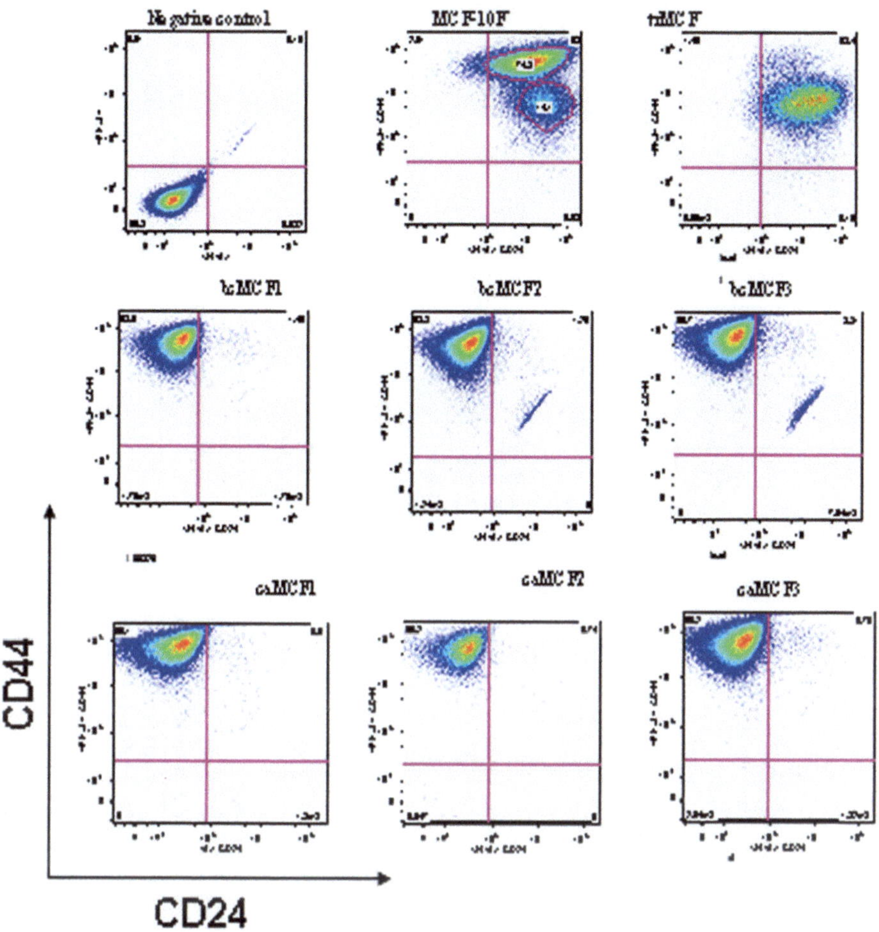

Fig. 7.5 Study of the expression of stem cell markers using FACS analysis. Expression of CD24 (*y*-axis) and CD44 (*x*-axis) (Reprinted with permission of: Yong Huang, Sandra V. Fernandez, Shirlean Goodwin, Patricia A. Russo, Irma H. Russo, Thomas R. Sutter, and Jose Russo. Epithelial to mesenchymal transition in human breast epithelial cells transformed by 17β-estradiol. Cancer Res 2007;67(23):11147–57)

the majority of CSCs, independent of marker, as compared to luminal breast cancer cell lines. In a study done by Wicha et al. [18], a group of 28 different breast cancer cell lines were sorted via the ALDEFLUOR assay for expression of ALDH. This group of cell lines contained 16 cell lines that are categorized as basal type cell lines and 12 cell lines categorized as luminal, cell types as determined by mRNA expression. Of the 16 basal cell lines tested, all were positive for some level of ALDH expression; of the 12-luminal type cell lines tested, only five expressed any detectable ALDH activity. In another seminal work, Ginestier et al. concluded that ALDH1 is a marker of normal and malignant human mammary stem cells and a predictor of poor clinical outcome. Indeed, in a series of 577 breast carcinomas, expression of ALDH1 detected by immunostaining correlated with poor prognosis [18].

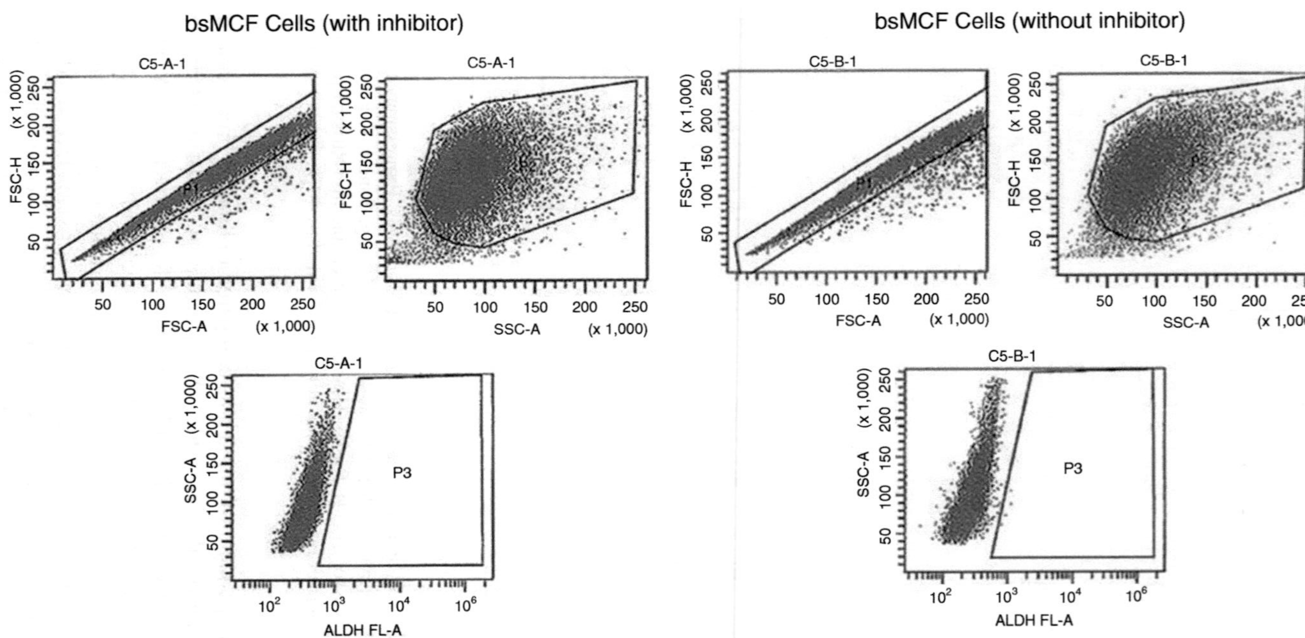

Tube: A-1

Population	#Events	%Parent	%Total
All Events	25,232		100.0
P1	24,618	97.6	97.6
P2	23,520	95.5	93.2
P3	0	0.0	0.0

Tube: B-1

Population	#Events	%Parent	%Total
All Events	20,000		100.0
P1	18,482	92.4	92.4
P2	17,565	95.0	87.8
P3	6	0.0	0.0

Fig. 7.6 Sorting of ALDEFLUOR+ C5 cells from an IL-8RA$^+$ (*right panel*) C5 cells using FACS. The *left panel* serves as a negative control

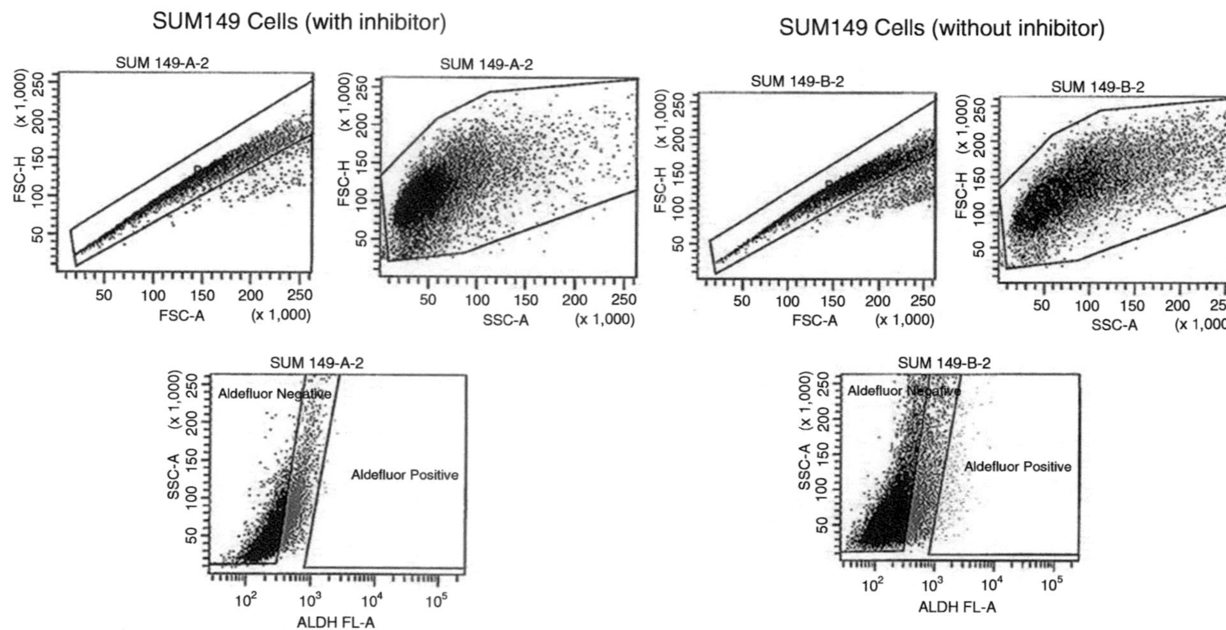

Population	#Events	%Parent	%Total
All Events	20,000		100.0
P1	19,691	98.5	98.5
P2	19,530	99.2	97.6
Aldefluor Positive	10	0.1	0.0
Aldefluor Negative	15,880	81.3	79.4

Population	#Events	%Parent	%Total
All Events	20,000		100.0
P1	18,518	92.6	92.6
P2	18,467	99.7	92.3
Aldefluor Positive	341	1.8	1.7
Aldefluor Negative	15,308	82.9	76.5

Fig. 7.7 Sorting of ALDEFLUOR+ SUM149 cells from an IL-8RA$^+$ (*right panel*) C5 cells using FACS. The *left panel* serves as a negative control

7.2.2 Separation of Stem Cells from Cell Lines

The choice of cell lineage when performing an in vitro–in vivo model is crucial to obtaining consistent and translatable results throughout the experiment. Cells of different origins, basal cells as opposed to luminal cell lines for example, and even cells of similar lineage can express many different phenotypes, leading to potentially different responses to certain compounds. When testing a compound, it is important to be aware that some compounds may only affect certain cell lineages or may affect them in different ways. For example, Wicha et al. [17] at the University of Michigan saw mass apoptosis and bulk tumor death when cells were treated with the anti-inflammatory drug Repertaxin, a novel low molecular weight inhibitor of the human CXCL8 receptor. This response however, was only observed in cell lines containing a PTEN mutation. Phenomena like this also make it important to test cells of different origin and lineage. For the same reason, consistency of cell line is crucial when performing both in vitro and in vivo models. When performing an in vivo study, the goal is generally to reproduce results in the animal model similar to those that were observed in an established cell line when treated with the compound being tested.

Currently in our lab, we are testing two novel compounds hypothesized to target breast CSCs though the IL-8 pathway that has been found to be highly expressed in ALDEFLUOR+ cells [18]. These compounds would be used in combination with existing treatments to increase their efficiency in targeting breast cancer CSCs and to prevent reoccurrence. Two different breast carcinoma cell lines have been sorted for CSCs using FCM: the cell line bsMCF, a highly invasive, triple negative cell line, and the cell line SUM149 (see Chap. 5), a primary inflammatory breast cancer cell line, both of which are basal type cell lines. In order to isolate CSCs from both populations, cells were sorted for ALDH activity using the ALDEFLUOR assay. In our lab, FCM was conducted using a FACstarPLUS and gates were established using propidium iodide to sort cells for viability and ALDEFLUOR fluorescent staining to sort for the presence of ALDH activity. bsMCF, the triple negative, highly invasive cell line surprisingly showed no presence of ALDH activity, instead the SUM149 cells displayed approximately 2 % of cells sorting positive for ALDH activity, data that is consistent with other published findings [18].

7.2.3 Validation of FCM with Immunofluorescence Microscopy

Data gathered from FCM can be confirmed using immunofluorescent (IF) microscopy.

In order to do this, a cell population is first cultured on a glass slide. Cells are then fixed using one of several solutions depending on cell type and antibody being used. This fixation renders the cells unviable. After fixation one of two methods are used, either primary (direct) staining or secondary (indirect) staining. For primary staining, an antibody specific to the target molecule, directly linked to a fluorochrome is used to stain cells. In secondary staining, a primary non-fluorescent antibody is used first which binds to the target molecule. Following this, a secondary antibody is used which recognizes the primary antibody. The benefit of secondary staining is that is allows for a more amplified signal. After staining, slides are viewed using a fluorescent microscope with the intensity of the fluorochrome proportional to the expression of targeted molecule.

7.2.4 Advantages and Disadvantages of FCM Sorting

Using FACS allows one to sort through a large population of cells [1–4, 6, 7] which provides a multitude of advantages. Because cell granularity can be measured during FCM, dead cells can be discarded from the analyses. Accurate quantitative counts of the collected population can be made and populations displaying small subgroups of cells with particular markers can be detected. Weak cell signals can be detected due to the amplified signals [7], and sorting based on multiple markers can be run due to the machine's ability to analyze cells individually (Fig. 7.1). This process can analyze in a few hours what would have previously taken weeks to complete. In addition, because FCM can be done in a manner that is not toxic to the cells, following sorting, the cells can be put back into culture and the population expanded based on the expression of particular biomarkers.

Although useful in many ways, FCM is not without its disadvantages. Some cells (especially primary tumor cells) are too fragile to successfully withstand the pressure and flow forces generated in the stream that is passed through the detection system. Because of the physical limitations of dispersion into single cells and the flow rate limits, it may take the machine long periods of time to identify and select very minor populations and, therefore, to select a sufficient number of cells. FCM is an expensive process, in terms of both equipment and interpretation of results. FCM requires a skilled operator in order to get accurate and reproducible results, especially when extremely sensitive measurements are called for [3]. Along with the expense of operation and equipment, biomarkers and fluorescent dyes can add significant cost. Another downfall to FCM, which comes as a result of the high price of machines, is that flow cytometers capable of sorting cells are not easily available. Often institutions share machines in effort to ease the price burden, which causes high demand resulting in the need to schedule appointments (and therefore experiments) weeks ahead of time. Because of this, flow cytometers may not be ideal for all experiments.

7.3 Use of Immunofluorescence Microscopy

Immunofluorescence microscopy (IF) is a technique used to visualize biological samples which have been labeled with antibodies conjugated with fluorophore [23]. IF is a powerful technique used to localize structures and expression within tissues, cells and subcellular compartments, such as in the nuclei or on the cell membrane, and is widely used both in clinical diagnostics and research. Immunofluorescence can be used alone or combined with other techniques, for example, with the use of immunocytochemical staining to evaluate the targets both in bright field and dark field; use of DAPI to stain DNA for the visualization of nuclei; and use of ISH for the assessment of RNA or DNA expression. The microscopes most widely used for analysis of immunofluorescence are epifluorescence microscope and confocal microscope [24].

There are two types of immunofluorescence techniques, primary (or direct) and secondary (or indirect). Direct immunofluorescence stain samples with primary antibody directly conjugated to fluorophore. Advantages of direct immunofluorescence include fewer staining steps in the procedure and lower background signal in some tissues with cross activity or nonspecificity, as well as the ability to use multiple antibodies raised from the same species. The disadvantages include low signal, since the fluorophore molecules conjugated to the primary antibody are limited, and a higher cost, since all the primary antibodies need to be labeled. In indirect immunofluorescence, the primary antibody is unlabeled; the secondary antibody carries fluorophore, recognizes primary antibody and binds to it. This technique is more sensitive than direct immunofluorescence because the fluorescent signal is amplified since one primary antibody can bind to more than one secondary antibody. Other advantages are lower cost and increased flexibility because one fluorophore conjugated secondary antibody can bind to different primary antibodies of the same species and the same primary antibody can be detected by the secondary antibodies conjugated with different fluorophores. Indirect immunofluorescence is more complex and time consuming compared to direct fluorescence, however. When performing multiple stainings in the same experiment using indirect immunofluorescence, it is important to find primary antibodies that are not raised in the same species or same isotype. For example, the primary antibody made in mouse, rabbit, and goat must be used separately. Another disadvantage is that some samples may have endogenous immunoglobulin, such as mouse tissues usually have mouse IgG and will result in a high background when the primary antibody is made in mouse. In this case, special block steps should be applied [25].

Tissues or cell samples used for immunofluorescence should be fixed. Fixation can effectively prevent autolysis and provide excellent preservation of tissue and cellular morphology. Ten percent of neutral buffered formalin (NBF) and 4 % paraformaldehyde (PFA) are the two fixatives most widely used for the preparation of samples for this technique. Cells are usually grown, treated, fixed, and stained

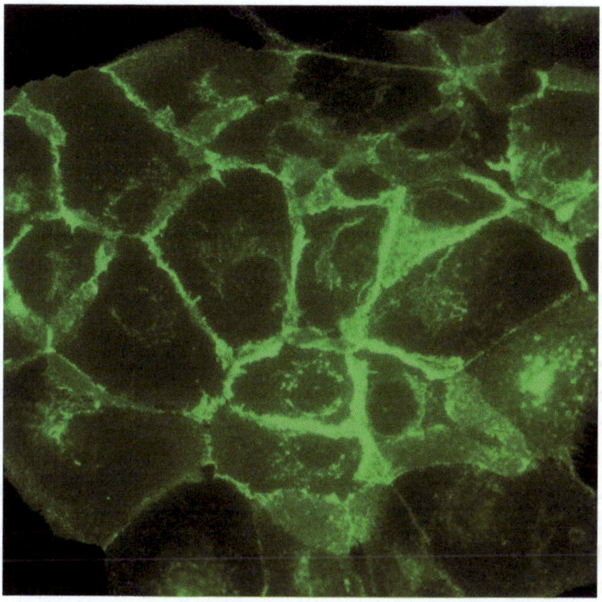

Fig. 7.8 Immunofluorescent staining of E-cadherin (*green*) in MCF10F cells

directly in multiwell plates, on chamber slides, and coverslips. Besides 10 % NBF and 4 % PFA, cells can also be fixed with methanol, acetone, or ethanol. Some antibodies work on different fixations, while some may only work on one particular fixation. Formalin-fixed paraffin-embedded sections need deparaffinization and an additional antigen retrieval step in order to break the protein crosslinks formed by formalin fixation. Antigen retrieve also exposes protein epitopes of interest to the primary antibody. A cell sample fixed with 10 % NBF or 4 % PFA usually needs permeabilization using 0.1–0.5 % Triton X-100 for the staining of cytoplasmic and nuclear protein.

The immunostaining is carried out by blocking the specimen in blocking buffer, and then incubating with primary antibody conjugated with fluorophore in direct immunofluorescence, or incubating with primary antibody and secondary antibody conjugated with fluorophore sequentially in indirect fluorescence. Washes should be extensive after antibody incubation to reduce the background signal. Coverslip slides with Prolong® Gold Antifade reagent; use the reagent with DAPI if the nuclear counterstain is required. For best results, allow mounting to cure overnight at room temperature and capture images under a fluorescent microscope as soon as possible. An example is shown in Fig. 7.8.

7.3.1 *Immunocytochemistry and Immunohistochemistry vs. Immunofluorescence Technique*

Immunocytochemistry (ICC) is a common technique that allows researchers to evaluate the expression of antigen in a particular cell sample, such as cultured cells on glass slides, cytospin, cell suspension, blood smears, and swabs. In contrast, immunohistochemistry (IHC) stains sections of biological samples, in which each cell is surrounded by tissue structure and other cells normally found in the intact tissue. Similar to immunofluorescence technique, both ICC and IHC use antibodies to detect proteins of interest. The difference is that the primary antibody or the secondary antibody is labeled with enzyme such as alkaline phosphatase (AP) and horseradish peroxidase (HRP); the corresponding substrate is then used to detect the signal. The common substrate for AP is nitroblue terrazolium (NBT)/5-bromo-4-chloro-3 indolyl phosphate (BCIP) which produces blue to purple color and fast red which produces red color. The substrate for HRP is 3,3′-diaminobenzidine (DAB) which produces a brown color. There are many methods for obtaining an immunological signal. The direct method uses antibody directly labeled with reporter. It is not used as frequently as it once was because of its low sensitivity. There are many indirect methods; of them polymetric methods are widely used, which permit binding of a large number of enzyme molecules such as HRP to a secondary antibody via the dextran backbone, thus increase sensitivity and minimize nonspecific background compared to the conventional Avidin-Biotin Complex (ABC) method. The Supersensitive™ polymer-HRP IHC detection system from Biogenex is one of the detection systems which uses this method. Modified by the use of super enhancer, it allows for excellent cell penetration resulting in intense nuclear, cytoplasmic, and membrane antigen staining. It has been used in our lab for routine IHC staining and has worked very well [26].

The benefits of ICC and IHC include high signal to noise ratio for intense stain, excellent sensitivity for weakly expressed antigens, and long storage time of stained slides, as well as time flexibility for taking pictures, especially important for large sample sizes. The disadvantages for these two methods are limited ability for colocalization study and inability to assess multiple proteins in the same sample. For this issue, immunofluorescence has an advantage of simultaneous visualization of multiple cellular targets. The disadvantages of immunofluorescence include high background in tissues, short preservation after staining, and the necessity of a fluorescence microscope. An example is shown in Fig. 7.9.

To some extent, IHC is a more general term that may encompass immunofluorescence. Both methods involve antigen-antibody reactions that enable the investigator to study tissues or cells. In IHC, the antibody is generally linked to the enzyme whereas it is linked to the fluorophore in immunofluorescence. Immunofluorescence is a type of IHC. The decision to use IHC or immunofluorescence depends on the purpose of the experiment.

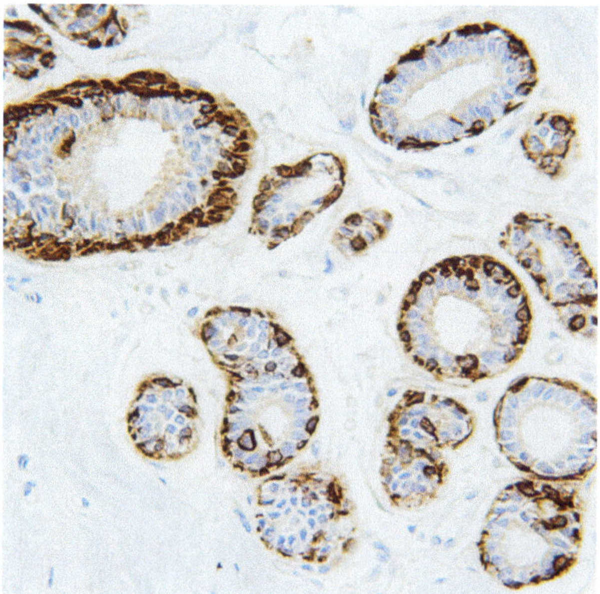

Fig. 7.9 Immunocytochemical staining of CK5 (*brown*) on breast tissue

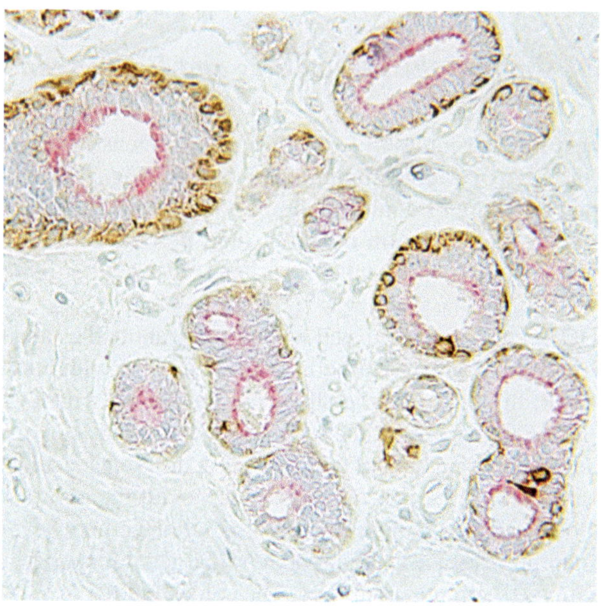

Fig. 7.10 Double immunohistochemical staining of CK5 (*brown*) and CK19 (*red*) on breast tissue

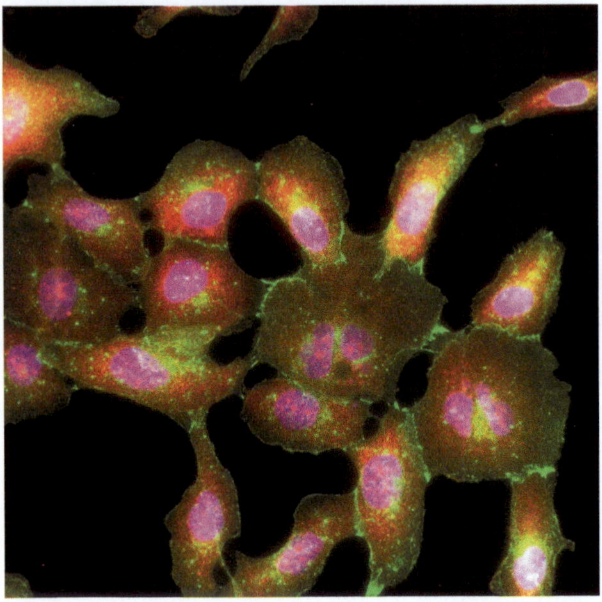

Fig. 7.11 Double immunofluorescent staining of E-cadherin (*green*) and claudin 5 (*red*) in transformed MCF10F cells

7.3.2 *Double Immunohistochemistry and Fluorescence*

In order to detect the distribution of two different antigens in the same sample, a double IHC (Fig. 7.10) or double immunofluorescence (Fig. 7.11) procedure can be carried out. The procedure of double staining is similar to single staining early on in the procedure. After the serum blocking (and peroxidase blocking for IHC), primary antibodies raised in different species can be used either in a mixture or in a sequential way.

In double IHC, sections are incubated with primary antibodies and then incubated sequentially with two secondary antibodies, one is labeled with HRP, and the other is labeled with AP; after washes the slides are developed with DAB for the first detection and then with fast red for the second detection. This enzyme based double staining is limited to the evaluation of two proteins at different locations. For example, one is a nuclear protein, the other a cytoplasmic or cell membrane protein. If two proteins are colocalized or too close to each other, this method will not be suitable. In double immunofluorescence staining, both primary and secondary antibodies can be incubated with the section in a mixture. The colocalization of two target proteins is not a problem in this method since it uses different lasers to detect the signals (Fig. 7.11). It is important that each primary antibody be tested prior to performing a double staining.

7.4 Importance of Validation of IHC Techniques

In 1998, the US Food and Drug Administration (FDA) classified IHC antibodies as "analyte-specific reagents" [27]. Within this defined group, most antibodies are deemed "Class I" medical devices, exempting them from premarket FDA notification, which entails providing evidence of safety and efficacy. A few IHC markers, including estrogen and progesterone receptors, are "Class II" medical devices, meaning they have associated guidance documents. The rationale behind Class I categorization is that IHC results are just one part of the final diagnostic interpretation by the pathologist. The flexibility provided by such a designation can be credited with rapid development of new antibodies. Predictive markers, such as human epidermal growth factor 2 (HER2/neu), whose results bear prognostic and treatment implications, have been the first targets of enhanced standardization [28, 29]. However, there is not a consistent validation system in IHC which results in a significant gap in test performance [30].

An example of how to evaluate an immunocytochemical reaction has been provided by the HercepTest (Dako Denmark A/S, Glostrup, Denmark), which classifies the HER2 membrane staining into four categories: 3+ is a positive result, with strong, complete membrane staining; 2+ is weakly positive or equivocal, with weak to moderate, complete membrane staining; 1+ is a negative result, with faint incomplete membrane staining; and 0 is negative, with no membrane staining. Only patients with 3+ score by IHC, and/or gene amplification, benefited [31]. Almost 10 years after the introduction of HER2 testing, joint guidelines were introduced by the American Society of Clinical Oncology (ASCO) and the College of American Pathologists (CAP) for the laboratory evaluation of HER2 assessment [32]. The new ASCO/CAP guidelines redefine a HER2-positive result as 3+ staining by IHC of greater than 30 % of invasive tumor cells, as compared with the 10 % cutoff used in the clinical trials leading to approval of trastuzumab [33–35]. The time it takes to score a HercepTest result using image analysis-assisted microscopy is significantly longer than it takes for manual scoring for most observers [36]. However, the use of image analysis did not impose an overall improvement or hindrance in the concordance of HER2 findings [36]. The benefit of digital microscopy to pathologist accuracy has been documented [37]. Atkinson reported that inter-observer agreement was significantly higher using image analysis-assisted microscopy compared to conventional microscopy [36].

7.5 In Situ Hybridization Techniques

ISH are techniques that use a labeled DNA or RNA probe to hybridize a specific DNA or RNA within a histological section, in cultured cells or the entire tissue (whole mount ISH). In fluorescence in situ hybridization (FISH), the probe can be detected by fluorochrome, which is conjugated to the nucleotides and then incorporated into the probe (direct method), or detected indirectly by the antibody conjugated to fluorochrome [38]. In chromogenic in situ hybridization (CISH), the incorporated nucleotides are detected by the antibody linked to enzymes, such as HRP, which are capable of reacting with substrates resulting in colored deposits and can be analyzed by light microscopy [39]. The advantage of FISH is the same as with other immunofluorescence techniques; i.e., simultaneous visualization of multiple targets. The disadvantages are similar as well. Compared to FISH, CISH has the advantage of revealing histological details more clearly, with concern about signal fading. DNA FISH is a powerful technique used in medical diagnostics to assess chromosomal integrity and gene amplification. For example, 5p minus syndrome, a rare genetic disorder caused by the absence of a part of chromosome 5, can be diagnosed by FISH. RNA FISH is usually used to measure and localize RNAs of interest within biological samples. This is particularly useful for the study of noncoding RNAs, some of which have been shown to play an important role in many cellular processes [40].

7.5.1 Choice of Probe

A probe is a fragment of DNA or RNA of variable length (usually 100–1,200 bases long) which is used to hybridize and detect DNA or RNA targets.

7.5.1.1 DNA Probes

Both single-stranded (ssDNA) and double stranded DNA (dsDNA) can be used as probes. ssDNA probes can be produced by reverse transcription of RNA or by amplified primer extension of a PCR-generated fragment in the presence of a single antisense primer. dsDNA probes can be produced by extraction, purification, and excision of plasmid DNA, or by PCR. dsDNA probes are generally less sensitive because of the tendency of the DNA strands to rehybridize to each other and are not widely used today [41].

7.5.1.2 RNA Probes

In general, RNA probes are more sensitive than DNA probes. RNA probes also have the advantage that RNA–RNA hybrids are more stable than RNA–DNA hybrids and are resistant to digestion by RNases. Therefore, the nonspecific signal can be removed by the post-hybridization digestion with RNase. The disadvantages of RNA probes are the amount of efforts required to make the probes and the degradation of probes by RNase. RNA probes are usually produced by in vitro transcription. The RNA probe is transcribed from a linear DNA template that is cloned and linearized from a vector. The plasmid vectors containing polymerase from bacteriophages T3, T7, or SP6 are used [42].

7.5.1.3 Oligonucleotide Probes

An oligonucleotide probe is a short piece of DNA that is complementary to a specific piece of DNA or RNA in the cell. These oligonucleotides are produced by an automated chemical synthesis. They are small, usually around 40–50 base-pairs. Regarding the advantages of oligonucleotide probes, besides the stability and availability, because of their small size they penetrate into cells and tissues better, and are faster and less expensive to use. The major disadvantage is their low sensitivity. For this reason, use of more than one piece of oligonucleotide probe can usually produce stronger signal [43].

7.5.2 Labeling of Probes

There are several labeling methods. In these methods, the label is attached to one of the four nucleotides and then incorporated into the probe.

7.5.2.1 PCR Labeling

PCR labeling is a very popular method of labeling dsDNA probes. When using this method, the primers that will amplify the sequence are to be used as probe, and a labeled dNTP should be included during PCR. This technique is especially useful for making highly labeled probes from a limited amount of DNA template [44].

7.5.2.2 Nick Translation

Nick translation is a DNA labeling method by which labeled nucleotides are incorporated into DNA using DNase I to introduce nicks in the DNA templates and DNA polymerase I to repair the nicks, thus forming the labeled DNA sequences. This procedure is good at making large, evenly labeled probes [45].

7.5.2.3 In Vitro Transcription

In vitro transcription is a method that generates sense or antisense RNA probes using T7, SP6 or T3 promoter and purified enzyme. Modified nucleotide containing isotopes like ^{32}P or detectable epitopes like DIG can be integrated into the RNA via the transcription [46].

7.5.2.4 End Labeling

End labeling is a labeling method for DNA and RNA probes. End labeling can be performed at 3′- or 5′-end. In 3′-end labeling method, terminal transferase is used to add a single labeled nucleotide to the 3′ ends of DNA/RNA or oligonucleotide. In 5′-end labeling method, T4 polynucleotide kinase is used to label DNA/RNA at 5′ end [47].

7.5.2.5 3′ Tailing Labeling

Tailing labeling is mainly used to label oligonucleotide probes. Terminal transferase is used to generate a tail of labeled nucleotides of 10–100 nucleotides in length at the 3′-end of oligonucleotides. This will increase the sensitivity of detection over probes labeled by 3′-end labeling [48].

7.5.3 Two Types of Labels: Radioactive Isotopes and Nonradioactive Labels

There are primarily two types of labels: radioactive isotopes and nonradioactive labels.

7.5.3.1 Radioactive Isotopes

Traditionally, ^{32}P, ^{35}S, and ^{3}H are used to label the probes for ISH. These probes are visualized by radioactive exposure of sections against photographic film which can then be developed. ^{35}S is the most commonly used because of its high sensitivity. However, the short shelf life of radioactive labels and the accompanying safety issues with waste disposal and contamination need to be considered [49].

7.5.3.2 Nonradioactive Labels

Fluorescent dye, biotin, and digoxigenin (DIG) have been used extensively for many years. Fluorescent dye can be used directly or indirectly. Biotin probes can be detected by streptavidin-peroxidase conjugate. DIG labeled probes may be detected with anti-DIG antibodies that are conjugated to peroxidase, alkaline phosphatase, fluorescein, and so on. Alternatively, conjugated secondary antibodies may be used [50].

The choice of label type is generally dependent on a number of considerations including sensitivity required, background in the particular tissue used, and availability of safety supply and measure for handling for radioactive labels and availability of a fluorescent microscope for fluorescent dye. Certain tissues such as liver and kidney contain high levels of endogenous biotin, thus the use of biotin probe for these tissues is not a good choice. Digoxigenin (DIG) is a steroid isolated from the digitalis plant and as the blossoms and leaves are the only known source of digoxigenin, the anti-DIG antibodies are not likely to bind to other biological material. In general, the DIG labeling and detection method is highly sensitive and applicable [51].

7.5.3.3 Preparation of Material for In Situ Hybridization (ISH)

The most common materials used for ISH are (1) paraffin tissue sections; (2) frozen tissue sections and (3) cells grown or cytospun on glass slides. The tissues or cells need to be fixed to preserve DNA or RNA for hybridization. As DNA is very stable, DNA for ISH can be easily preserved by fixing the tissues or cells with formalin or paraformaldehyde, as it is usual for IHC. However, preserving RNA is much more difficult due to presence of RNase. RNase can be found almost everywhere including glassware and reagents and will cause the degradation of the RNA in cells and RNA probes. Therefore, users must make sure they use RNase free solutions and sterile techniques to avoid RNase contamination. Use of DEPC-treated water, buffer, and disposable plastic ware, as well as diligent use of gloves, are common requirements.

Details of DIG Labeled Oligonucleotide Probe In Situ Hybridization Protocol (FISH)

Slide Preparation

Cells are grown on chamber slides. At the end of treatment, cells are washed once with PBS and then fixed with 4 % paraformaldehyde for 10 min at room temperature, followed with three washes with PBS, 5 min for each wash at room temperature.

Permeabilization

Treat cells with PBS containing 0.4 % Triton X-100 for 10 min at room temperature. Wash slides three times with PBS.

Acetylate Slides

Plate slides on a rocking platform and incubate slides with 0.1 M triethanolamine (TEA) buffer, pH 8.0, containing 0.25 % (v/v) acetic anhydride (*Note*: acetic anhydride is quite unstable, add acetic anhydride to TEA buffer immediately before incubation). Then wash slides twice with 2× SSC, 5 min for each wash.

Prehybridization

Use Enzo ISH buffer (1.25×, Cat# ENZ-33808) to make 1× prehybridization buffer containing 100 μg/mL of Poly(A). Incubate slides at 37 °C for 2 h with prehybridization buffer in a humidified chamber. Hybrislips coverslips from EMS

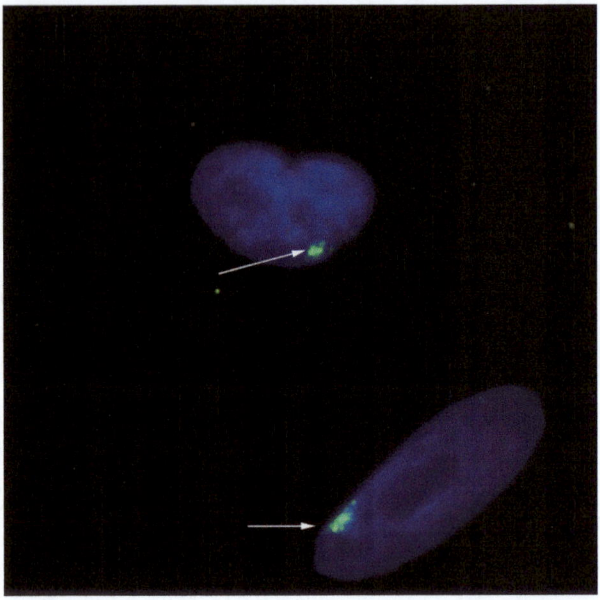

Fig. 7.12 In situ hybridization of XIST in MCF10F cells. *Arrows* indicate the signal of XIST RNA

(Cat# 70329-22) are used to cover the samples with prehybridization buffer. Remove coverslips and wash slides once in 2× SSC for 5 min at room temperature following prehybridization.

Hybridization

Drain 2× SSC from slides and apply sufficient hybridization buffer (containing 20 pmol/mL of DIG labeled 50-mer oligonucleotide probes) onto the slides and cover with coverslips. Sense probe control and negative control that use only prehybridization buffer must be used. Seal the coverslips with rubber cement. Incubate the slides at 37 °C for 16–20 h.

Stringency Washes

Remove coverslips by immersing slides in 2× SSC. Wash slides in 2× SSC for 5 min at room temperature. In a shaking water bath at 37 °C, wash slides as follows: twice with 2× SSC, 10 min for each wash; twice with 1× SSC, 10 min for each wash; twice with 0.2× SSC, 10 min for each wash.

Immunological Detection

(1) Wash slides once with washing buffer (PBS containing 0.1 % tween 20); (2) Block slides with blocking solution (10 % goat serum in washing buffer) at room temperature for 30 min; (3) Incubate slides with primary antibody (mouse monoclonal antibody to digoxigenin, 1:200 dilution in washing buffer) at room temperature for 60 min; (4) Wash slides three times with washing buffer, 10 min for each wash; (5) Incubate slides with secondary antibody (Alexa fluor 488, goat anti-mouse IgG, 1:400 dilution in washing buffer) for 60 min at room temperature; (6) Wash slides three times with washing buffer, 10 min for each wash; (7) Dehydrate slides: 70 % ethanol, 5 min; 100 % ethanol, 5 min. (8) Mount slides with Prolong Gold Antifade Reagents with DAPI. Leave slides flat in the folder and keep at room temperature to cure overnight. For long-term storage, seal the coverslip to the slide with clear nail polish on the following day and store slides in a covered slide box at −20 °C.

Image Acquisition

Pictures must be taken under a fluorescent microscope as soon as possible to avoid fading of the signal. An example of FISH used for detection of XIST RNA signal in MCF10F treated in presence and in absence of the hormone chorionic gonadotropin is shown in Fig. 7.12.

7.6 Comparative Genomic Hybridization (CGH)

Chromosomal instability has long been recognized as a hallmark of cancer [52–58]. Aberrations in chromosomal copy numbers are among the most common of the number and types of chromosomal aberrations that may occur as a result of chromosomal instability. The clear difference in chromosomal copy number between benign and malignant tumors was the basis for introducing, in 1992, a novel technique known as comparative genomic hybridization (CGH) [59]. This technique allowed for evaluation of the entire genome of a tumor of interest for copy number aberrations in one experiment. In the traditional method, tumor DNA and reference DNA are labeled with different fluorochromes and are then simultaneously hybridized to immobilized and denatured chromosome metaphase spreads [59, 60]. The fluorescence ratio along the axis of each chromosome can then be measured, indicating the presence of DNA gains or losses from the tumor sample at each chromosomal locus. The development of an array platform for CGH has since greatly increased the magnitude of resolution, allowing for significantly more precise definition of the area of chromosomal loss or gain [59, 60]. We have successfully used this technique (Fig. 7.13) to evaluate gain and losses in MCF10F cells transformed with 17 beta estradiol [61, 62]. Gain or losses are progressive at the different stages of the in vitro–in vivo model; there are few gains or losses in trMCF, and more in the invasive bsMCF and tumorigenic caMCF cells [62] (Fig. 7.12).

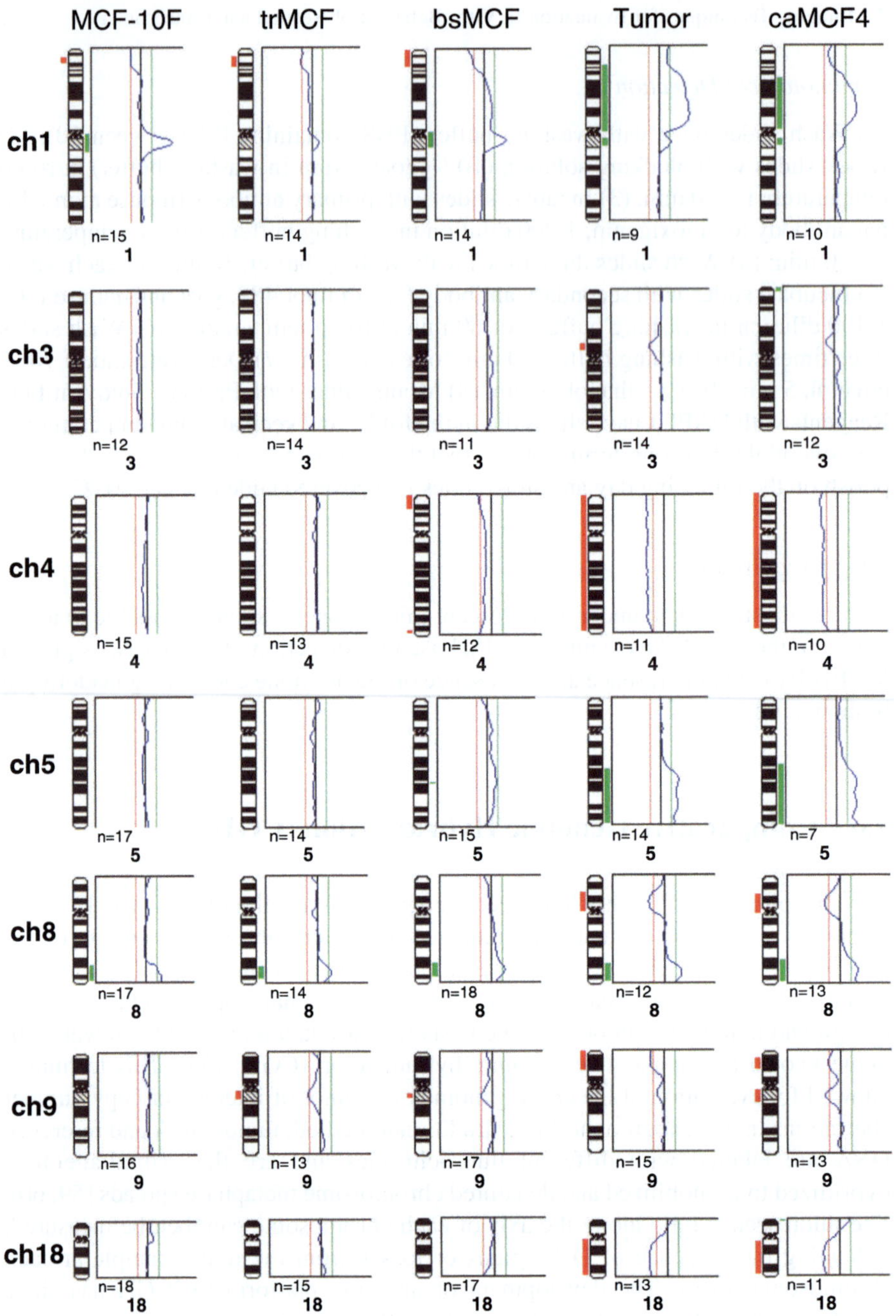

Fig. 7.13 Comparative genomic hybridization (CGH) analysis of the in vitro–in vivo model of cell transformation induced by 17β-estradiol (E2). The threshold was set at 0.8 and 1.2 for losses and gains, respectively; the mean values of individual ratio profiles were calculated from at least seven metaphase spreads, and averaged values were plotted as profiles alongside individual chromosome ideograms. The chromosomal imbalances are shown in a histogram, which represent the DNA gains (in *green*) and losses (in *red*) as an incidence curve along each chromosome. Only chromosomes with changes are represented. Also, the gains and losses found in the tumor (Tumor) that gave origin to caMCF4 are represented (Printed with permission from: Fernadez, S.V. and Russo, J. Estrogen and xenoestrogens in breast cancer. Toxicologic Pathology, 38: 110–122, 2010)

Only small chromosome gains were observed in trMCF, and most alterations were observed in bsMCF and caMCF cells (Fig. 7.13). Chromosomal amplifications were found in 1p36.12-pter, 5q21.1-qter in both bsMCF, and caMCF. Losses of the complete chromosome 4 and 8p11.21-23.1 were found in bsMCF and caMCF cells (Fig. 7.13). In caMCF cells, additional losses were found in 3p12.1-14.1, 9p22.1-pter, and 18q11.21-qter. Also, a chromosomal amplification in 13q21.31-qter was found in caMCF using high-density SNP arrays [62]. Functional profiling of dysregulated genes revealed progressive changes in the integrin signaling pathway, inhibition of apoptosis, acquisition of tumorigenic cell surface markers, and epithelialmesenchymal transition (EMT). In tumorigenic cells, the levels of E-cadherin, EMA, and various keratins were low and CD44E/CD24 was negative, whereas SNAI2, vimentin, S100A4, FN1, HRAS, TGFb1, and CD44H levels were high.

7.7 Fluorescence In Situ Hybridization (FISH) for Detecting Chromosomal Abnormalities

FISH has emerged as a preferred molecular technique to interrogate chromosomal abnormalities [63]. FISH has proven its utility as a diagnostic adjunct in a number of lymphoid and solid tumors including melanoma. FISH uses short DNA fragments (FISH probes) that are hybridized to a 5 mm-thick, formalin-fixed, paraffin-embedded section of a tumor of interest. Overlapping wavelength spectrums of the currently available fluorochromes limit the number of probes that can be concurrently hybridized in a single experiment to a maximum of four. Hence, in contrast to CGH, a maximum of four chromosomal targets can be evaluated in a single FISH test. Nonbound, fluorescently labeled DNA is washed away [64]. The section can then be examined under a fluorescence microscope. FISH offers the possibility of examining individual tumor cells in different areas of a lesion and of correlating the results with tumor morphology under conventional light microscopy. Each fluorescently labeled fragment of DNA that hybridizes to a tumor cell nucleus will appear as a distinct fluorescent dot. Each dot identifies a single copy of the chromosomal locus with a homologous DNA. As discussed in the literature [64], in an idealized scenario, each diploid nucleus will bind to two probes, resulting in two dots. If a DNA region of interest is gained, more than two dots will be observed. If a DNA region of interest is deleted, less than 2 (1 or 0) signals will be identified. In a real-life scenario, for many perfectly normal diploid cells, probes will be outside the plane of the section or overlap, resulting in less than two dots per nucleus. Imperfect hybridization can also result in false-positive or false-negative signals. The number of dots per nucleus with a specific fluorescent color can then be detected either manually or with software designed for automated analysis. Because of intrinsic variability of FISH signals, a sufficient number of cells must be examined in each experiment and strict quality control measures must be implemented. The data must be presented as a percentage of nuclei containing more than two copies of a particular locus or as a percentage of cells showing an increase or loss of signals in comparison to a probe directed against a centromere region of a chromosome. The signal cutoffs—that is, a percentage of cells that must fulfill a particular FISH criterion for a result to be considered abnormal—have to be determined and validated empirically for each probe, with appropriate controls. Having rigorously interrogated cutoff values is critical to the robustness of a FISH assay. Increasing the cutoff values will increase the specificity of a test at the expense of the sensitivity and vice versa [64].

FISH applied to standard 5 mm–thick, paraffin-embedded sections allows for histopathologic correlation. Specific area or areas of tumor examined by FISH and using as the background fluorescence of DAPI (6-diamidino-2-phenylindole) allows visualization of the tumor's silhouette, and with some experience one can easily make correlations between the histopathologic appearance and the chromosomal aberrations.

FISH can be used to mark specific genes in a given chromosome. For example, allele loss in breast cancer has been reported in chromosome 17p [65, 66], suggesting that genes located in that chromosome arm might be likely targets for this event. We have searched for a gene located in the telomeric portion of chromosome 17p, on 17p13.3, distal to p53 that has been observed to be host in primary breast cancer [67]. LOH of this region, using the probe p144D6, has been found in 75 % of primary breast cancers and in cells that are progesterone receptor negative. LOH of 17p has been associated with high cell proliferation in primary breast tumors, and with a shorter doubling time in the tumorigenic cell line BP1-E [68, 69]. A number of breast tumors have been reported to lose the tip of 17p without affecting the *p53* gene, suggesting that a second gene, distinct from *p53*, may be involved in breast cancer. With this concept in mind we performed DNA analysis for allelic loss on chromosome 17p 13.3 using probes pYNZ22 and p144D6 for detecting variable number of tandem repeats (VNTR) [69]. The Southern blots showed that the cell lines MCF-10M and MCF-10F are heterozygous for probe pYNZ22. Clones D3, D3-1, BP1 and BPI-E, derived from DMBA and BP treated cells, respectively, showed the same heterozygous restriction length polymorphism pattern (RFLP). RFLP analysis with probe pl44D6 revealed a heterozygous pattern in MCF-10M, MCF-10F cells and in clones D3, D3-1, and BPI. Only the cell line BP1-E exhibited LOH, which probed with p144D6 [69]. This phenomenon was demonstrated by FISH analysis (Fig. 7.14) [68]. Conventional banding analysis did not reveal any alterations of 17p in either MCF-10F or BPI-E cells. FISH analysis showed that both pl44D6 and the chromosome 17 centromeric probe hybridized to both copies of chromosome 17 in MCF-10F cells (Fig. 7.14). In these cells, the hybridization signals were parallel in both chromatids of both homologues (Fig. 7.14). In transformed BPI-E cells, the 17 centromeric probe was positive, but the p144D6 probe did not hybridize both chromatids in the two homologues, but only two chromatids of one homologue were positive (Fig. 7.14), indicating that LOH in the telomeric position of chromosome 17p was observed only in the cell line expressing the pre-tumorigenic phenotype [68] that resulted in a pre-tumorigenic cell line that evolved to a tumorigenic one.

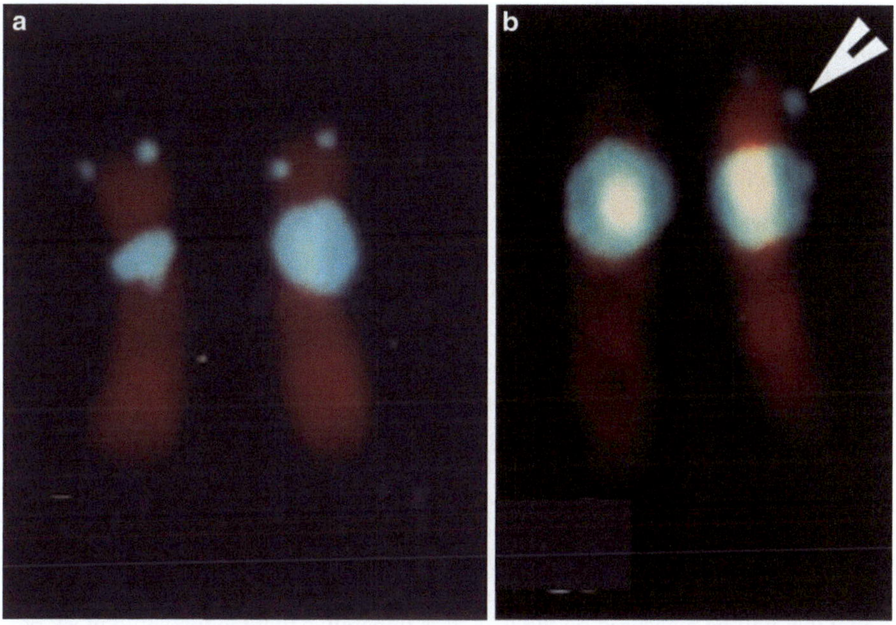

Fig. 7.14 (**a**) FISH analysis showed that both pl44D6 and the chromosome 17 centromeric probe hybridized to both copies of chromosome 17 in MCF-10F cells. The chromosome 17 shows the hybridization signals in parallel in both chromatids of both homologues. (**b**) In transformed BPI-E cells, the 17 centromeric probe was positive, but the p144D6 probe did not hybridize both chromatids in the two homologues. Two chromatids of one homologue were positive (*arrow*), indicating that LOH in the telomeric position of chromosome 17p. With permission from: Xiaoqi Yang, Irma H. Russo, Yajue Huang, and Jose Russo. Microsatellite instability on chromosome 17 is associated with progression of breast cancer. International J. of Oncology 11:41–46, 1997

7.8 Validation of Fluorescence In Situ Hybridization (FISH)

As the fields of cancer cytogenetics and molecular cytogenetics continue to expand, new tests are being developed to identify an increasing number of chromosome abnormalities that are important for the diagnosis, prognosis, and monitoring of neoplastic diseases. Molecular cytogenetic methods, such as FISH, are more sensitive than routine metaphase chromosome analysis because they can detect a specific abnormality in both dividing and nondividing cells and have a lower threshold for detecting small populations of abnormal cells. Thus, FISH testing has become part of routine clinical practice in many laboratories that deal with neoplastic specimens. Introduction of a new FISH test requires assessment of performance characteristics of the assay and validation of the test by the laboratory. If the FISH test has been approved by the US FDA, the laboratory must independently verify the performance characteristics of the assay. However, most FISH probes used in clinical testing are not US FDA–approved but instead are analyte-specific reagents. Although the technical performance of these probes has been evaluated by the manufacturer prior to marketing, they must be validated as a laboratory-developed test prior to use in clinical testing. General guidelines on validation of FISH tests are available and provide an excellent introduction to issues that are particular to this methodology.

The validation of FISH probes includes several steps: familiarization and planning; assessment of performance characteristics (accuracy, precision, sensitivity and specificity, reference ranges, and reproducibility), and implementation of the test.

7.9 Quantitation of Immunocytochemistry and Immunofluorescence

An example of how to evaluate immunocytochemical reaction has been provided by the HercepTest (Dako Denmark A/S, Glostrup, Denmark) (see previous sections). Our experience is based on the use of Metamorph v7.7.8.0. We have found that an important step is to select the threshold of the image and select the range of intensity of pixels you are willing to accept. For example, creating a threshold mask for the DAPI stain, making sure that the mask does not envelope two or more nuclei or cover area outside of the DAPI signal. The region that the mask overlays is the region of the signal that will be measured by the ICC. It is imperative that the threshold mask be a good fit and any joined objects be separated. The threshold allows for the creation of a mask composed of binary 8-bit images, selecting pixels in a specific intensity range. Auto threshold can be used, but manual is the best. If threshold is used in any picture, the threshold range should be maintained throughout all fields and pictures. For repetitive tasks such as thresholding, transferring, and creation of regions, one can create what Metamorph calls a Journal, also known as a Macro. A journal is programming script that tells metamorph how to process a particular task, and in what order to do it. The ease of doing this will compress most of this protocol into just a few short clicks which can be easily added to a taskbar.

7.10 Conclusions and Future Directions

Technology used in the field of breast cancer research has been increasing exponentially over the last few decades and has shown few signs of slowing down. As we advance further and deeper into the biological mechanisms underlying oncogenesis and metastases we will come ever closer to understanding breast cancer and the mechanisms underlying it. Methods used today will soon be out of date and replaced with new technology, making forms of cell imaging and sorting faster, more efficient, more precise, and cheaper. Machines used for FACS analysis, while becoming cheaper, smaller, and more user friendly may soon be regularly found in labs. Different subtypes of biology are discovered constantly, first on the cellular level, then on the molecular level, and more recently on the epigenetic level. Without a doubt there will be another field discovered and more pathways and components adding to the complexity of breast cancer. We can easily imagine a new way of going about research with an emphasis on bioengineering. With the ability to alter molecules, pathways, cells, and a multitude of other biological components, bioengineering may represent the way of the future, and with great hope, be the key to understanding of breast cancer.

The need to understand and visualize molecular processes in living cells will require a combination of multiple techniques in one instrument. For example, the combination of fluorescence life time imaging microscopy (FUM) in combination with atomic force microscopy. FUM provides optically encoded information about processes in live cells, and the atomic force microscope provides nanometer-resolved surface topography and mechanical information [70]. Characteristics of surfaces on a single molecule level can be examined using both confocal fluorescence microscopy and atomic force microscopy. The techniques analyze different aspects of the sample. For example, one can investigate molecular dynamics with time-resolved fluorescence spectroscopy and characterize surface topography with atomic force microscopy. Thus, combining both techniques into a single instrument would enhance the methodology and open up new investigation schemes for live cell imaging, high-resolution molecular scale imaging, and both force and mechanical studies [70].

Literature Cited

1. Smith C (2013) Buyers guide: flow cytometer. Am Lab:17–19
2. Jahan-Tigh RR, Ryan C, Obermoser G, Schwarzenberger K (2012) Flow cytometry. J Invest Dermatol 132(10):282
3. Givan AL, (ed) (2001) The sorting of cells. In: Flow cytometry: first principles. Wiley, London, pp 159–174
4. Gilman-Sachs A (1994) Flow cytometry. Anal Chem 66(13):700A–707A
5. Givan AL, (ed) (2001) Seeing the light: lasers, fluorochromes, and filters. In: Flow cytometry: first principles. Wiley, London, pp 59–80
6. Delude RL (2005) Flow cytometry. Crit Care Med 33(12 Suppl):S426–S428
7. Jordan CT, Guzman ML, Noble M (2006) Cancer stem cells. N Engl J Med 355(12): 1253–1261
8. Reya T, Morrison SJ, Clarke MF, Weissman IL (2001) Stem cells, cancer, and cancer stem cells. Nature 414(6859):105–111
9. Blacking TM, Waterfall M, Samuel K, Argyle DJ (2012) Flow cytometric techniques for detection of candidate cancer stem cell subpopulations in canine tumour models. Vet Comp Oncol 10(4):252–273
10. Moreb JS (2008) Aldehyde dehydrogenase as a marker for stem cells. Curr Stem Cell Res Ther 3(4):237–246
11. Alfaro LA, Dick SA, Siegel AL, Anonuevo AS, McNagny KM, Megeney LA, Rossi FM (2011) CD34 promotes satellite cell motility and entry into proliferation to facilitate efficient skeletal muscle regeneration. Stem Cells 29(12):2030–2041
12. Naor D, Nedvetzki S, Golan I, Melnik L, Faitelson Y (2002) CD44 in cancer. Crit Rev Clin Lab Sci 39(6):527–579
13. Gil da Costa RM, Matos E, Rema A, Lopes C, Pires MA, Gartner F (2007) CD117 immunoexpression in canine mast cell tumours: correlations with pathological variables and proliferation markers. BMC Vet Res 3:19

14. Tang KH, Ma S, Lee TK, Chan YP, Kwan PS, Tong CM, Ng IO, Man K, To KF, Lai PB, Lo CM, Guan XY, Chan KW (2012) CD133(+) liver tumor-initiating cells promote tumor angiogenesis, growth, and self-renewal through neurotensin/interleukin-8/CXCL1 signaling. Hepatology 55(3):807–820
15. Marcato P, Dean CA, Pan D, Araslanova R, Gillis M, Joshi M, Helyer L, Pan L, Leidal A, Gujar S, Giacomantonio CA, Lee PW (2011) Aldehyde dehydrogenase activity of breast cancer stem cells is primarily due to isoform ALDH1A3 and its expression is predictive of metastasis. Stem Cells 29(1):32–45
16. Sheridan C, Kishimoto H, Fuchs RK, Mehrotra S, Bhat-Nakshatri P, Turner CH, Goulet R Jr, Badve S, Nakshatri H (2006) CD44+/CD24– breast cancer cells exhibit enhanced invasive properties: an early step necessary for metastasis. Breast Cancer Res 8(5):R59
17. Ginestier C, Liu S, Diebel ME, Korkaya H, Luo M, Brown M, Wicinski J, Cabaud O, Charafe-Jauffret E, Birnbaum D, Guan JL, Dontu G, Wicha MS (2010) CXCR1 blockade selectively targets human breast cancer stem cells in vitro and in xenografts. J Clin Invest 120(2):485–497. doi:10.1172/JCI39397
18. Charafe-Jauffret E, Ginestier C, Iovino F, Wicinski J, Cervera N, Finetti P, Hur MH, Diebel ME, Monville F, Dutcher J, Brown M, Viens P, Xerri L, Bertucci F, Stassi G, Dontu G, Birnbaum D, Wicha MS (2009) Breast cancer cell lines contain functional cancer stem cells with metastatic capacity and a distinct molecular signature. Cancer Res 69(4):1302–1313
19. Li C, Heidt DG, Dalerba P, Burant CF, Zhang L, Adsay V, Wicha M, Clarke MF, Simeone DM (2007) Identification of pancreatic cancer stem cells. Cancer Res 67(3):1030–1037
20. Fillmore CM, Kuperwasser C (2008) Human breast cancer cell lines contain stem-like cells that self-renew, give rise to phenotypically diverse progeny and survive chemotherapy. Breast Cancer Res 10(2):R25
21. Wang N, Shi L, Li H, Hu Y, Du W, Liu W, Zheng J, Huang S, Qu X (2012) Detection of circulating tumor cells and tumor stem cells in patients with breast cancer by using flow cytometry: a valuable tool for diagnosis and prognosis evaluation. Tumour Biol 33(2):561–569. doi:10.1007/s13277-011-0303-1
22. Gupta V, Zhang QJ, Liu YY (2011) Evaluation of anticancer agents using flow cytometry analysis of cancer stem cells. Methods Mol Biol 716:179–191
23. Coons AH, Kaplan MH (1950) Localization of antigen in tissue cells; improvements in a method for the detection of antigen by means of fluorescent antibody. J Exp Med 91(1):1–13
24. Paddock SW (1999) Confocal laser scanning microscopy. Biotechniques 27(5):992–996, 998–1002, 1004
25. Lu QL, Partridge TA (1998) A new blocking method for application of murine monoclonal antibody to mouse tissue sections. J Histochem Cytochem 46(8):977–984
26. Ramos-Vara JA (2005) Technical aspects of immunohistochemistry. Vet Pathol 42(4): 405–426
27. Taylor CR (1999) FDA issues final rule for classification and reclassification of immunohistochemistry reagents and kits. Am J Clin Pathol 111(4):443–444
28. Wolf AC, Hammond EH, Schwartz JN et al (2007) American Society of Clinical Oncology/College of American Pathologists guideline recommendations for human epidermal grown factor receptor 2 testing in breast cancer. Arch Pathol Lab Med 131(1):18–43
29. Fitzgibbons PL, Murphy DA, Hammond ME et al (2010) Recommendations for validating estrogen and progesterone receptor immunohistochemistry assays. Arch Pathol Lab Med 134(6):930–935
30. Hardy LB, Fitzgibbons PL, Goldsmith JD, Eisen RN, Beasley MB, Souers RJ, Nakhleh RE (2013) Immunohistochemistry validation procedures and practices a College of American Pathologists survey of 727 laboratories. Arch Pathol Lab Med 137:19–25
31. Slamon DJ, Leyland-Jones B, Shak S et al (2001) Use of chemotherapy plus a monoclonal antibody against HER2 for metastatic breast cancer that overexpresses HER2. N Engl J Med 344(11):783–792
32. Hicks DG, Kulkarni S (2008) HER2+ breast cancer: review of biologic relevance and optimal use of diagnostic tools. Am J Clin Pathol 129(2):263–273

33. Vang Nielsen K, Jorgensen JT, Schonau A, Oster A (2007) Human epidermal growth factor receptor 2 testing in breast cancer. J Clin Oncol 25(25):4020; author reply 4021–4023
34. Yaziji H, Taylor CR (2007) Begin at the beginning, with the tissue! The key message underlying the ASCO/CAP task-force guideline recommendations for HER2 testing. Appl Immunohistochem Mol Morphol 15(3):239–241
35. Chibon F, de Mascarel I, Sierankowski G et al (2009) Prediction of HER2 gene status in Her2 2+ invasive breast cancer: a study of 108 cases comparing ASCO/CAP and FDA recommendations. Mod Pathol 22(3):403–409
36. Atkinson R, Mollerup J, Lænkholm A-V, Verardo M, Hawes D, Commins D, Engvad B, Correa A, Ehlers CC, Nielsen KV (2011) Effects of the change in cutoff values for human epidermal growth factor receptor 2 status by immunohistochemistry and fluorescence in situ hybridization: a study comparing conventional brightfield microscopy, image analysis-assisted microscopy, and interobserver variation. Arch Pathol Lab Med 135(8):1010–1016
37. Bloom K, Harrington D (2004) Enhanced accuracy and reliability of HER-2/neu immunohistochemical scoring using digital microscopy. Am J Clin Pathol 121(5):620–630
38. Langer-Safer PR, Levine M, Ward DC (1982) Immunological method for mapping genes on Drosophila polytene chromosomes. Proc Natl Acad Sci U S A 79(14):4381–4385
39. Zhao J, Wu R, Au A, Marquez A, Yu Y, Shi Z (2002) Determination of HER2 gene amplification by chromogenic in situ hybridization (CISH) in archival breast carcinoma. Mod Pathol 15(6):657–665
40. Sirchia SM, Tabano S, Monti L, Recalcati MP, Gariboldi M, Grati FR, Porta G, Finelli P, Radice P, Miozzo M (2009) Misbehaviour of XIST RNA in breast cancer cells. PLoS One 4(5):e5559
41. Chantratita W, Henchal EA, Yoosook C (1989) Rapid detection of herpes simplex virus DNA by in situ hybridization with photobiotin-labelled double-stranded DNA probes. Mol Cell Probes 3(4):363–373
42. Schaeren-Wiemers N, Gerfin-Moser A (1993) A single protocol to detect transcripts of various types and expression levels in neural tissue and cultured cells: in situ hybridization using digoxigenin-labelled cRNA probes. Histochemistry 100(6):431–440
43. Lewis ME, Sherman TG, Burke S, Akil H, Davis LG, Arentzen R, Watson SJ (1986) Detection of proopiomelanocortin mRNA by in situ hybridization with an oligonucleotide probe. Proc Natl Acad Sci U S A 83(15):5419–5423
44. Schowalter DB, Sommer SS (1989) The generation of radiolabeled DNA and RNA probes with polymerase chain reaction. Anal Biochem 177(1):90–94
45. Macgregor HC, Mizuno S (1976) In situ hybridization of "nick-translated" 3H-ribosomal DNA to chromosomes from salamanders. Chromosoma 54(1):15–25
46. Young ID, Ailles L, Deugau K, Kisilevsky R (1991) Transcription of cRNA for in situ hybridization from polymerase chain reaction-amplified DNA. Lab Invest 64(5):709–712
47. Harwood AJ (2002) End-labeling of DNA probes. Methods Mol Biol 187:17–22
48. Collins ML, Hunsaker WR (1985) Improved hybridization assays employing tailed oligonucleotide probes: a direct comparison with 5′-end-labeled oligonucleotide probes and nick-translated plasmid probes. Anal Biochem 151(2):211–224
49. Simmons DM, Arriza JL, Swanson LW (1989) A complete protocol for *in situ* hybridization of messenger RNAs in brain and other tissues with radio-labeled single-stranded RNA probes. J Histotechnol 3:169–181
50. Bauman JG, Wiegant J, Borst P, van Duijn P (1980) A new method for fluorescence microscopical localization of specific DNA sequences by in situ hybridization of fluorochromelabelled RNA. Exp Cell Res 128(2):485–490
51. Komminoth P (1992) Digoxigenin as an alternative probe labeling for in situ hybridization. Diagn Mol Pathol 1(2):142–150
52. Sandberg AA, Chen Z (2002) Cytogenetics and molecular genetics of human cancer. Am J Med Genet 115(3):111–112
53. Boland CR, Komarova NL, Goel A (2009) Chromosomal instability and cancer: not just one CINgle mechanism. Gut 58(2):163–164
54. Cheung AL, Deng W (2008) Telomere dysfunction, genome instability and cancer. Front Biosci 13:2075–2090 [review]

55. Ganem NJ, Godinho SA, Pellman D (2009) A mechanism linking extra centrosomes to chromosomal instability. Nature 460(7252):278–282
56. Murnane JP (2010) Telomere loss as a mechanism for chromosome instability in human cancer. Cancer Res 70(11):4255–4259 [review]
57. Ye CJ, Liu G, Bremer SW et al (2007) The dynamics of cancer chromosomes and genomes. Cytogenet Genome Res 118(2–4):237–246 [review]
58. Albertson DG, Collins C, McCormick F et al (2003) Chromosome aberrations in solid tumors. Nat Genet 34(4):369–376 [review]
59. Pinkel D, Albertson DG (2005) Array comparative genomic hybridization and its applications in cancer. Nat Genet 37(Suppl):S11–S17 [review]
60. Albertson DG, Snijders AM, Fridlyand J et al (2006) Genomic analysis of tumors by array comparative genomic hybridization: more is better. Cancer Res 66(7):3955–3956
61. Russo J, Fernandez SV, Russo PA, Fernbaugh R, Sheriff FS, Lareef HM, Garber J, Russo IH (2006) 17 beta estradiol induces transformation and tumorigenesis in human breast epithelial cells. FASEB J 20:1622–1634
62. Fernandez SV, Russo J (2010) Estrogen and xenoestrogens in breast cancer. Toxicol Pathol 38:110–122
63. Dongsong N, Zhou JY (2010) FISH is more sensitive than Southern analysis at identifying increased levels of cyclin D1 gene amplified in breast cancer cell lines. Mol Biol Rep 37(7):3473–3480
64. Gerami P, Zembowicz A (2011) Update on Fluorescence In Situ Hybridization in Melanoma. Arch Pathol Lab Med 135:830–837
65. Callahan R, Campbell A (1989) Mutations in human breast cancer: an overview. J Natl Cancer Inst 81:1780–1786
66. Devilee P, Cornelisse CJ (1990) Genetics of human breast cancer. Cancer Survey 9:605–630
67. Kirchwerger R, Zellinger R, Schneeberger C, Speiser P, Lovason G, Theillet C (1994) Patterns of allele losses suggest the existence of five distinct regions of LOH on chromosome 17 in breast cancer. Int J Cancer 6:193–199
68. Barnabas N, Bell D, Calaf G, Moraes RCB, Testa J, Russo J (1993) Loss of heterozygosity on chromosome 17p loci in transformed *in vitro* human breast epithelial cells treated with chemical carcinogens. Proc Am Assoc Cancer Res 34:649a
69. Yang X, Huang Y, Russo IH, Balsara BR, Barret JC, Russo J (1999) Functional roles of chromosomes 11 and 17 in the transformation of human breast epithelial cells *in vitro*. Int J Oncol 15:629–638
70. BRUKER NANO SURFACES (2013) Combining FLIM and AFM for enhanced multiparameter cen imaging. Microsc Anal, SPM, p 16

Chapter 8
In Vivo Studies of Breast Cancer Cells

Jose Russo, Ricardo Lopez de Cicco, Thomas J. Pogash, and Irma H. Russo

Keywords Xenograft model • 17 beta estradiol • HCG on malignant human breast epithelial cells • Labeling of cancer cells • Imaging procedure • TAIL vs. intraventricular injection • Metastatic foci in liver and lung

8.1 Introduction

The process of transplanting a living tissue, cells, or organs from one species to another is known as xenotransplantation and, the tissue, cells, or organs transplanted are known as xenografts. Xenotransplantation is possibly most well known for the transplant of organs originating from animals such as baboons and pigs into humans. The first and possibly most famous case of a xenotransplantation is that of Fae, an infant girl who successfully received a baboon heart in 1984. Despite the potential future use of this method to replace organs, cell xenotransplantation is a much more common procedure utilized in cancer research. This process most commonly involves implantation of human tumor cells into immunodeficient mice and is used to test the efficiency of compounds and their interactions with pathways within the body. Because human tumor cells can relatively easily be injected into mice and tumor growth can be routinely established, these models are commonly used for testing compounds. In a paper published by Baselga et al. xenografts were used to

J. Russo, M.D. (✉) • R.L. de Cicco, Ph.D. • T.J. Pogash, B.S. • I.H. Russo, M.D.
Irma H Russo MD Breast Cancer Research Laboratory, Fox Chase Cancer Center, Cottman Avenue 333, Philadelphia, PA 19111, USA
e-mail: jose.russo@fccc.edu; Irma.Russo@fccc.edu

J. Russo and I.H. Russo, *Techniques and Methodological Approaches in Breast Cancer Research*, DOI 10.1007/978-1-4939-0718-2_8,

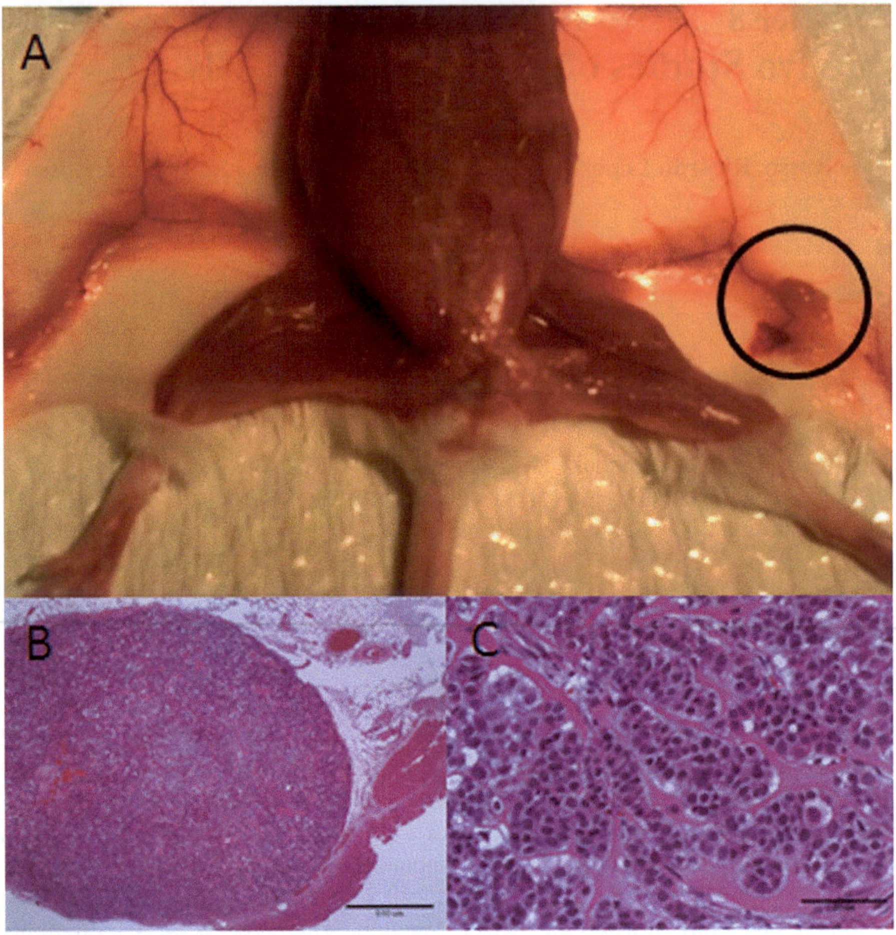

Fig. 8.1 (**a**) SUM149 xenograph-induced tumor on left fat pad of SCID mouse. Mice were administered subcutaneous injections of two million SUM149 cells suspended in 100 μL matrigel. Tumor was collected three weeks after implantation. (**b**) Histological section of xenographed-induced tumor 40×. Section was stained with Hemotoxylin and Eosin. (**c**) Poorly differentiated xenographed-induced tumor. Histological section of xenographed-induced tumor 400×. Section was stained with Hemotoxylin and Eosin

determine the effects of two anticancer drugs paclitaxel and doxorubicin in combination with an anti-HER2 antibody on breast cancer cell growth [1]. We have used xenografts to test the tumor formation of the SUM149 cell line in SCID mice. Mice were given subcutaneous injections of two million cells suspended in matrigel into the posterior quadrant of animals' abdomens. Tumors are observed 3 weeks after the implantation (Fig. 8.1a), and the histology reveals a well-defined poorly differentiated tumor (Fig. 8.1b, c).

8.2 The Use of a Xenograft Model for Testing the Tumorigenic Potential of MCF7 Cells

An important criterion of malignancy is the ability of transformed cells to grow in an adequate heterotransplantation system [2]. Immunologically depressed athymic mice (*nu/nu*) [3–6] have the striking capability of discriminating between normal and neoplastic cells. Normal cells do not induce tumors [5], whereas malignant cells do. MCF-7 cells, cultured as previously described [7], were removed from the culture vessel by trypsinization, suspended in phosphate buffer saline (PBS) (1×10^6 cells per 0.05 mL) and transplanted in 21-day-old Balb/c (*nu/nu*) mice into the mammary gland fat pad which was cleared according to the method of DeOme [8]. The first experiment demonstrated that MCF 7 cells were unable to grow neither in female nor in male athymic mice [9]. The gross examination of the area of cell inoculation and the histological study revealed a complete absence of the inoculated cells; only disorganization of the fat and some fibrosis were observed. Only those mice that have received a transplant of pituitary glands or ovaries from syngenic mice induced the growth of MCF-7 cells. Nine of the 11 (82 %) inoculated female mice that received pituitary grafts developed palpable tumors within 12–18 days after inoculation. The tumors adhered to the skin and underlying muscle. No macroscopic metastatic growths were observed. Eight of the 13 (61.5 %) inoculated female mice that received ovary grafts developed palpable tumors within 12–18 days. Tumors were attached to the skin and underlying muscles; no metastatic growths were observed. The tumors were small, oblong masses of 1.5–2.5 mm at their largest diameter. They adhered to the dermis of the skin and to the muscle of the abdominal wall. The tumors were firm, of a rubbery consistency, and presented resistance to sectioning. The tumor's vascular bed was well developed. The area of the tumor was easily distinguished from the scar produced by the cauterization and the incision made during the transplant procedure. The histological pattern of the 17 tumors studied was identical. The tumors were composed of nests of cells arranged in either clusters or single- or double-row strands. The inoculated epithelial cells were surrounded by a dense stroma formed by collagen fibers and fibroblasts. Blood vessels were scarce in the central portion of the tumor and more abundant in the periphery and in areas of invasion. The cells presented a considerable degree of pleomorphism and atypia. The nucleus was oval with few indentations. The nucleoplasm was pale, and a thin layer of heterochromatin was observed on the inner side of the nuclear envelope. More than two nucleoli per nucleus were frequently observed. Intracellular lumina with cellular detritus within were present in some cells. When stained with toluidine blue, the cytoplasm of most cells appeared strongly basophilic. A few cells with pale cytoplasm were also observed. Similar epithelial cells were also observed in the dermis of the skin overlying the inoculation site and among muscular fibers of the abdominal wall. The intense fibrous reaction observed at the inoculation site and in the dermis was not observed around cells invading skeletal muscle. Mitoses were frequently observed in areas of invasion. No metastases were found in any of the tissues studied; however, clusters of cells attached to the adventitia of blood vessels or adjacent to the perineurium were observed in the periphery of the tumor. Invasion of blood vessels or nerves by neoplastic cells was not observed in serial sections. The tumors observed in mice isografted with pituitary glands or ovaries were indistinguishable.

The successful heterotransplantation of human tumors [6, 10] and cultured human malignant cells [3, 4] into nude mice has proven to be an excellent model for the study of neoplastic tissue and an effective diagnostic tool for differentiating malignant from benign cells [5]. The growth of MCF-7 cells as tumors in nude mice might be predicted by the malignant nature of the tumor of origin and by the demonstration of several transformation markers [9]. However, MCF-7 cells did not form tumors in all inoculated mice but only in those receiving pituitary or ovarian grafts, thus suggesting a hormone dependency for in vivo growth. The fact that more tumors were observed in mice receiving pituitary grafts (82 %) than in those receiving ovarian grafts (61.5 %) suggested that some pituitary hormone could be involved in the development of these tumors. The inoculation of MCF-7 cells into nude mice induces tumors morphologically similar to the tumor of origin. This property of malignant cells has been described for other cell lines maintained for almost 100 passages in vitro [5] and transplanted into nude mice, and for human tumors transplanted into the anterior chamber of the guinea pig eye [11]. MCF-7 cells develop a histological pattern in the nude mice similar to that observed in the tumor of origin. The tumor of origin was an infiltrative ductal carcinoma with productive fibrosis. This same pattern of epithelial cells surrounded by a dense stroma is observed in the mouse, suggesting that it is the neoplastic epithelial cell that elicits a stromal response in the host. This observation was also supported by results obtained in an experimental model developed for the study of scirrhous carcinoma [12].

The absence of tumors in untreated animals could be explained by an inadequate hormonal milieu for the growth of MCF-7 cells. The fact that the original tumor from which MCF-7 cells were derived was responsive to hormones and that MCF-7 cells still retain specific high-affinity estradiol and progesterone receptors after more than 160 passages in culture, supports this explanation. The utilization of hormonal supplementation in the growth of MCF-7 cells in 1976 [13], suggested the replacing the isografts by hormone pellets [14]. We found out that the use of castrated male, estrogen-supplemented, was also suitable for the growth of MCF 7 cells [9]. The removal of the uterus and supplementation with estradiol either as pellets or silastic tube containing 5 mg of 17-β-estradiol in female mice is also a standard procedure [9]. The removal of the uterus avoids the swelling and accumulation of fluid in this organ due to the estrogenic stimulation.

8.3 The Use of Xenografts for Testing the Tumorigenicity of Chemically Transformed Cells

In the experimental system of human breast epithelial cells (HBEC) transformed with chemical carcinogens the SCID mice as heterologous hosts for testing tumorigenicity has been used. SCID mice have an autosomal dominant recessive defect that impairs the rearrangement of antigen receptor genes in both T and B lymphocyte progenitors. SCID mice also lack functional T and B cells, which make these animals a suitable host for heterotransplantation. Tumorigenesis in heterologous hosts is considered to be the final reliable criterion for assessing complete transformation of human cells [11, 15], even though the validity of this model has been questioned, since many human malignancies or cell lines derived from them are not tumorigenic in the nude mouse [16]. The adequacy of SCID mice for testing tumorigenicity has been validated by the observations that 100 % of the mice inoculated with either T24 or MCF-7 cells developed tumors with a short latency period. SCID mice proved to be more adequate as a host for MCF-7 cells than nude mice, since in our model they did not require estrogen supplementation [13]. The tumors developed in inoculated SCID mice were proven to be of human origin by determination of Alu sequences [17], which showed that they were derived from the cells inoculated and not a host reaction. T24 and BPI-E, the two cell lines expressing tumorigenesis in SCID mice [9], also exhibited the highest colony formation in agar–methocel. BP1-E cells also exhibited a higher colony efficiency (CE) and, larger colony size and higher colony number than the other non-tumorigenic clones, which indicated that these phenotypes need to be sequentially expressed prior to the manifestation of the tumorigenic phenotype. Anchorage independent growth is considered to be a predictor of tumorigenesis in other cell systems [11, 18]. Nonetheless, in the experimental model neither anchorage independence nor any of the other parameters served, when considered individually, as an indicator of cell tumorigenic potential. These observations are supported by results on HBEC obtained from milk, which have been reported to acquire anchorage independent growth and immortalization after SV40 infection, but do not elicit tumorigenesis in nude mice [11, 18–20] and by the observation that the extended lifespan induced in HBEC treated with B[*a*]P in vitro was not accompanied by the expression of anchorage independence or tumorigenicity [21]. Tumorigenesis in a heterologous host emerges in chemically treated immortal cells as a consequence of clonal expansion, in which phenotypes indicative of neoplastic transformation are cumulatively expressed through successive processes of selection over long periods of time. During the process of neoplastic transformation no chromosomal changes were detected. The molecular events that are operational in each phase of the transformation process indicate that each carcinogen induces different degrees of point mutations in codons 12 and 61 of the c-Ha-ras oncogene [22] as an event detectable by the 10th passage post-carcinogen treatment. Other genes such as p53, Rb, and erbB2 were differently expressed by the various clones derived from carcinogen-treated cells [23, 24].

8.4 Ha-ras Enhances the Tumorigenic Phenotype of Chemically Transformed Cells

As indicated above, the levels and localization of *ras* expression in normal and malignant breast tissues have been examined and quantitated by analyzing breast tissue samples for the expression of *ras*-related mRNA and p21 ras protein which has been found to be expressed in biopsies of both normal and malignant breast tissues [25]. However, whether the *ras* oncogene is a causative agent of human breast cancer has not been proved as yet. Therefore, one way to evaluate the contribution of *ra*s genes in the development of the tumorigenic phenotype is to introduce this gene into suitable acceptor cells. Transfection of the non-tumorigenic cell lines, clones D3-1 and BP1, derived from the carcinogen-treated MCF-10F cell lines, and the MCF-10F cell line with the cHa-ras oncogene not only enhanced colony formation in agar–methocel and invasiveness but also induced tumorigenicity with a short latency period in SCID mice [9].

The MCF-10F cells, DMBA or BP-treated cells and the clones D3-1 and BPI did not exhibit tumorigenicity in SCID mice. We have already shown that the subclone BP1-E derived from BP1 expressed the tumorigenic phenotype after 101 days of inoculation whereas the MCF-10F-Tras had lower tumorigenicity because it took 99 days to induce tumors in 3/14 animals, the clones D3-1-Tras and BPI-Tras were highly tumorigenic and the tumors appeared between 47 and 60 days post-inoculation, in four out of four animals and 11 out of 14 animals, respectively [9]. All the tumors derived from D3-1-Tras and BP1-Tras cells were poorly differentiated adenocarcinomas. They were immunocytochemically positive for keratin whereas the human milk fat globule membrane antigen (HMFGMA) was frankly expressed only in tumors induced by BP1-E cells; tumors derived from *c-Ha-ras* transfected cells showed either a notably reduced expression of this antigen, as observed in D3-1-Tras-induced tumors, in which only 10 % of the tumor cells were positive, or complete abolishment of HMFGMA reactivity, as in the BP1-Tras-induced tumor cells. Tumor cell lines derived from the tumors thus originated have been an important resource for understanding the molecular basis of mammary carcinogenesis [9].

Altogether the data presented in this chapter summarizes an in vitro model of breast cancer in which the final malignant phenotype of tumorigenesis can be induced in HBEC by carcinogen treatment alone or in combination with *ra*s oncogene, provided that the treated cells are previously immortalized and enough time and clone selection are allowed for its expression. This model provided clones of cells expressing different stages of progression to malignant transformation, which are useful for determining whether specific phenotypes are the result of specific genotypic alterations.

8.5 Tumorigenicity of 17 Beta Estradiol Transformed Human Breast Epithelial Cells

The invasive bsMCF cells were plated at low density and cell colonies were isolated using cloning rings. The cells were cultured in Dulbecco's modified Eagle medium (DMEM): F12 medium containing 1.05 mM calcium, antimycotics, hormones, growth factors, and equine serum [26]. After trypsinization and plating, the clones obtained were identified as bcMCF-1, bcMCF-2, bcMCF-3, bcMCF-4, bcMCF-5,

Table 8.1 Histological and immunocytochemistry of the bcMCF clones

Clones	Tumor type	AE1	CAM5.2	EMA	Vimentin
bcMCF-1 bcMCF-4	Invasive poorly differentiated spindle cell type	–	–	–	+++
bcMCF-2 bcMCF-6 bcMCF-7	Invasive poorly differentiated epithelial cell type	±	±	±	++
bcMCF-3 bcMCF-5	Invasive poorly differentiated with mix features of spindle and epithelial type	–	±	±	++

The mouse monoclonal antibodies anti-human cytokeratin of low molecular weight (AE1, Biogenex, San Ramon, CA), cytokeratin peptides 7 and 8 (CAM5.2, Ventana, Tucson, AZ), epithelial membrane antigen (EMA) clone E29 and vimentin, clone V9, both from DakoCytomation Inc., Fort Collins, CO were used. Negative (–), weak (±), moderate (++), or strong (+++)

bcMCF-6, and bcMCF-7. The tumorigenic ability of the bcMCF subclones was tested by injecting them into the mammary fat pad of 45-day-old female SCID mice (Table 8.1).

MCF-10F cells that after treatment with 17β-estradiol (E_2) expressed high colony efficiency (CE) and loss of ductulogenic capacity in collagen-matrix represented the first level of in vitro transformation. Cells expressing these two paramcters were classified as transformed (trMCF), which after further selection for invasiveness in a Matrigel invasion chamber originated the second level of transformation: the invasive (bsMCF) and the cloned (bcMCF) cells [27]. The bsMCF cells formed tumors in SCID mice from which four cell lines, caMCF, were derived. By ring cloning, seven subclones were isolated from the invasive bsMCF cells: bcMCF-1, bcMCF-2, bcMCF-3, bcMCF-4, bcMCF-5, bcMCF-6, and bcMCF-7. All the bcMCF subclones produced invasive poorly differentiated tumors in SCID mice with different morphological phenotypes: spindle cell type (bcMCF-1 and bcMCF-4), epithelial cell type (bcMCF-2, bcMCF-6, and bcMCF-7) and, with mix features of spindle and epithelial type (bcMCF-3 and bcMCF-5) (Table 8.1). As it was previously reported, MCF-10F cells were seeded on Boyden chamber as control; cells that passed through the membrane were selected, expanded, and injected in SCID mice; these cells did not produced tumors [26].

8.6 Tumoristatic Effect of HCG on Malignant Human Breast Epithelial Cells Transplanted in Heterologous Host

The observation that hCG had an inhibitory effect on chemically induced rat mammary carcinomas led us to test whether this hormone had an effect on the in vivo growth of malignant HBEC. For these purposes, MCF-7 cells, a cell line derived from a metastatic breast carcinoma, were injected to Balb/c nude mice (nu/nu). The animals were divided into five groups: the animals of four groups had implanted a silastic tube containing 5 mg 17-β-estradiol in the interscapular region 5 days after castration, and one group was castrated but did not receive the estrogen supplementation. The cells were injected in the mammary fat pad of mice in all the groups at a concentration of 1×10^6 cells. HCG was administered to the group of animals that did not receive the estrogen at a dose of 1,000 IU/day, and to the three estrogen-supplemented groups at the doses of 10, 100, or 1,000 IU/day. The animals that received estradiol pellets alone had an incidence of 85 % tumor formation. The group of animals injected with hCG alone did not develop tumors [9]. Animals that received estradiol pellets and also hCG exhibited a reduction in both tumor incidence and tumor size which were dose-dependent. These studies led us to conclude that the treatment with hCG abrogates the effect of the estrogen growth dependency of MCF-7 cells in a heterologous hosts (For more details on MCF7 growth in athymic mice see Sect. 8.2).

8.7 The Labeling of Cancer Cells for an In Vivo Imaging System

Many times, it is necessary to follow the growth and migration of cancer cells while the animal is still living [28]. There is a plethora of reporter systems that are used to introduce and select for gene target in cells. These systems have been historically used to study protein expression, interaction between proteins, and function of the proteins in the cells. These reporter genes confer drug resistance, bioluminescence, or fluorescence properties into the cells they are introduced. Typical reporter studies link reporter genes directly to a promoter region of interest, the function of which can be monitored by the reporter activity. Tagging of fused proteins is used to detect intracellular localization, degradation, protein–protein interactions, etc. The most common fusion tags used in research are fluorescent proteins (e.g., eGFP) or small protein epitopes (e.g., FLAG, Myc HA) which can be detected by fluorescence FACS or western blots [29]. One method of in vivo imaging involves in vitro labeling of cancer cells prior to injection in animals. The labeling of cells can be done using a vector that encodes the luciferase reporter gene isolated from *Photinus pyralis*. This gene has been optimized for mammalian expression. These vectors often contain selection markers for prokaryotes and eukaryotes. Cells can be transfected with the plasmid through different techniques: electroporation, lipid-based transfection, calcium phosphate transfection, etc. Once transfected, cells that incorporated the vector are selected with the appropriate antibiotic. Because the gene is incorporated into the genome, as cells divide, the daughter cells will also contain the gene allowing for complete tracking of tumor and cancer cell growth (Fig. 8.2). Luciferase positive cells are then injected in animals. Noninvasive in vivo study of experimental tumor formation and/or metastasis with bioluminescent imaging allows the study of tumor distribution, growth, and regression in individual animals. This technology permits repeated measurements over an extended time period in living animals without the need to sacrifice animals at preset time points [30].

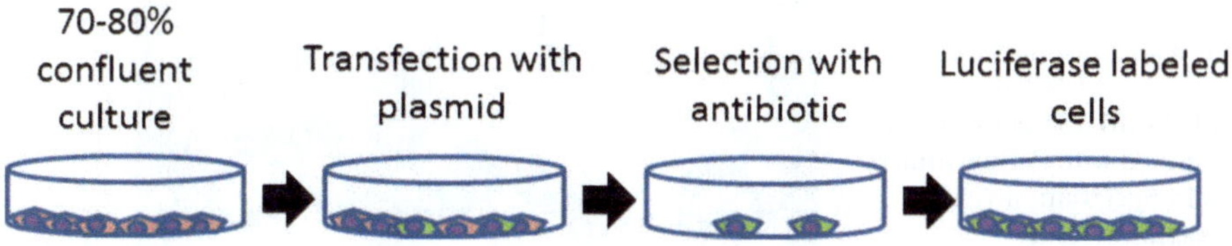

Fig. 8.2 Transfection of luciferase on breast cancer cells. Cells are plated to a 70–80 % confluence. Transfection is performed by the addition of pGL4.51 [luc2/CMV/Neo] plasmid. After transfection selection over the course of 10–14 days with Neomycin is necessary to select cells that are successfully transfected. After selection cells are expanded. *Pink cells* represent cells not transfected with plasmid, *green cells* represent transfected cells

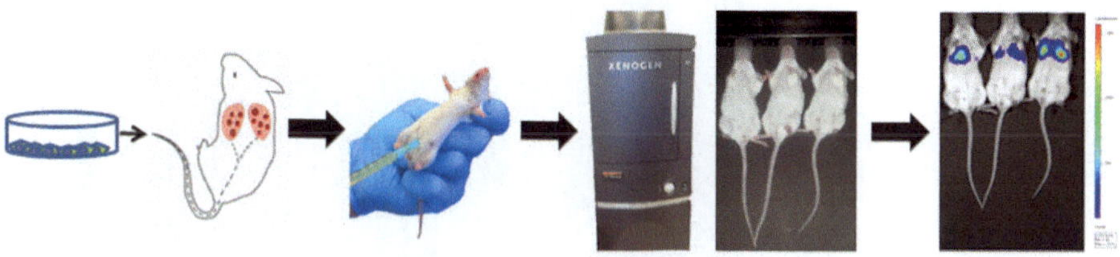

Fig. 8.3 Luciferase-labeled MDA-MB-231 cells are injected into the tail vein of SCID mice. Cells circulate throughout the body and metastasize in various organs. At the time of imaging, animals are injected with 4 mg of Luciferin dissolved in 200 μL PBS. Animals are anesthetized in an isoflurane chamber and are transferred to imaging station where isoflurane is delivered to animals during imaging. Imaging is conducted 10 min after Luciferin injections. Location of metastasis is confirmed by fluorescent cells. MDA-MB-231 cells metastasis occurred in the lungs of the animal. Animals were imaged weekly for 6 weeks at which sacrifice was performed

All imaging is performed using an imaging system, such as the "Perkin-Elmer IVIS Spectrum in vivo imaging system". This system is designed to image bioluminescent or fluorescent signals in animals. To visualize injected cancer cells in vivo, animals are anesthetized and then transferred to the imaging system, where the animals are imaged. For animals bearing tumors expressing firefly luciferase, an injection of the appropriate Luciferin substrate dissolved in phosphate-buffered saline at a physiological concentration is injected intraperitoneally. Then the animals are placed in the imaging system, and the light emitted by the tumors is measured. Animals bearing tumors that produce fluorescent markers (enhanced green fluorescent protein, red fluorescent protein, etc.) may be imaged, but injection of substrates is not needed. However, animals will be illuminated at the excitation frequency of the fluorophore for brief periods (generally a few seconds). Since the animals are under a general anesthetic they suffer no stress or discomfort due to the substrate injections or illumination (Fig. 8.3).

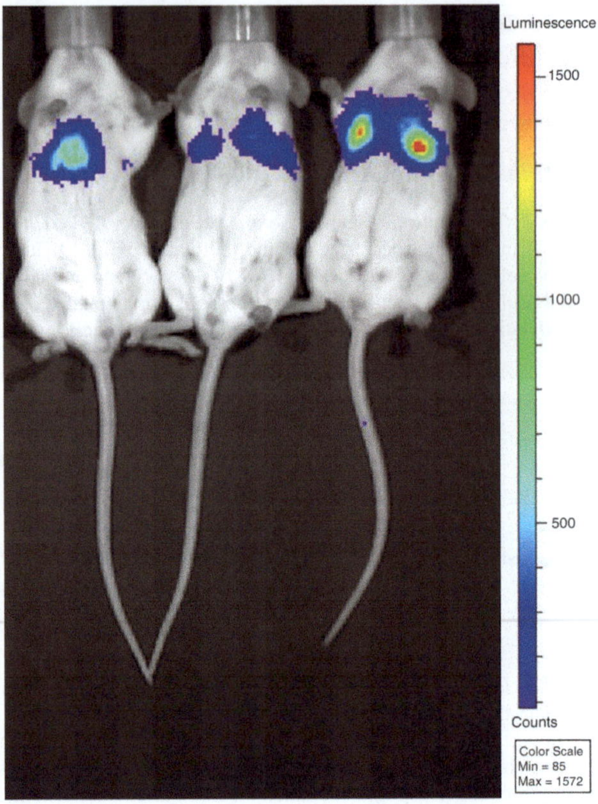

Fig. 8.4 Lung metastasis of SCID mice injected with Luciferase-labeled MDA-MB-231 cells. Image acquired with Perkin-Elmer IVIS Spectrum in vivo imaging system

We have used MDB.MA-231 cells transfected by Lipofectamine/Plus Reagent from Life Technologies. The plasmid used for the transfection was pGL4.51 [luc2/CMV/Neo] from Promega. Transfected cells were selected occurred over a period of 10–14 days using Neomycin (Invitrogen). After selection, cells were allowed to expand (Fig. 8.2). To ensure the presence of luciferase in the cells, a Luciferase Assay (Promega) was performed using the EnVision Workstation plate reader. Two million (2×10^6) cells suspended in PBS were injected into the lateral tail vein using a 26 gauge needle. The animals were followed over a period of 3 weeks to determine the location of the MDA-MB-231 cells (Fig. 8.4). Weekly inspection of the animals allows to follow up closely the process of nesting the metastatic cells as it is shown in Fig. 8.5.

8.7.1 Injection of Cells in the Vein Tail of Mice

For small animals, the procedure basically consists in injecting 2×10^6 cells suspended in 200 μL Dulbecco's PBS (calcium and magnesium free) per animal. The animals must be weighed and marked with an individual identifying mark. For

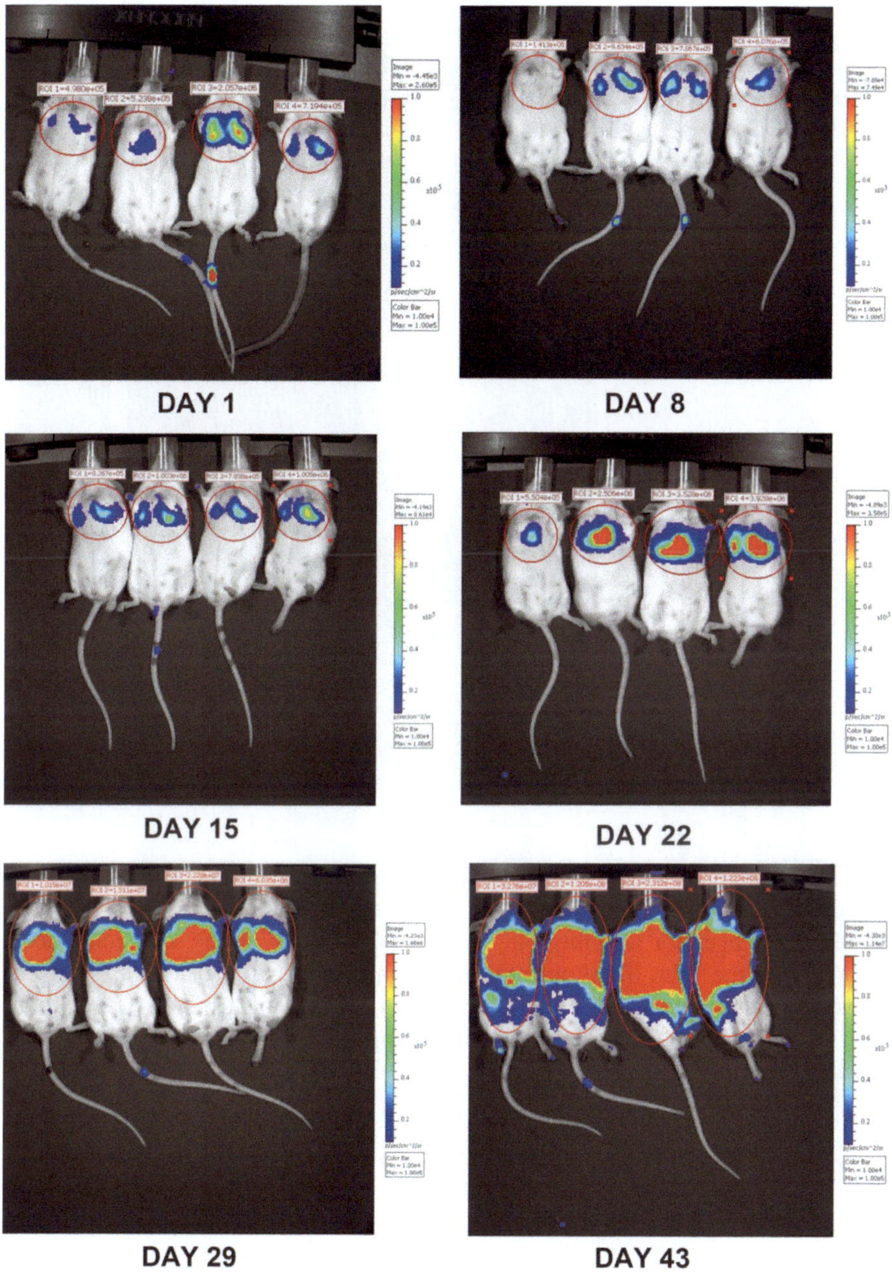

Fig. 8.5 Lung metastasis of SCID mice injected with Luciferase-labeled MDA-MB-231 cells. Image acquired with Perkin-Elmer IVIS Spectrum in vivo imaging system at different time up to 43 days post injection

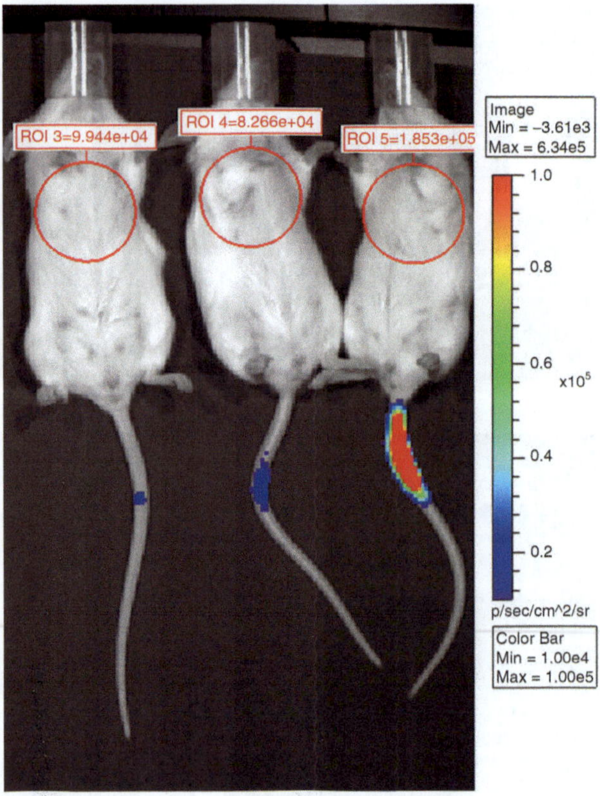

Fig. 8.6 On some occasion the operator does not inject the cells in the vein and they extravasate and grow in the tail

making the vein tail injection must be considered that the mouse tail is shaped like a cube, with a vein on each side. Use one of the lateral veins, using a 26-gauge needle. If the operator is in the vein the injected material will be seen going down the length of the vein. If the operator misses the vein it will get a "blib." In some cases, the cells remain there and the tumor is formed in the tail of the animals (Fig. 8.6). The largest volume that should be injected is 5 cm^3, but the mouse may go into shock.

8.7.2 Imaging Procedure Using the IVIS Spectrum Imaging System

The imaging procedure can be separated in several steps:

Step 1: Prepare Luciferin Solution. Luciferin is stored in the −80° freezer and the solution should be prepared in a 4 mg-200 μL PBS (per animal) concentration. Always prepare extra, you will lose some solution in the base of the syringe. You will waste less if you prepare the suspension in 1.7 mL Eppendorf tubes.

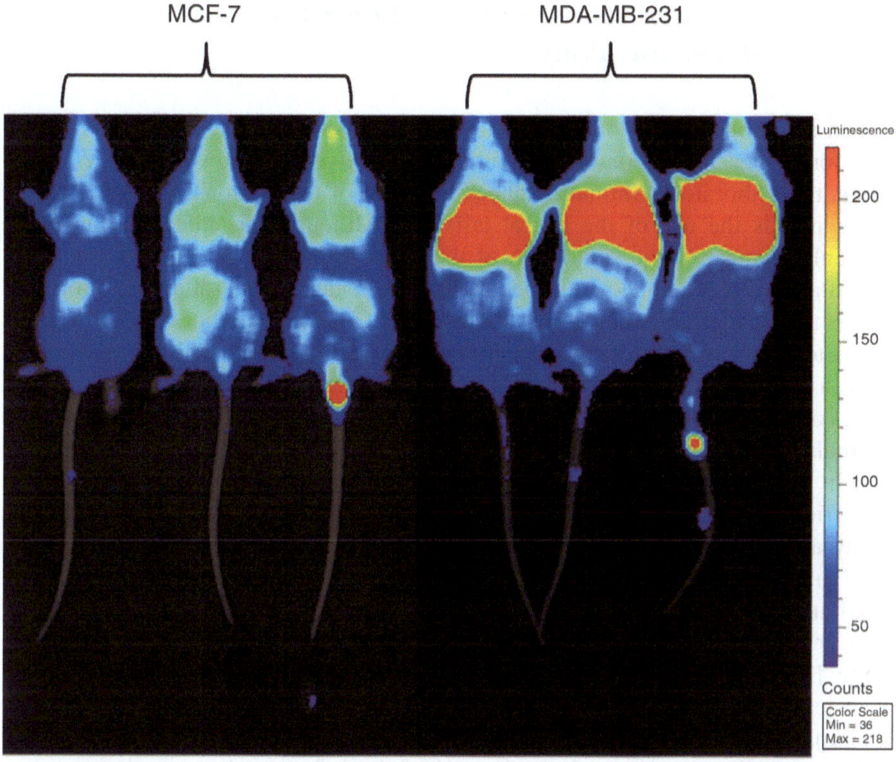

Fig. 8.7 SCID mice injected with Luciferase-labeled MCF-7 and MDA-MB-231 cells. Metastatic cells localize in the lungs of the MDA-MB-231 injected mice. Circulating cells can be seen in the head, neck, lungs, and liver of the MCF-7 injected mice. Image acquired with Perkin-Elmer IVIS Spectrum in vivo imaging system

Step 2: Animal Imaging. The operator needs to turn the IVIS flow and the oxygen "on" and at the same time adjusts the isoflurane dial. A piece of black art paper is placed under the area where the mice will be placed. The mice are placed into the anesthesia chamber. While the mice are "falling asleep," the operator prepares the syringes with 200 μL Luciferin. Once the mice are asleep the operator injects the Luciferin via IP injection. Once all mice are injected with Luciferin they place them in the IVIS chamber. Mice should be placed on their backs with their noses securely in the nose tubes (isoflurane will be released through the tube and it will keep the animal asleep).

Step 3: Living Image is done by following the software of the instrument. It is important to save the image and two have one bitmap image and one csv file for each group. To prepare the graph, take the average flux (p/s) for each group and plot the average flux per group. Include the standard deviation. In Fig. 8.7 is compared two different cells lines MCF7 and MBA-MD-231, whereas the latter one nest in the lung relatively fast as can be fully visualized at 43 days the MCF7 cells remain alive and circulating in the animal for longer period of time without nesting.

8.8 Tail vs. Intra-ventricular Injection Sites for Metastatic Assays

When using mice as a model for metastasis assays, the site of injection is important for proper absorption or distribution of additive. For intravenous (IV) injections, the Laboratory Animal Sciences Program of the National Cancer Institute recommends the use of the lateral tail vein. Although this injection can prove difficult because of the small size of the area of injection, it is readily used in metastasis assays and shown to be an effective means of cell administration, and this is the standard model for metastatic assays [31, 32]. Another option for injection site to observe metastasis is an intra-ventricular injection. This method is used less in literature due to the location of the injection site [32]. In order to inject drugs or molecules to be tested in the lateral ventricles of the brain of experimental animals, a cannula is implanted several days before the experiment, under anesthesia. In most of the protocols, the 3D location of the cannula is reported (antero-posterior, lateral, and depth). The drug or vehicle is injected at a preset schedule using a microinjector at a high rate or at a low rate. Special attention should be paid to the volume of fluid injected, as it will directly depend on the type of animal used and the age of the animal. For example, Passini et al. injected a volume of 2 μL into each lateral ventricle of neonate mice with a finely drawn glass micropipette [33]. When an experiment calls for multiple injections into the lateral ventricle/s of the brain, subcutaneously implanted osmotic minipump connected to an implanted indwelling lateral ventricle cannula similar to that used for the acute infusion studies can be used. At the end of the infusion, the cannula is closed, the infusion pump is removed under anesthesia, and the animals remain alive until the end of the experiment, time at which they are sacrificed and the tissue dissected and processed [34].

8.9 Evaluation of the Metastatic Foci in Liver and Lung

Evaluation of metastatic foci in the lungs and liver is a relatively simple, yet a laborious procedure. After dissection, the lungs and liver are washed in PBS and are placed in Bouin's fixative for a minimum of 24 h. Metastatic foci will appear white whereas the normal tissue will stain yellow (Fig. 8.8a). Quantification of surface metastasis can be done by counting the number of surface metastatic foci under a stereo microscope or by the naked eye. The number of foci can be expressed as a total sum, tumor volume, percentage, or metastatic score [35–39]. The presence of microscopic metastatic foci can be histologically examined by embedding the organ in a paraffin block and staining histological sections with Hematoxylin and Eosin (Fig. 8.8b, c) [35–37].

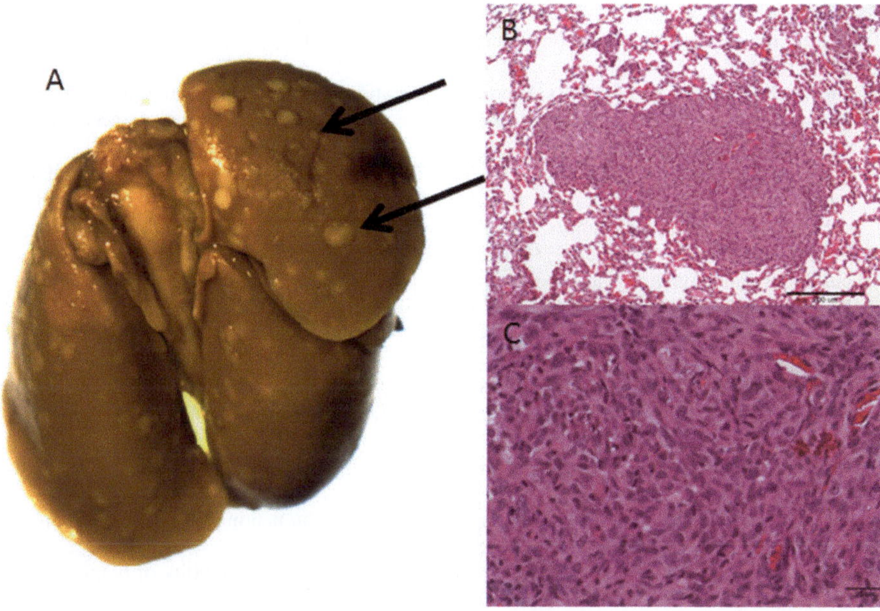

Fig. 8.8 (**a**) Lung metastasis of SCID mice injected with Luciferase-labeled MDA-MB-231 cells. Number of surface metastasis foci is counted under a stereo microscope or by the naked eye. *Arrows* indicate metastatic foci. (**b**) Histological section of metastatic tumor foci 10×. Section was stained with Hematoxylin and Eosin. (**c**) Histological section of metastatic tumor foci 100×. Section was stained with Hematoxylin and Eosin

8.10 Future Directions

Novel techniques for imaging biological processes in live animals are being offered by many companies and could eventually be further developed based on the demands of researchers that are pushing the limits of technology. These state-of-the-art procedures will definitely change our view and understanding of the development of diseases such as cancer. The possibilities are infinite, from the understanding of how cells work in the organism to the monitoring of physiological processes and the effects of drugs and small molecules.

Literature Cited

1. Baselga J, Norton L, Albanell J, Mee Kim Y, Mendelsohn J (1998) Recombinant humanized anti-HER2 antibody (Herceptin™) enhances the antitumor activity of paclitaxel and doxorubicin against HER2/*neu* overexpressing human breast cancer xenografts. Cancer Res 58:2825m. http://cancerres.aacrjournals.org/content/58/13/2825#fn-3#fn-3
2. Sanford KK (1974) Biologic manifestation of oncogenesis in vitro: a critique. J Natl Cancer Inst 53:1481–1485

3. Giovanella BC, Stehlin JS, Williams LJ Jr (1972) Development of invasive tumors in the "nude" mouse after injection of cultured human melanoma cells. J Natl Cancer Inst 48: 1531–1533
4. Giovanella BC, Stehlin JS (1973) Heterotransplantation of human malignant tumors in "nude" thymusless mice. I. Breeding and maintenance of "nude" mice. J Natl Cancer Inst 51: 615–619
5. Giovanella BC, Stehlin JS, Williams LJ Jr (1974) Heterotransplantation of human malignant tumors in "nude" thymusless mice. II. Malignant tumors induced by injection of cell cultures derived from human solid tumors. J Natl Cancer Inst 52:921–930
6. Povlsen CO, Fialkow PJ, Klein E (1973) Growth and antigenic properties of a biopsy-derived burkitt's lymphoma in thymusless (nude) mice. Int J Cancer 11:30–39
7. Russo J, Soule HD, McGrath C, Rich MA (1976) Re-expression of the original tumor pattern by a human breast carcinoma cell line (MCF-7) in sponge cultures. J Natl Cancer Inst 56:279–282
8. DeOme KB, Faulkin LJ Jr, Bern HA, Blair PB (1959) Development of mammary tumors from hyperplastic alveolar nodules transplanted into gland-free mammary fat pads of female C3H mice. Cancer Res 19:350–359
9. Russo J, Russo IH (2004) Biological and molecular basis of breast cancer. Springer, Heidelberg, Germany
10. Rygaard J, Povlsen CO (1969) Heterotransplantation of a human malignant tumor to "nude" mice. Acta Pathol Microbiol Scand 77:758–760
11. Greene HSN (1952) The significance of the heterologous transplantability of human cancer. Cancer 5:24–44
12. Russo J, McGrath CM (1975) Scirrhous carcinoma in the mouse: a model for human mammary carcinoma. Excerpta Medica, Amsterdam, p 488
13. Russo J, McGrath CM, Russo IH, Rich MA (1976) Tumoral growth of a human breast cancer cell line (MCF-7) in athymic mice. In: Nieburgs HE (ed) III international symposium on detection and prevention of cancer, New York, pp 617–626
14. Shafie SM, Giartham FH (1981) Role of hormones in the growth and regression of human breast cancer cells (MCF-7) transplanted into athymic mice. J Natl Cancer Inst 67:51–56
15. Ura H, Bonfil RD, Reich R et al (1989) Expression of type IV collagenase and procollagen genes and its correlation with the tumorigenic, invasive and metastatic abilities of oncogene-transformed human bronchial epithelial cells. Cancer Res 49:4615–4621
16. Smith HS, Wolman SR, Hackett AJ (1984) The biology of breast cancer at the cellular level. Biochim Biophys Acta 738:103–123
17. Cooper CS, Blair DG, Oskarsson MK, Tainsky MA, Eader LA, Vande Woude GF (1984) Characterization of human transforming genes from chemically transformed teratocarcinoma, and pancreatic carcinoma cell lines. Cancer Res 44:1–10
18. Strange R, Aguilar-Cordova E, Young UT, Billey HT, Dandekar S, Cardiff R (1989) Harvey-ras mediated neoplastic development in the mouse mammary gland. Oncogene 4:309–315
19. Ozzello L (1971) Ultrastructure of the human mammary gland. In: Summers SC (ed) Pathology annual. Appleton-Century-Crofts, New York, pp 1–59
20. Soule HD, Maloney TM, Wolman SR, Peterson WD, Brenz R, McGrath CM, Russo J, Pauley RJ, Jones RF, Brooks SC (1990) Isolation and characterization of a spontaneously irmno human breast epithelial cell line, MCF-10. Cancer Res 50:6075–6086
21. Stampfer MR, Bartley JC (1984) Induction of transformation and continuous cell lines from normal human mammary epithelial cells after exposure to benzo[*a*]pyrene. Proc Natl Acad Sci U S A 82:2394–2398
22. Zhang PL, Calaf G, Russo J (1992) Point mutation in codons 12 and 61 of the c-Ha-ras gene in carcinogen-treated human breast epithelial cells (HBEC). Proc Am Assoc Cancer Res 33:669a
23. Abarca-Quinones J, Calaf G, Estrada S, Barnabas-Sohi N, Zhang PL, Garcia M, Russo J (1992) Phenotypic progression of human breast epithelial cells HBEC transformed with chemical carcinogen. Proc Am Assoc Cancer Res 33:670a

24. Calaf G, Russo J (1992) Emergence of progressive neoplastic phenotypes of human breast epithetial (HBEC) treated with chemical carcinogens *in vitro*. Proc Am Assoc Cancer Res 33:1141a
25. Rochlitz CF, Scott GK, Dodson JM, Liu E, Dollbaum CH, Smith HS, Benz CH (1989) Incidence of activating ras oncogene mutations associated with primary and metastatic human breast cancer. Cancer Res 49:357–360
26. Russo J, Fernandez SV, Russo PA et al (2006) 17-Beta-estradiol induces transformation and tumorigenesis in human breast epithelial cells. FASEB J 20:1622–1634
27. Huang Y, Fernandez S, Goodwin S, Russo PA, Russo IH, Sutter T, Russo J (2007) Epithelial to mesenchymal transition in human breast epithelial cells transformed by 17-beta-estradiol. Cancer Res 67:11147–11157
28. Zelmer A, Ward TH (2013) Noninvasive fluorescence imaging of small animals. J Microsc 252(1):8–15. doi:10.1111/jmi.12063
29. Liu C, Billadeau DD, Abdelhakim H, Leof E, Kaibuchi K, Bernabeu C, Bloom GS, Yang L, Boardman L, Shah VH, Kang N (2013) IQGAP1 suppresses TbetaRII-mediated myofibroblastic activation and metastatic growth in liver. J Clin Invest 123(3):1138–1156. doi:10.1172/JCI63836
30. Edinger M, Cao YA, Hornig YS, Jenkins DE, Verneris MR, Bachmann MH, Negrin RS, Contag CH (2002) Advancing animal models of neoplasia through in vivo bioluminescence imaging. Eur J Cancer 38(16):2128–2136
31. Vines DC, Green DE, Kudo G, Keller H (2011) Evaluation of mouse tail-vein injections both qualitatively and quantitatively on small-animal PET tail scans. J Nucl Med Technol 39(4):264–270. doi:10.2967/jnmt.111.090951
32. Ray DE, Holton JL, Nolan CC, Cavanagh JB, Harpur ES (1998) Neurotoxic potential of gadodiamide after injection into the lateral cerebral ventricle of rats. AJNR Am J Neuroradiol 19:1455–1462
33. Passini MA, Watson DJ, Vite CH, Landsburg DJ, Feigenbaum AL, Wolfe JH (2003) Intraventricular brain injection of adeno-associated virus type 1 (AAV1) in neonatal mice results in complementary patterns of neuronal transduction to AAV2 and total long-term correction of storage lesions in the brains of β-glucuronidase-deficient mice. J Virol 77:7034–7040
34. Allen RM, Uban KA, Atwood EM, Albeck DS, Yamamoto DJ (2007) Continuous intracerebroventricular infusion of the competitive NMDA receptor antagonist, LY235959, facilitates escalation of cocaine self-administration and increases break point for cocaine in Sprague-Dawley rats. Pharmacol Biochem Behav 88(1):82–88
35. Becker JC et al (1996) Eradication of human hepatic and pulmonary melanoma metastases in SCID mice by antibody-interleukin 2 fusion proteins. Proc Natl Acad Sci U S A 93(7):2702–2707
36. Bijian K et al (2012) Synthesis and biological activity of novel organoselenium derivatives targeting multiple kinases and capable of inhibiting cancer progression to metastases. Eur J Med Chem 48:143–152
37. Bugge TH et al (1997) Growth and dissemination of Lewis lung carcinoma in plasminogen-deficient mice. Blood 90(11):4522–4531
38. Gillies SD et al (1998) Antibody-IL-12 fusion proteins are effective in SCID mouse models of prostate and colon carcinoma metastases. J Immunol 160(12):6195–6203
39. Louie E et al (2010) Identification of a stem-like cell population by exposing metastatic breast cancer cell lines to repetitive cycles of hypoxia and reoxygenation. Breast Cancer Res 12(6):R94

Chapter 9
RNA Sequencing in the Human Breast

Julia Santucci-Pereira, Maria Barton, Irma H. Russo, and Jose Russo

Keywords Sequencing • Next-generation sequencing (NGS) • DNA sequencing • RNA sequencing • Target-enrichment • Library preparation • Library validation • Cluster generation • Library preparation adjustments

9.1 Introduction

Breast cancer has an intrinsic cellular and molecular heterogeneity [1], and, traditionally, the treatment decisions have been based mainly on clinical–pathological characteristics such as size of the tumor, lymph node involvement, histological classification, and the presence of markers such as estrogen, progesterone, and epidermal growth factor receptors [2]. In order for treatment and prevention to improve, a better understanding of the molecular mechanisms which lead to breast cancer is needed by studying the normal breast in different physiological conditions like pregnancy for example. New tools have been developed to help us to better understand these molecular mechanisms, and high-throughput technologies, such as microarrays and more recently next-generation sequencing (NGS), have offered better opportunities to explore the molecular complexity of the breast tissue. In this chapter, we describe NGS and how we have applied RNA sequencing in our studies to understand the normal human breast.

J. Santucci-Pereira, Ph.D. (✉) • M. Barton, M.S. • I.H. Russo, M.D. • J. Russo, M.D. (✉)
Irma H Russo MD Breast Cancer Research Laboratory, Fox Chase Cancer Center, Cottman Avenue 333, Philadelphia, PA 19111, USA
e-mail: julia.pereira@fccc.edu; Irma.Russo@fccc.edu; jose.russo@fccc.edu

J. Russo and I.H. Russo, *Techniques and Methodological Approaches in Breast Cancer Research*, DOI 10.1007/978-1-4939-0718-2_9,

9.2 Sequencing

Sequencing is the process of determining the order of nucleotides within a DNA/cDNA. The first DNA sequences were obtained in the early 1970s [3, 4]. In 1977, Sanger et al. described the method using DNA polymerase with chain-terminating inhibitors [5]. In this technique, the template DNA is incubated with a primer, DNA polymerase, the normal deoxynucleosidetriphosphates (dNTPs), and one of the nucleotides modified dideoxynucleosidetriphosphates (ddNTPs). In this procedure it is necessary to make four reactions, one for each modified nucleotide (ddATP, ddCTP, ddGTP, ddTTP). The modified ddNTP does not allow the DNA polymerase to extend the chain because it does not contain the 3′-hydroxyl group, so termination occurs specifically at positions where the specific nucleotide is incorporated. It results in four mixtures of DNA fragments of different sizes, all fragments having the same 5′ and with ddNTP residues at the 3′ ends. These mixtures are then parallelly fractionated in a gel by electrophoresis and the patterns of bands show the distribution of the ddNTPs [5] (Fig. 9.1). In 1985, Smith et al. introduced colored fluorophore to each of the bases A, C, G, and T. Therefore, the reactions could be combined in one tube and when electrophoresed, the separated fluorescent bands of DNA could be detected near the bottom of the gel and the fluorescence could be read by a computer, allowing partial automation of DNA sequence analysis [6, 7]. In the following years, many technical improvements were incorporated to Sanger sequencing, however its limitations gave space for new and improved technologies to emerge, these new methods are often called NGS [8].

9.3 Next-Generation Sequencing (NGS)

The NGS systems use massively parallel sequencing to generate tens of millions of short (36- to 400-bp) DNA reads that can be aligned to the genome or used for de novo assembly. In 2004, the Roche (454) GS FLX sequencer was the first commercially available sequencer [9]. Today there are several platforms (Table 9.1) available, each with a specific technology; however, all of them follow the same basic steps for the sequencing DNA or RNA (Fig. 9.2). In the first step, the DNA/cDNA of interest is used to prepare a library. During the library preparation, the DNA or cDNA is randomly fragmented in specific sizes and has adapter sequences bound to the ends of the fragments. These fragments are then amplified (some techniques do not require amplification), sequenced, and the data is analyzed.

As consequence of the progress of this technology and the introduction of different platforms into the market, there has been significant reduction of the cost of the sequencing (Fig. 9.3) [10].

The detailed information resulting of NGS can be used in a broad of applications. NGS has being applied in a variety of research areas [8, 11] such as, characterization of ancient genomes [12], sequencing of different species [13], identification of

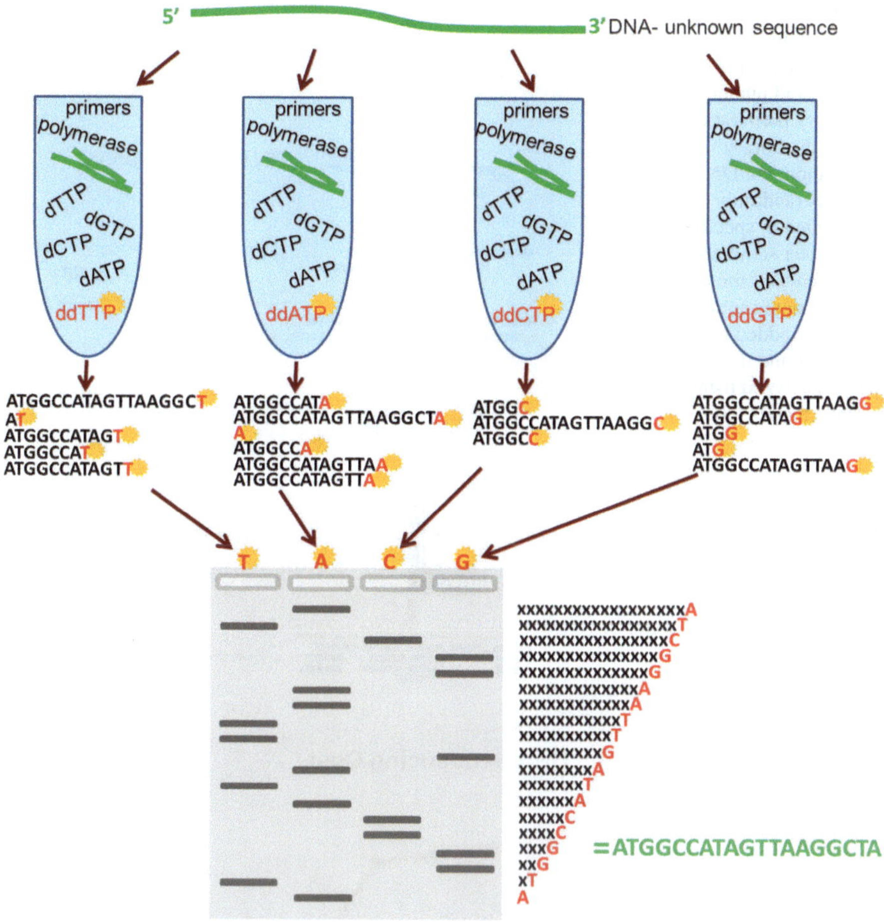

Fig. 9.1 Representation of Sanger sequencing method. The unknown DNA is incubated with primers, DNA polymerase, normal dNTPs (dTTP, dATP, dCTP, dGTP), plus one of the nucleotides modified—ddNTPs. One reaction per modified nucleotide is prepared (four reactions). Because the ddNTPs do not contain the 3′-hydroxyl group, the reaction is terminated where the modified nucleotide is incorporated, resulting in four mixtures of DNA fragments of different sizes. All fragments have the same 5′ with ddNTP residues at the 3′ ends. These mixtures are then parallelly fractionated by electrophoresis and the patterns of bands show the distribution of the ddNTPs

Table 9.1 Sequencing technologies available

Platform	Company	Amplification	Sequencing
454	Roche	Emulsion PCR	Synthesis
SOLiD	Applied Biosystems (Life Technologies)	Emulsion PCR	Ligation
Illumina	Illumina	Bridge	Synthesis
PGM/proton	Ion Torrent (Life Technologies)	Emulsion PCR	H+ detection
RS	Pacific Biosciences	None	Synthesis
GridION/MinION	Oxford Nanopore Technologies	None	Ion detection

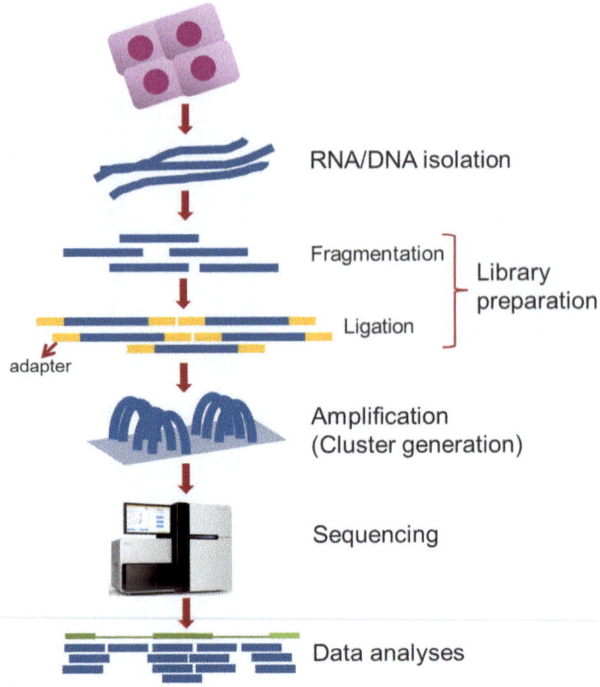

Fig. 9.2 Diagrammatic representation of the sequencing workflow. The isolated acid nucleic is used for library preparation. During the library preparation, the DNA or cDNA is randomly fragmented in specific sizes and adapter sequences bound to both ends of the fragments. The fragments are then amplified, added to the sequencer and the resulting data is analyzed using Bioinformatics tools

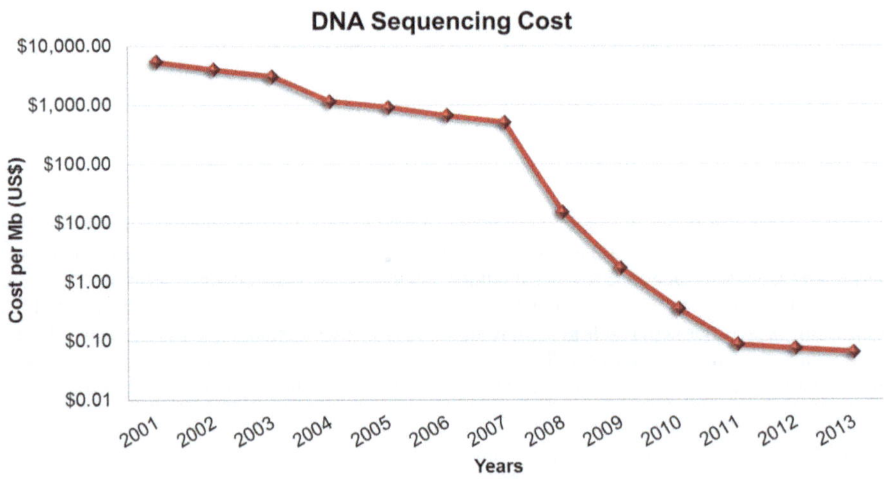

Fig. 9.3 Graph of cost of DNA sequencing per one million of bases—Megabase (Mb). Notice the drop in cost after 2007. The data plotted was extracted from http://www.genome.gov/sequencingcosts/ [10]

noncoding RNAs [14], risk assessment of genetic diseases [15], and molecular diagnosis of various diseases including cancer [16, 17] among others, all leading to a road for personalized medicine [18].

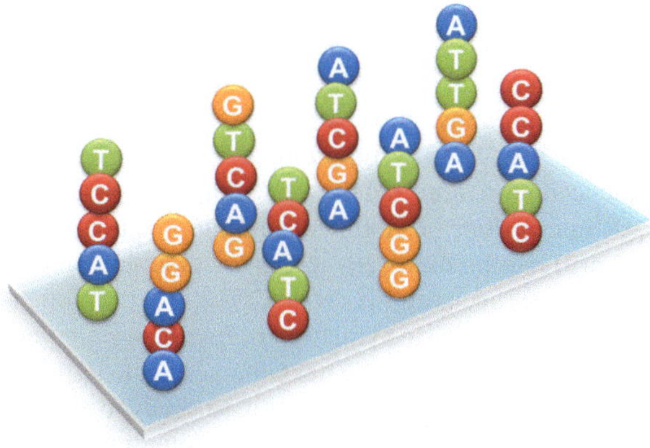

Fig. 9.4 Representation of an oligo-microarray slide. In the oligo-microarrays, several previously described sequences are spotted to a slide or chip and the samples are hybridized to the oligos. This technology is limited to sequences that are already known

9.3.1 Sequencing vs. Microarrays

Previous to the NGS, oligo-microarrays (Fig. 9.4) was the high-throughput technology mostly used to analyze the transcriptome and also genome variations. This technology is relatively inexpensive, however because it is based on hybridization approach, oligo-microarrays rely on existing knowledge about the genome being studied [19], which is the main difference when compared to NGS. Due to the hybridization approach, oligo-microarrays can have cross-hybridization, high levels of background, and limited range of gene expression levels detection because of the background or saturation of the oligos [19]. While oligo-microarrays are limited to what is already known, sequencing has the potential for new discoveries; NGS can be used to study organisms which the genome have not yet been described. Also offer the potential for identification of novel transcripts, splice junction areas, and level of expression of different isoforms. NGS allows the identification of single-nucleotide polymorphisms (SNP) or other type of genome variations.

9.3.2 DNA Sequencing

With NGS sequencing, entire genomes became more accessible. It allows identifying mutations and other structural variations such as SNP, copy number variation, insertions, and deletions that might be fundamental to understand and potentially treat genetic diseases. The use of sequencing also can be used to sequence new organism genomes, such as pathological agents, which could facilitate the drug targets against them.

Another advantage of the DNA sequencing is the possibility to study the interactions between DNA and proteins, such as histones and transcription factors which are important for gene expression regulation, controlling of DNA availability for transcription and replication, and other processes [9].

The study of the interaction of DNA with transcription factors or other proteins can be performed using chromatin immunoprecipitation (ChIP) prior to the sequencing (ChIP-Seq). In ChIP, first, the cells are fixed to preserve the interaction between DNA and proteins. Next, an antibody is used to separate the protein of interest bound to the DNA by immunoprecipitation, and lastly, the DNA is purified and sequenced (for more details, see Chap. 10, Sect. 10.4.2). In the case of histone–DNA interactions, a modified immunoprecipitation technique called Native-ChIP (N-ChIP) can be used [20]. Other methodologies to evaluate the chromatin status are nucleases accessibility (NA-seq) [20, 21], which uses endonucleases and FAIRE (formaldehyde-assisted isolation of regulatory elements) [22]; these methodologies are further discussed in the next chapter Sect. 10.4.2.

In addition, NGS can be used to analyze the methylated DNA, providing a detailed picture of the methylome. DNA methylation plays important roles in adverse changes of gene expression, in differentiation of the cells, and it is not only related to cancer but also to other age-related and neuropsychiatric diseases [23]. There are three main ways to separate the methylated DNA from the un-methylated DNA, (a) bisulfite conversion, (b) use of restriction endonucleases, and (c) use of affinity properties [23], these techniques are more detailed in the next chapter, Sect. 10.4.1.

9.3.3 RNA Sequencing

NGS sequencing was a large step for understanding the transcriptome of the cell. Transcriptome is defined as the complete set of transcripts in the cell [19], which not only includes the mRNAs, but also rRNA, tRNA, small RNA, and other noncoding RNAs. The comprehension of the transcriptome is fundamental to understand development and diseases.

NGS permits to obtain the gene expression levels, which can be extracted from the number of reads that a specific gene had normalized by the size of the gene [19]. In addition, information about the different transcripts and splice regions can be extracted from the data. Due to the independency of previous knowledge, novel transcripts can be found, as well, other regulatory RNAs, such as microRNA, or additional noncoding regions can now be explored. The advances of NGS and computational tools made possible the amplification of studies of long noncoding RNA [24, 25], as well as the description of some of the properties of this class of RNAs that has been shown to be important for gene regulation [26], especially in development [27].

9.3.4 Target-Enrichment

It is not always feasible or of interest for the laboratory to sequence the whole genome to get the information it is looking for. The sequencing of large genomes,

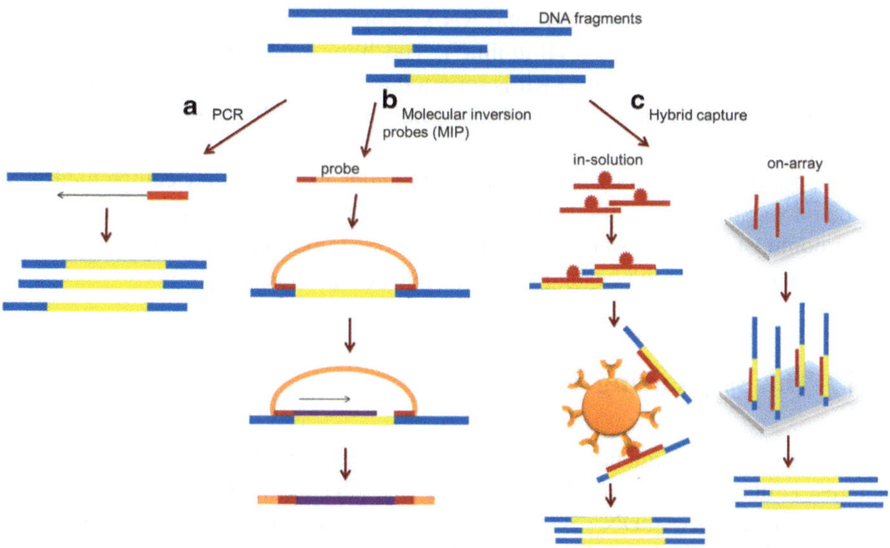

Fig. 9.5 Target enrichment approaches. (**a**) PCR, primers are designed to target the regions of interest and PCR is used for amplification of these areas. It is possible to generate single amplicons or perform multiplexed PCR. (**b**) Molecular inversion probes (MIP), probes that recognize the target sequence and become circularized. The gap between the probes is filled by a ligase. All the uncircularized fragments are digested by exonucleases, and then the circularized species are amplified. (**c**) Hybrid capture, the areas of interest of genomic DNA libraries are hybridized to probes that can be on an array, or can be captured by beads (on-solution). The unwanted DNA, which is not hybridized, is washed out, and the DNA of interest is eluted and amplified

with a good coverage level, can be costly, and requires time for data analyses; therefore, methods to select specific areas of the genome to be sequenced have been developed—target-enrichment. The current target enrichment approaches available are based on (a) PCR, (b) molecular inversion probes (MIP), or (c) hybrid capture (Fig. 9.5) [28]. In the PCR approach, primers are designed to detect the region of interest. It is possible to generate single amplicons or perform multiplexed PCR to obtain several amplicons per reaction. In the MIPs technique, probes, consisting of a common linker flanked by target-specific sequences, anneal to the target sequence and become circularized. The uncircularized fragments are digested by exonucleases while the circularized species (target sequences) are amplified [28]. The hybrid capture can be performed on-array or in-solution. In both techniques, the genomic DNA libraries are hybridized to probes specific for the areas of interest, the DNA that doesn't hybridize is washed away, while the DNA of interested is eluted [28].

In order to perform target RNA-seq, some modifications can be applied for the techniques used for targeting DNA. Targeted RNA-seq methods have been used to detect fusion transcripts, allele-specific expression, mutations, and RNA-editing events [29]. When there is interest in building RNA-seq in which both poly-A and non-poly-A transcripts are maintained, one can use techniques which remove the abundant ribosomal RNA from total RNA sample [19], for example using Ribo-Zero™ rRNA Removal Kits (Illumina®). The library prepared after treatment of RNA with the Ribo-Zero Gold Kit is reported to enrich in reads aligning to exomic and nonexomic sequences that are biologically relevant RNA species, species that are lost when using poly-A enrichment [30].

The performance of target enrichment strategies vary, and they can be measured based on sensitivity, specificity, uniformity, reproducibility, cost, practicability, and the amount of required starting material [28], all important characteristics to consider when choosing the best approach for one's experiment.

9.4 The Sequencing Technique

For RNA or DNA sequencing, the workflow is the same. Once the acid nucleic of interest is obtained, it is used to prepare a DNA/cDNA library. During the library preparation, the DNA/cDNA is randomly fragmented in specific sizes and has adapter sequences bound to the ends of the fragments. These fragments are then amplified (some current techniques do not require amplification)—cluster generation, and then the samples are sequenced and the data resulting is analyzed (Fig. 9.2). As mentioned previously, there are several platforms available (Table 9.1); however, some of the sequencers mostly used are manufactured by Illumina® (Fig. 9.6). In this section, we describe the methodology used in our laboratory to perform RNA sequencing following Illumina® TruSeq™ RNA Sample Preparation Kit protocol [31].

9.4.1 Library Preparation

Illumina® recommends the use of 0.1–4 μg of total RNA and use PCR plates with a magnetic plate stand for the library preparation [31]. To process 48 samples or less,

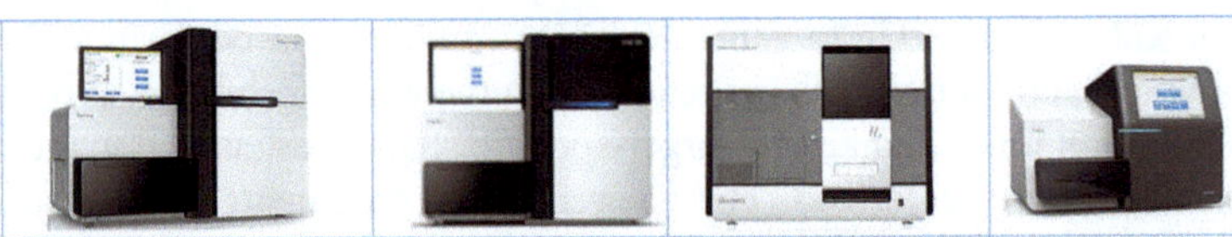

Name	HiSeq 2500		HiSeq 1000	Genome Analyzer IIx	MiSeq
Mode	High Output	Rapid Run	High Output	High Output	Benchtop
Output	600 Gb	120 Gb	300 Gb	95Gb	15 Gb
Run Time	~ 11 days	~27 hours	~8.5 days	~ 14 days	~65 hours
Paired-End reads	6 Billion	1.2 Billion	3 Billion	640 million	50 Million
Read Length	2 × 100 bp	2 × 150 bp	2 × 150 bp	2 × 150 bp	2 x 300 bp

Fig. 9.6 Some of the Illumina® sequencers and their capabilities

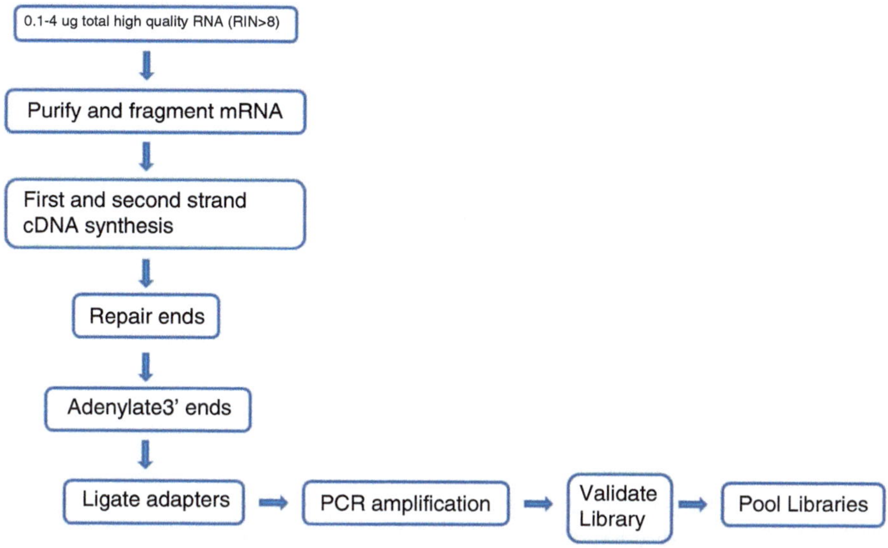

Fig. 9.7 Steps for library preparation. Steps for preparation of the samples for sequencing following the low-throughput (LT) protocol from TruSeq™ RNA Sample Preparation Kit protocol (Illumina®)

Illumina® recommends the low-throughput (LT) protocol [31]. Figure 9.7 illustrates the process of the LT TruSeq RNA sample preparation to prepare a template composed of 12 pooled indexed libraries. In the first step of the library preparation, the mRNA (containing the poly-A tail) is separated from the total RNA using poly-T oligo attached to magnetic beads. Then the purified RNA is fragmented and cDNA is prepared. Once the cDNA is prepared, its ends are repaired into blunt ends. A single "A" nucleotide is added to the 3′ ends of the blunt fragments to prevent them from ligating to one another, and then adapters are ligated, preparing the cDNA fragments for hybridization onto a flow cell [31].

Once the library is ready, it is necessary to amplify it through PCR. It is important to adjust the amount of cycles in the PCR depending on the amount of starting material to not overamplify the samples. Overamplifying the samples lead to a skewed representation of the samples' transcriptome. For example, if the amount of starting material is 0.1–0.5 μg, the amount of cycles should be minimized to 10–12 cycles (instead of 15) to get a fair representation of the transcriptome after PCR sequencing.

Validating the library is essential for quality control on the sample library and quantification of the cDNA library templates [31]. In order to achieve the highest quality of data on Illumina® sequencing platforms, it is important to go for the best cluster densities across every lane in every flow cell. In order to obtain the best cluster densities, accurate quantification of the DNA library templates is essential. Validation of the library is further discussed in the next section (Sect. 9.4.2).

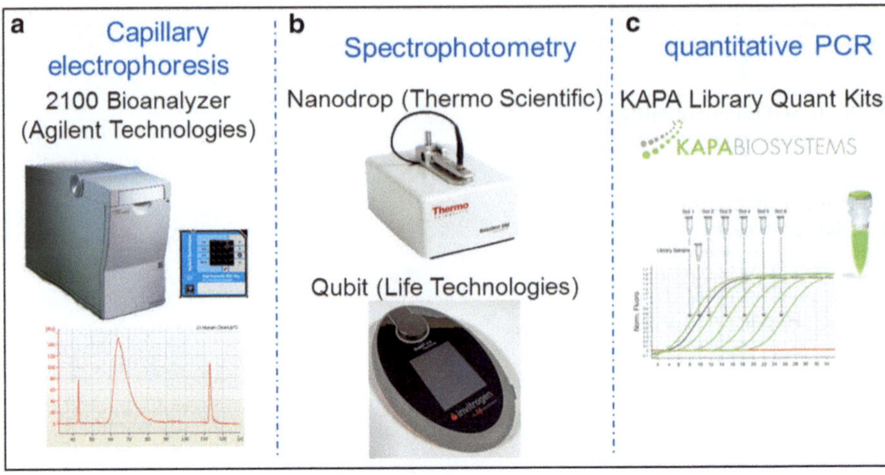

Fig. 9.8 Methods for library validation. (**a**) 2100 Bioanalyzer (Agilent Technologies) can check quality and concentration of the libraries. The electropherogram should present a peak correspondent to fragments' size. (**b**) NanoDrop (Thermo Scientific) uses UV/VIS spectrophotometry, Qubit (Life Technologies) is a fluorometer for quantitation of DNA, RNA, and protein, using fluorescence-based quantitation assays. (**c**) Quantitative PCR (Kapa Biosystems) specifically quantifies PCR-competent DNA molecules. Kapa Biosystems' kits include sets of serially diluted DNA concentration standards and qPCR reagents

Each lane of the flow cell may contain three to six samples pooled together. The samples are tagged with different indexes, also known as barcodes, added together with the adapters during the library preparation. The fewer amount of samples per lane, the best coverage for each sample. Depending on the size of the genome (*Saccharomyces cerevisiae* (*S. cerevisiae*) vs. human for example), one can pool more or less samples in each lane. A small genome like *S. cerevisiae* will allow for more samples to be pooled in a lane with a very good coverage for each sample.

Multiplexed DNA libraries are normalized to 10 nM in the DCT (Diluted Cluster Template) plate and then pooled in equal volumes in the PDP (Pooled DCT plate) [31].

Once the DNA templates are pooled, they can be applied to cluster generation (Fig. 9.2).

9.4.2 Library Validation

To achieve the highest quality of data sequencing, it is important to obtain the best cluster densities across every lane in every flow cell. In order to obtain the best cluster densities, accurate quantification and good quality of the DNA library are essential. To check the quality, Illumina® suggests the use of Agilent Technologies 2100 Bioanalyzer [31]; the quantification can be done using different approaches: (a) capillary electrophoresis,

Bioanalyzer (Agilent Technologies); (b) high sensitivity spectrophotometry, Nanodrop (Thermo Scientific) or Qubit (Life Technologies); and (c) quantitative PCR (Kapa Biosystems) (Fig. 9.8).

(a) Capillary electrophoresis: Using the High Sensitivity DNA kit from the 2100 Bioanalyzer (Agilent Technologies), it is possible to check quality and concentration of the libraries, although for quantification, other more precise methods are recommended. When checking for quality of the library, the graph obtained after running 1 μL of the library template should yield results of purity and fragment size. One should observe the peak of the sample with an appropriate size (approximately 260 bp for using Illumina® TruSeq™ RNA Sample Preparation Kit) [31] (Fig. 9.8).

(b) High sensitivity spectrophotometry: The NanoDrop 1000 Spectrophotometer (Thermo Scientific) uses UV/VIS spectrophotometry to measure 1 μL of sample without any dilution (up to 3,700 ng/μL for dsDNA) [32]. The Qubit 2.0 Fluorometer is a fluorometer for quantitation of DNA, RNA, and protein, using highly sensitive and accurate fluorescence-based Qubit™ quantitation assays [33]. This quantization system relies on dyes that only fluoresce when bound to specific molecules, such as dsDNA, ssDNA, or RNA [33]. The assay is highly selective and common contaminants, such as salts, free nucleotides, solvents, detergents, or protein, do not interfere with the molecule of interest measurement [34] (Fig. 9.8).

(c) Quantitative PCR: Kapa Biosystems is a sensitive way to quantify low concentration libraries [35]. It is a qPCR method that specifically quantifies only PCR-competent DNA molecules. KAPA Library Quantification Kits are optimized for different platforms, including Illumina®. The kits include sets of serially diluted DNA concentration standards and qPCR reagents. The DNA polymerase specifically amplifies long templates with SYBR® Green I-tolerant technology [35] (Fig. 9.8).

9.4.3 Cluster Generation

After the library preparation, the DNA fragments contain 5′ and 3′ adapters and they are ready for the cluster generation. During cluster generation, template molecules are attached to a flow cell and then amplified locally to form clonal clusters (Fig. 9.9). The flow cells containing the clusters can then be sequenced.

The Illumina® cBlot system provides automation of a complex process that involves the clonal amplification of each fragment of cDNA by a series of extensions, bridge amplification [36]. In the clonal cluster, DNA polymerase generates multiple DNA copies (clusters), each representing the single molecule that initiated the cluster amplification. Each cluster contains approximately one million copies of the original RNA/cDNA fragment [36].

Most imaging systems need amplified templates to detect the fluorescent nucleotide. The ratio of the primers to the template on the support defines the surface density of the amplified clusters. Solid-phase amplifications can produce millions of

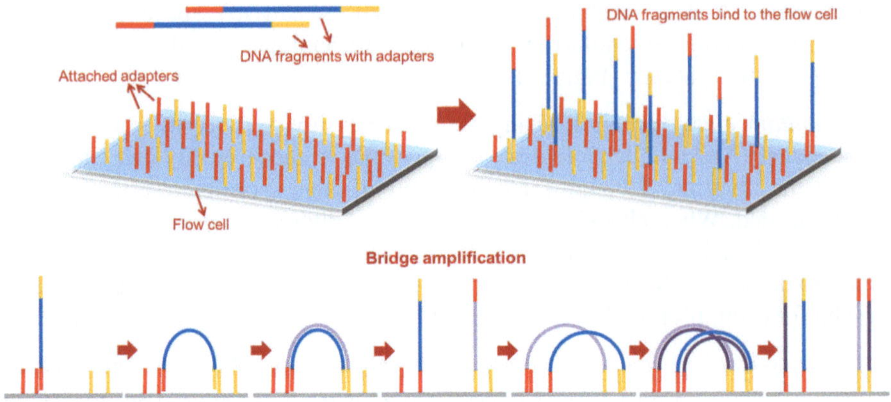

Fig. 9.9 Cluster generation. During cluster generation, the DNA fragments containing the adapters attach to the flow cell. The free end of the template hybridizes to the adjacent primer, forming a bridge, which is amplified. These bridges are amplified several times, creating a dense cluster

template clusters spatially separated on the support, providing free ends to which a universal sequencing primer can be hybridized to initiate the NGS reaction [8].

The cluster generation requires very little preparation and all the steps (except the DNA template preparation) are performed on the cBlot [36]. The steps required are: (1) preparation of the reagent plate; (2) dilution, denature, and then load of the library into tubes; and (3) load of the reagent plate and flow cell. Once the pre-run check is performed, the run can begin.

Creating optimum cluster densities across every lane of the flow cell is essential to achieve the highest quality data on Illumina® sequencing platforms. This requires accurate quantification of template DNA libraries. If the concentration is too low, fewer clusters are generated and this results in a low sequencing yield. If the concentration is too high, clusters are too dense and this may lead to poor data analysis.

During the cluster generation, the following steps are taken (Fig. 9.9):

1. The DNA fragment library is diluted, denatured, and introduced into the lanes of the flow cell.
2. The templates are captured by the oligonucleotides attached to the surface of the flow cell.
3. Templates bound to the primers are 3′-extended.
4. The double-stranded molecule is denatured, and the original template is washed away.
5. Free ends of the bound templates hybridize to adjacent lawn primers to form U-shaped bridges.
6. The DNA bridge is copied from the primer to create a double-stranded DNA bridge.
7. The bridge is amplified 35 times to create a dense cluster of more than 2,000 molecules.

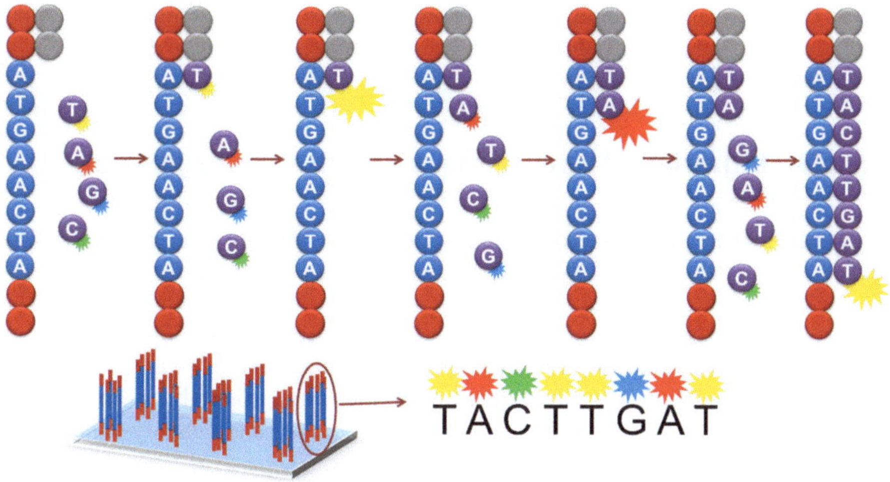

Fig. 9.10 Sequencing. The sequencing performed by the Illumina® systems is based on sequencing-by-synthesis approach using four fluorescent-labeled nucleotides. The DNA polymerase and all four nucleotides are added simultaneously to the flow cell. The four bases compete to bind to the template, and when one nucleotide binds to the template, it stops the incorporation of others because the 3′-OH group is chemically blocked. After each round of synthesis, the clusters are excited by a laser, emitting the color of the added base and an image is taken. Then, the fluorescence and the blocking group are removed to allow the next base to bind. This cycle repeats several times, depending on the experiment

8. The reverse strands in the cluster are removed by cleavage at the reverse strand-specific primers, leaving a cluster with forward strands only.
9. The free 3′-OH ends are blocked to prevent nonspecific priming.
10. Sequencing primers are hybridized to the free ends of the DNA templates.

The flow cell with the synthesized clonal clusters is now ready to be sequenced.

9.4.4 Sequencing

The sequencing performed by the Illumina® systems is based on sequencing-by-synthesis approach. All four nucleotides and the DNA polymerase are added simultaneously to the flow cell. The nucleotides carry a distinctive fluorescent label in each nucleotide and the 3′-OH group is chemically blocked such that each incorporation is a unique event. An image is taken after each step and then the block at the 3′end is chemically removed so that the next fluorescently labeled nucleotide can be incorporated [9]. This series of steps continues for a specific number of cycles depending on the experiment's needs (Fig. 9.10).

Clonal amplification results in a population of identical templates, each of which has undergone the sequencing reaction. The image is a combination of signals added to identical templates for a given cycle (Fig. 9.10). Thus, the addition process has to be highly efficient and precise. After NGS reads have been generated, they are aligned to a known reference sequence for comparison and analysis.

9.5 Optimization of Small Amount of RNA for Sequencing

To obtain a better quality and final concentration of cDNA that can be analyzed by sequencing methods after library preparation some adjustments have to be made to the TruSeq™ RNA Sample Library Preparation Protocol from Illumina® [31]. The adjustments are essentially concentrated in the ligation step. Basically, the objective is to minimize the amount of cDNA lost during clean-up steps and reduce the amount of clean ups necessary after enrichment of sample if a considerable dimer peak is observed when running the Bioanalyzer for cDNA quality and quantity control.

9.5.1 Quality Control

Determining the quality of the samples by Bioanalyzer before starting to prepare the library is essential. Only high quality RNA samples should go through the process of sequencing and thus no library preparation of poor quality RNA should be made. All RNA RINs should be eight or more (Fig. 9.11).

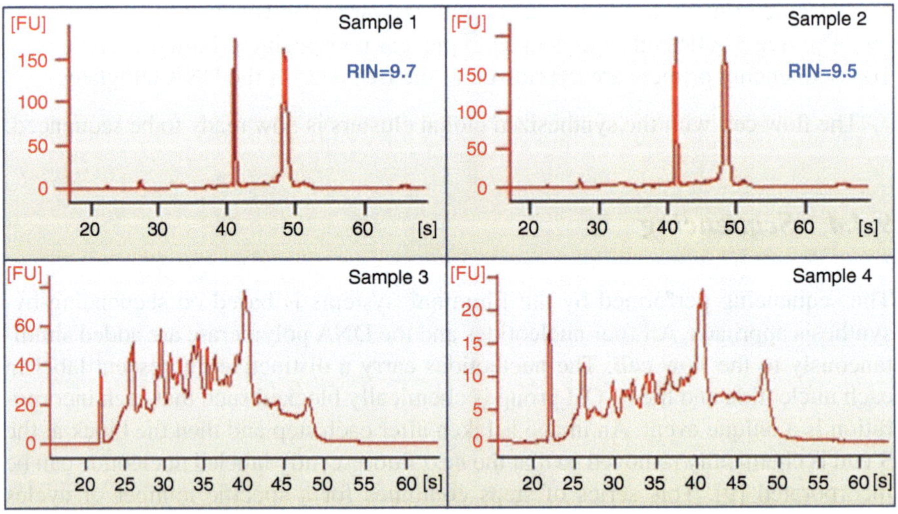

Fig. 9.11 Examples of Bioanalyzer's electropherograms obtained in our laboratory. Both electropherograms at the top show good quality RNA (RIN > 8.0), while the samples at the bottom have low quality; therefore, they are not adequate for RNA-seq

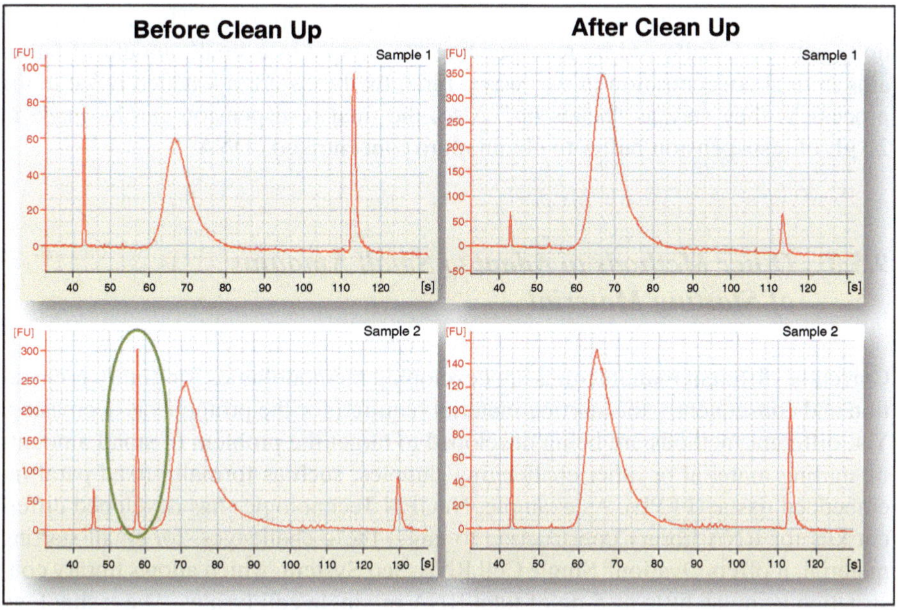

Fig. 9.12 Bioanlyzer's electropherograms of samples before and after clean up to reduce dimers formed by the adapters (dimer circled in green - sample2)

9.5.2 Library Preparation Adjustments

The protocol is followed as indicated by Illumina® with some adjustments. First, during the ligation step, you should add a 5-min at 70 °C step after the 30-min incubation at 37 °C. This additional step decreases the amount of dimers formed by adapters binding to each other. Also, a dilution of 1:5 or 1:10 of the ligate adapters is suggested to decrease the amount of dimers formed in this step. The decrease or total elimination of a dimer peak after library preparation ensures that no follow up clean-up steps are needed (Fig. 9.12). Each clean-up step not only reduces the amount of dimer peak formed but also the amount of sample peak and when working with low concentration of starting RNA this can be the difference between having enough material for sequencing or having no sample left. The clustering and sequencing stages are not affected by these changes.

The other suggested modification is doing a "beads on protocol" for clean up after the ligation step, instead of the double clean up suggested by Illumina®. This adjustment helps spare as much cDNA as possible as we do not lose sample when doing the clean up twice. Leaving the same beads on the tube and precipitating the cDNA between clean ups with polyethylene glycol (PEG) helps preserve the little amount of cDNA at this point. The PEG is used as a precipitant after the cDNA is eluted with resuspension buffer (RB) following the protocol. This precipitation done with PEG followed by resuspending the sample in a small amount of buffer aids in final concentration in two ways. First, the beads are not added twice in the process as they stay as "beads on." Also, the final resuspension can be made in 20 μL of resuspension buffer to obtain more concentrated cDNA.

9.5.3 *Other Methods to Adapt to Small Amounts of Starting Material*

Currently, different manufactures are developing methodologies for library preparation and consequently the starting material required for the protocols is decreasing. Yet, different methods are being developed to target the problem of small amounts of starting material or other challenging samples, such as formalin-fixed paraffin-embedded tissue (FFPE). As example, NuGEN Technologies has developed different kits for RNA library construction to target these challenges, for small starting material, it offers Ovation® Single Cell RNA-Seq System, which allows library construction using as little as 5 pg of total RNA or single cell lysates. The technology used in this protocol enriches for non-ribosomal RNA during cDNA synthesis through the use of proprietary whole transcriptome primers [37]. Also from NuGEN, Ovation® Human FFPE RNA-Seq Multiplex System can be used for library preparation of FFPE samples using 100 ng. The Insert Dependent Adaptor Cleavage (InDA-C) technology provides targeted depletion of unwanted high abundance transcripts and results in a reduction in sequencing reads derived from rRNA [37].

When choosing which method to use for low starting material or other challenging samples is important to consider if the RNA contained in the sample represents the transcriptome to be evaluated, otherwise the information resulting from the sequencing will be repeats of reads that are not representative or meaningful.

9.6 Conclusions and Future Perspectives

The development of new technologies, such as NGS, and methods to improve the capabilities of this technology has been a revolution for the study of genomics and transcriptomics. NGS opens doors for progress in a variety of biology fields, including biomedical research. NGS allows the sequencing of the whole genome and transcriptome in massive scale, accessible price, and it is not limited, as for example oligo-microarrays, which are limited to what was already known. Using sequencing, more details are extracted from the transcriptome, which provide us additional tools to understand the mechanisms of different diseases, including breast cancer. The advantages that NGS have already been brought to science will lead to better diagnosis, better treatments, and even, personalized treatments.

Literature Cited

1. Perou CM, Sorlie T, Eisen MB, van de Rijn M, Jeffrey SS, Rees CA, Pollack JR, Ross DT, Johnsen H, Akslen LA, Fluge O, Pergamenschikov A, Williams C, Zhu SX, Lonning PE, Borresen-Dale AL, Brown PO, Botstein D (2000) Molecular portraits of human breast tumours. Nature 406:747–752
2. Cavallaro S, Paratore S, de Snoo F, Salomone E, Villari L, Buscarino C, Ferrau F, Banna G, Furci M, Strazzanti A, Cunsolo R, Pezzino S, Gangi S, Basile F (2012) Genomic analysis: toward a new approach in breast cancer management. Crit Rev Oncol Hematol 81:207–223
3. Sanger F, Coulson AR (1975) A rapid method for determining sequences in DNA by primed synthesis with DNA polymerase. J Mol Biol 94:441–448
4. Sanger F, Donelson JE, Coulson AR, Kossel H, Fischer D (1973) Use of DNA polymerase I primed by a synthetic oligonucleotide to determine a nucleotide sequence in phage fl DNA. Proc Natl Acad Sci U S A 70:1209–1213
5. Sanger F, Nicklen S, Coulson AR (1977) DNA sequencing with chain-terminating inhibitors. Proc Natl Acad Sci U S A 74:5463–5467
6. Smith LM, Fung S, Hunkapiller MW, Hunkapiller TJ, Hood LE (1985) The synthesis of oligonucleotides containing an aliphatic amino group at the 5′ terminus: synthesis of fluorescent DNA primers for use in DNA sequence analysis. Nucleic Acids Res 13:2399–2412
7. Smith LM, Sanders JZ, Kaiser RJ, Hughes P, Dodd C, Connell CR, Heiner C, Kent SB, Hood LE (1986) Fluorescence detection in automated DNA sequence analysis. Nature 321:674–679
8. Metzker ML (2010) Sequencing technologies—the next generation. Nat Rev Genet 11:31–46
9. Mardis ER (2008) The impact of next-generation sequencing technology on genetics. Trends Genet 24:133–141
10. Wetterstrand KA (2013) DNA sequencing costs: data from the NHGRI genome sequencing program (GSP). [cited 2013 11/01/2013]. Available from http://www.genome.gov/sequencingcosts/
11. Mardis ER (2008) Next-generation DNA sequencing methods. Annu Rev Genomics Hum Genet 9:387–402
12. Green RE, Krause J, Ptak SE, Briggs AW, Ronan MT, Simons JF, Du L, Egholm M, Rothberg JM, Paunovic M, Paabo S (2006) Analysis of one million base pairs of Neanderthal DNA. Nature 444:330–336
13. Hudson ME (2008) Sequencing breakthroughs for genomic ecology and evolutionary biology. Mol Ecol Resour 8:3–17
14. Ponting CP, Oliver PL, Reik W (2009) Evolution and functions of long noncoding RNAs. Cell 136:629–641
15. Gonzalez-Garay ML, McGuire AL, Pereira S, Caskey CT (2013) Personalized genomic disease risk of volunteers. Proc Natl Acad Sci U S A 110:16957–16962
16. Normanno N, Rachiglio AM, Roma C, Fenizia F, Esposito C, Pasquale R, La Porta ML, Iannaccone A, Micheli F, Santangelo M, Bergantino F, Costantini S, De Luca A (2013) Molecular diagnostics and personalized medicine in oncology: challenges and opportunities. J Cell Biochem 114:514–524
17. Zhang J, Ding L, Holmfeldt L, Wu G, Heatley SL, Payne-Turner D, Easton J, Chen X, Wang J, Rusch M, Lu C, Chen SC, Wei L, Collins-Underwood JR, Ma J, Roberts KG, Pounds SB, Ulyanov A, Becksfort J, Gupta P, Huether R, Kriwacki RW, Parker M, McGoldrick DJ, Zhao D, Alford D, Espy S, Bobba KC, Song G, Pei D, Cheng C, Roberts S, Barbato MI, Campana D, Coustan-Smith E, Shurtleff SA, Raimondi SC, Kleppe M, Cools J, Shimano KA, Hermiston ML, Doulatov S, Eppert K, Laurenti E, Notta F, Dick JE, Basso G, Hunger SP, Loh ML, Devidas M, Wood B, Winter S, Dunsmore KP, Fulton RS, Fulton LL, Hong X, Harris CC, Dooling DJ, Ochoa K, Johnson KJ, Obenauer JC, Evans WE, Pui CH, Naeve CW, Ley TJ, Mardis ER, Wilson RK, Downing JR, Mullighan CG (2012) The genetic basis of early T-cell precursor acute lymphoblastic leukaemia. Nature 481:157–163
18. Kolata G (2012) In treatment for leukemia, glimpses of the future. The New York Times. 07/08/2012; Sect. A1

19. Wang Z, Gerstein M, Snyder M (2009) RNA-Seq: a revolutionary tool for transcriptomics. Nat Rev Genet 10:57–63
20. Cullum R, Alder O, Hoodless PA (2011) The next generation: using new sequencing technologies to analyse gene regulation. Respirology 16:210–222
21. Gargiulo G, Levy S, Bucci G, Romanenghi M, Fornasari L, Beeson KY, Goldberg SM, Cesaroni M, Ballarini M, Santoro F, Bezman N, Frige G, Gregory PD, Holmes MC, Strausberg RL, Pelicci PG, Urnov FD, Minucci S (2009) NA-Seq: a discovery tool for the analysis of chromatin structure and dynamics during differentiation. Dev Cell 16:466–481
22. Shendure J, Lieberman Aiden E (2012) The expanding scope of DNA sequencing. Nat Biotechnol 30:1084–1094
23. Zuo T, Tycko B, Liu TM, Lin HJ, Huang TH (2009) Methods in DNA methylation profiling. Epigenomics 1:331–345
24. Cabili MN, Trapnell C, Goff L, Koziol M, Tazon-Vega B, Regev A, Rinn JL (2011) Integrative annotation of human large intergenic noncoding RNAs reveals global properties and specific subclasses. Genes Dev 25:1915–1927
25. Guttman M, Garber M, Levin JZ, Donaghey J, Robinson J, Adiconis X, Fan L, Koziol MJ, Gnirke A, Nusbaum C, Rinn JL, Lander ES, Regev A (2010) Ab initio reconstruction of cell type-specific transcriptomes in mouse reveals the conserved multi-exonic structure of lincRNAs. Nat Biotechnol 28:503–510
26. Rinn JL, Chang HY (2012) Genome regulation by long noncoding RNAs. Annu Rev Biochem 81:145–166
27. Guttman M, Donaghey J, Carey BW, Garber M, Grenier JK, Munson G, Young G, Lucas AB, Ach R, Bruhn L, Yang X, Amit I, Meissner A, Regev A, Rinn JL, Root DE, Lander ES (2011) lincRNAs act in the circuitry controlling pluripotency and differentiation. Nature 477:295–300
28. Mamanova L, Coffey AJ, Scott CE, Kozarewa I, Turner EH, Kumar A, Howard E, Shendure J, Turner DJ (2010) Target-enrichment strategies for next-generation sequencing. Nat Methods 7:111–118
29. Ozsolak F, Milos PM (2011) RNA sequencing: advances, challenges and opportunities. Nat Rev Genet 12:87–98
30. Benes V, Blake J, Doyle K (2011) Ribo-Zero Gold Kit: improved RNA-seq results after removal of cytoplasmic and mitochondrial ribosomal RNA. Nat Methods 8:iii–iv
31. Illumina Inc. (2010) TruSeqTM RNA sample preparation guide. Illumina® Inc.
32. Thermo Fisher Scientific Inc. (2008) NanoDrop 1000 Spectrophotometer V3.7 User's Manual Thermo Fisher Scientific Inc.
33. Life Technologies (2010) Qubit® 2.0 Fluorometer User Manual. Life Technologies
34. Simbolo M, Gottardi M, Corbo V, Fassan M, Mafficini A, Malpeli G, Lawlor RT, Scarpa A (2013) DNA qualification workflow for next generation sequencing of histopathological samples. PLoS One 8:e62692
35. Kapa Biosystems (2013) KAPA library preparation kits. Kapa Biosystems. [cited 2013 11/19/2013]. Available from http://www.kapabiosystems.com/products/name/kapa-library-preparation-kits
36. Illumina Inc. (2012) cBlot user guide. Illumina® Inc.
37. NuGEN Technologies Inc. (2013) Challenging samples. NuGEN Technologies Inc. [cited 2013 11/19/2013]. Available from http://www.nugeninc.com/nugen/index.cfm/products/cs/

Chapter 10
Methodological Approaches for Understanding the Epigenetic Landscape of the Human Breast and Its Implications in Cancer and Prevention

Julia Santucci-Pereira, Steven G. Doll, Ryan R. Smalley, Colleen O'Malley, Irma H. Russo, and Jose Russo

Keywords DNA methylation • The histone code • Epigenomic events • Targeting methylated DNA • TFF1 • HDAC2 • NSD1 and PRMT6 • NCOA3

10.1 Introduction

Breast cancer is among the most prevalent of cancers diagnosed within women worldwide [1]. To this end, much effort has been directed towards the research and development of viable methods for treating the disease. While treatment of breast cancer is both important and necessary, investigation of means for preventing the disease is of equal significance. In order to prevent the development of breast cancer, it is important that we understand the normal breast. The differentiation of the mammary gland during pregnancy is a key event that provides protection from neoplasia [2, 3]. At the heart of lobular differentiation is a specific pattern of gene expression induced by pregnancy [4–8]. Here, we report a methodological approach to understand the unique methylation profiles within the parous and nulliparous postmenopausal breast. We consider that an epigenomic signature of pregnancy, combined with our knowledge of the transcriptomic signature, may provide a comprehensive understanding of the molecular basis of mammary differentiation. In this chapter, we discuss some of the paradigms in order to provide a better understanding of our data on the epigenetic basis of breast cancer prevention.

J. Santucci-Pereira, Ph.D. • I.H. Russo, M.D. • J. Russo, M.D. (✉)
Irma H Russo MD Breast Cancer Research Laboratory, Fox Chase Cancer Center, Cottman Avenue 333, Philadelphia, PA 19111, USA
e-mail: jose.russo@fccc.edu; julia.pereira@fccc.edu

S.G. Doll • R.R. Smalley • C. O'Malley
Drexel University at the Irma H Russo Breast Cancer Research Laboratory, Fox Chase Cancer Center, Philadelphia, PA, USA

J. Russo and I.H. Russo, *Techniques and Methodological Approaches in Breast Cancer Research*, DOI 10.1007/978-1-4939-0718-2_10,

10.2 Historical Background and Significance of DNA Methylation

During the 1940s, Conrad Waddington proposed a theory of "epigenetics" that would explain the development of complex organisms as the result of interacting networks of genes, with different gene interactions mediating development on a temporal scale. Phenotypic variation need not be the result of altered gene sequence, rather the result of different genes being expressed at different times; therefore, phenotypic variation does not reflect genotypic variation [9]. Waddington referred to the sum of the gene networks that shaped development as the "epigenotype" [10]. It was the epigenotype as well as the genotype that created the pheonotype. A more modern definition of the epigenotype states that it is a pattern of differential gene expression that exists between cell types [11]. Evidence would later accumulate, some of which will be discussed here, and that would establish epigenetics as a study of differential gene expression and the underlying biochemical mechanisms.

Chemists had detected the presence of methylated cytosine (Fig. 10.1) within mammalian DNA well before the enzymes responsible for DNA methyltransfer had been characterized. In 1951, Cohn detected a novel nitrogenous base in the deoxycytidylic acid fraction separated from total thymus DNA [12]. This base absorbed light in a similar pattern as synthetic 5-methyl cytosine and produced the same fluorescence pattern when separated by paper chromatography. Several other chemical tests also demonstrated the uncanny likeliness of this base to that of 5-methyl cytosine. Cohn concluded that he had detected a deoxynucleotide that included 5-methylcytosine within thymus DNA. However, only a very small proportion of the total deoxycytosine fraction contained this base [12, 13]. In 1962, Doskocil and Sormova determined the distribution of 5-methylcytosine in DNA isolated from various mammalian tissues. Out of the total DNA, the greatest percentage of 5-methylcytosine occurred in the sequence CpPu (cytosine-phosphate-purine). While they could not directly determine the exact nature of the purine, Doskocil and

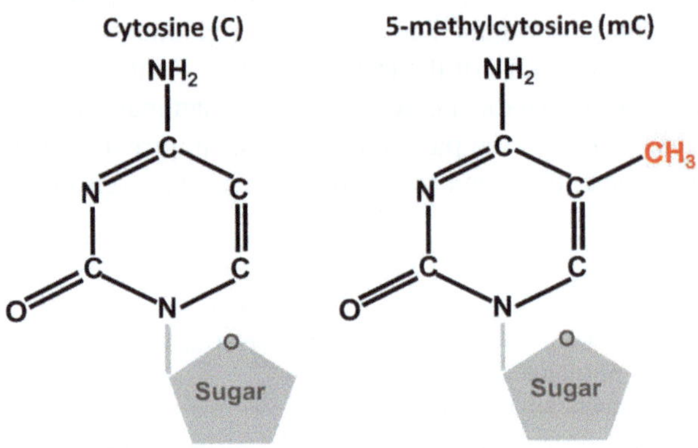

Fig. 10.1 Formulas of unmethylated (*left*) and the methylated (*right*) cytosine. The methyl group is added to the carbon in position 5

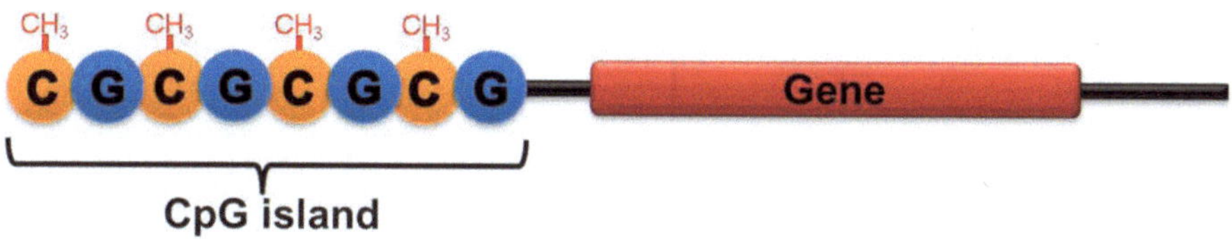

Fig. 10.2 Representation of CpG island. Methylation of cytosine often happens in regions of the DNA with high content of guanine and cytosine

Sormova deduced it to be guanine, because they did not detect pMpA (phosphate-methylated base-phosphate-adenosine) in the DNA [14, 15]. It was from these results that the concept of CpG methylation originated. CpG islands (CGIs) are regions of DNA with high content of guanine and cytosine and with a high frequency of CpG dinucleotides relative to the genome (Fig. 10.2) [16].

In 1975, Holliday and Pugh were among the first to propose a mechanism for eukaryotic DNA methylation based upon observations in bacterial systems [13, 17]. Bacteria possess modified bases and site-specific enzymes capable of such modification. Holliday and Pugh thus assumed a level of symmetry between bacterial and mammalian DNA base modification, they proposed that methylation of cytosine at CpG sequences alters gene expression and that CpG methylation is enzyme mediated. Central to this theory was the idea that base modifications would create a pattern of gene expression reflected in the cells phenotype. Holliday and Pugh determined that there must be "modification enzymes," the first of which methylates the parental strand, and the second of which replicates the methylation pattern to the nascent strand [13, 17]. In this way DNA methylation may be transmitted mitotically. Such a maintenance methylase hypothesis would not be proven entirely incorrect, as we shall further explore.

Gardiner-Garden and Frommer, in 1987, found CpG dinucleotide enrichment in the promoter regions of tissue-specific genes [16] (Fig. 10.2). Methylation at these CGIs inhibits transcription (Fig. 10.3), as seen through work on the tissue-specific expression of hSERPINB5. hSERPINB5 is differentially expressed between tissues, which correlates with the magnitude of cytosine methylation at the CGI within the genes promoter. In cells that normally express hSERPINB5, no methylation is detected within the CGI in either allele. Conversely, cells that do not express hSERPINB5 are heavily methylated at the CGI in both alleles [18]. Such findings not only indicate that CGI methylation at gene promoters results in gene inactivation (Fig. 10.3); they also suggest differential methylation patterns between tissues. The distribution of CpG methylation near the transcription start site of hSERPINB5 gene in various tissues highlights an important concept: *CpG methylation is not maintained at the level of the individual dinucleotide. Individual CpG methylation will change across cell generations; however, the overall methylation state of the CGI will remain consistent* [19].

Most promoter regions within the mammalian genome include CGIs, and nearly all are unmethylated. Interestingly, CGIs lack many of the elements needed to

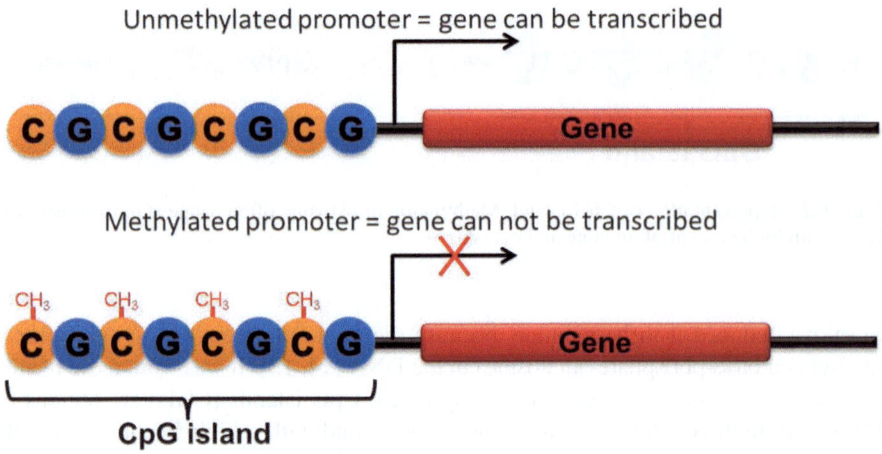

Fig. 10.3 Representation of CpG island and inhibition of transcription. When the CpG island is located in the promoter region or close to the promote region of a gene, its methylation inhibits the transcription of this gene

recruit the transcriptional machinery, yet are capable of promoting transcription nonetheless [20]. Certain transcription factors recognize and bind CpG rich promoters, and thus CGIs may represent an adaptive advantage by which genes with CGI-containing promoters were transcribed and selected for as necessary components of the genome through the virtue of nonspecific transcription factor binding [20]. Regardless, the CGIs of some promoters become methylated, a modification crucial to embryonic development and tissue differentiation.

The methylation patterns of the mature gametes are erased, or reset, upon fertilization and development of the zygote to the blastocyst [21]. In 2012, Smith et al. determined methylation levels within mouse gametaszygote, and throughout the preimplantation development of the embryo using reduced representation bisulfite sequencing [22]. Genomic methylation within the sperm was far greater than that of either the oocyte or preimplantation embryo (at all stages). By the two-cell stage, the paternal genome had been drastically demethylated. Genome-wide methylation remained low until gastrulation of the post-implantation embryo at which point de novo methylation occurred (Fig. 10.4). Particular focus was given to the methylation status of differentially methylated regions (DMRs) between paternal and maternal alleles [22]. DMRs represent imprinted genes, although the concept of genomic imprinting will not be discussed here (for review, see [21]). Smith et al. found that the majority of maternal DMRs included CGIs, which became demethylated leading up to implantation. Methylation within the paternal genome occurred primarily within retrotransposons, and methylation of these regions was not erased [22]. Thus maternal DMRs are most likely involved in a genomic signature of the preimplantation embryo while paternal DMRs prevent the transcription of transposable elements [22, 23].

DNA methylation is established and maintained within the genome through the action of the de novo methyltransferases, Dnmt3a and Dnmt3b, and the maintenance methyltransferase Dnmt1 (Fig. 10.5) [19, 24]. Dnmt1 localizes the replicating

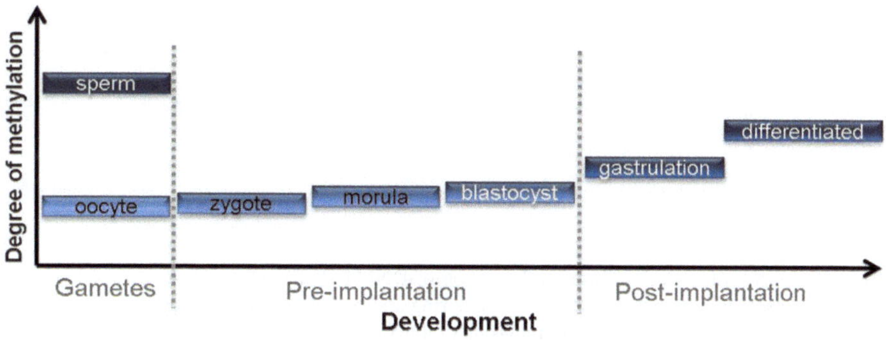

Fig. 10.4 Changes in methylation are observed during the embryonic development and tissue differentiation. The sperm presents a higher level of methylation, while the oocyte has lower levels of methylation. After fertilization, the zygote keep methylation levels similar to the oocyte, and throughout the development, an increased overall level of methylation is observed

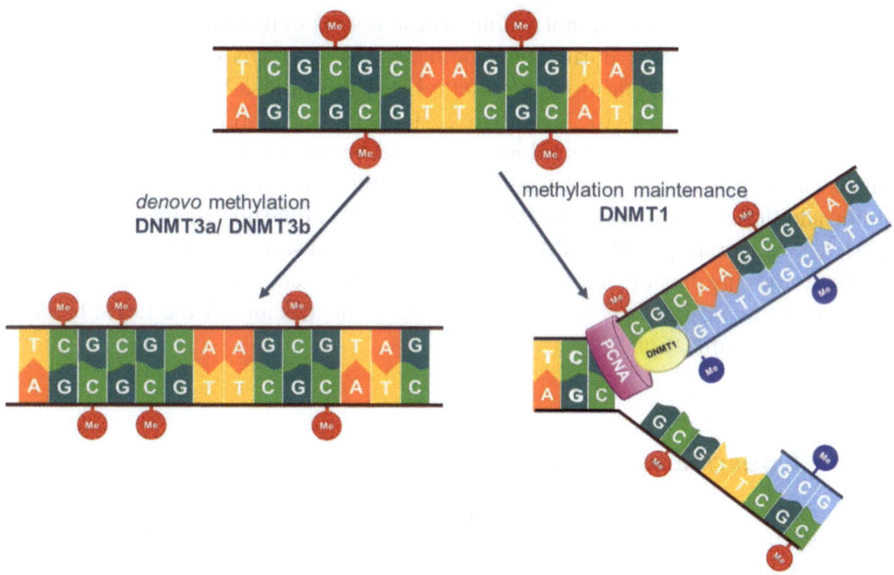

Fig. 10.5 DNA methylation is established and maintained within the genome through the action of DNA methyltransferases (DNMT). Dnmt3a and Dnmt3b are responsible for de novo methylation, while Dnmt1is responsible for the maintenance of the methylation during replication, and it is associated to PCNA

DNA when associated with proliferating cell nuclear antigen (PCNA) in vitro, suggesting that Dnmtl functions as a maintenance methylase that confers methylation from the parental DNA strand to the daughter strand [25]. Dnmt3a and Dnmt3b may modify cytosine residues without a preexisting methylation template [24, 26]. These methyltransferases are necessary for establishing methylation patterns within the

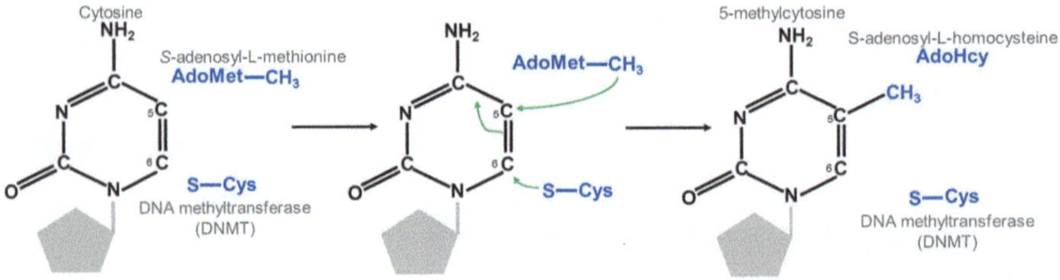

Fig. 10.6 DNA methylation through the action of DNMT. The catalytic cysteine residue of the DNMT disrupts the aromaticity of the cytosine ring and positions a double bond between C4 and C5. This double bond attacks the methyl group of the *S*-adenosyl-L-methionine, which is converted to *S*-adenosyl-L-homocysteine, and the methyl group is added to the C5 of cytosine

embryo as demonstrated by Okano et al. [27]. Retroviral DNA that was normally methylated in mouse embryos was unmethylated in both the blastocyst and embryo of Dnmt3a/Dnmt3b double negative homozygous mice. Furthermore, the double negative homozygotes were not viable. These results demonstrate the importance of de novo methylation during development [28]. The DNA methyltransferases (DNMT) catalyze the addition of a methyl group from *S*-adenosyl-L-methionine to C5 of cytosine (Fig. 10.6). A catalytic cysteine residue disrupts the aromaticity of the cytosine ring and positions a double bond between C4 and C5. This double bond attacks the methyl group of the *S*-adenosyl-L-methionine substrate through a nucleophilic addition, resulting in methylated cytosine at C5 after subsequent deprotonation [28]. The bacterial type-II DNMT HhaI rotates its substrate cytosine in a way that positions the ring outside of the double helix and allows access to active site residues. Such a finding describes how the three dimensional structure of the DNMT allows for cytosine methylation [29].

10.3 Understanding the Histone Code

The nucleosome comprises a histone octamer, with two of each of histones H2A, H2B, H3, and H4 forming the core particle (Fig. 10.7) [30, 31]. The core particle may be subdivided into H3–H4 and H2A–H2B heterodimers. The two H3–H4 heterodimers form a tetramer through a four-helix bundle between the H3 subunits, with both H2A–H2B heterodimers conjugated to the H3–H4 tetramer though four-helix bundles. Core histones H3, H4, H2B, and H2A comprise a half-core particle that binds 73 bp of supercoiled DNA (Fig. 10.7). The other half-core particle consisting of H3′, H4′, H2B′, and H2A′ binds 72 bp of supercoiled DNA, allowing the entire octamer to bind approximately 145–147 bp of DNA. From each core histone protrudes an N-terminal region or "tail" [31, 32]. A short segment of each tail associates with the DNA superhelix, by intercalation through the minor groove or hydrogen bonding to the outer backbone. The majority of the N-terminal tails extends away

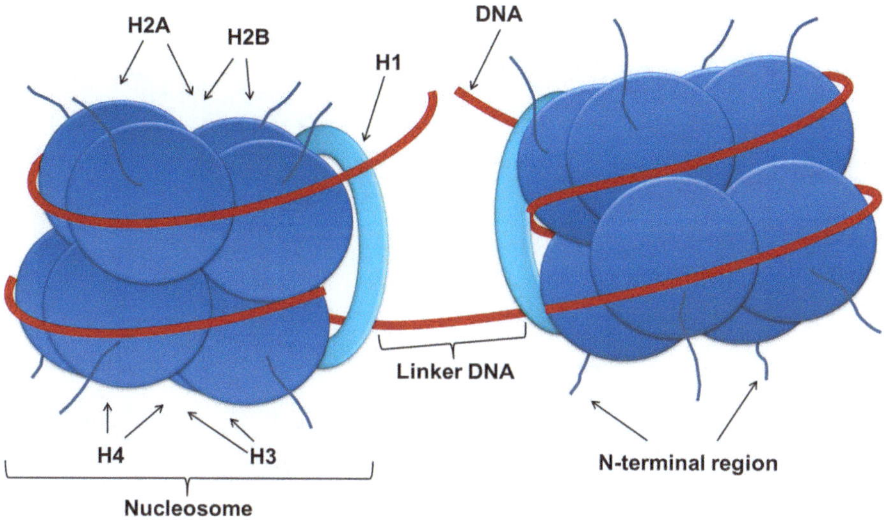

Fig. 10.7 The nucleosome core particle. The core particle is subdivided into two H3–H4 and two H2A–H2B heterodimers. The entire octamer binds to approximately 145–147 bp of DNA. From each core histone protrudes an N-terminal region or "tail"

from the core particle and only weakly interacts with the DNA (Fig. 10.7). A negatively charged region of the H4 N-terminal tail is capable of hydrogen bonding to an acidic region of the H2A–H2B heterodimer of nearby nucleosomes [31]. It is possible that disruption of such interaction through histone N-terminal tail modification alters the higher-order structure of the chromatin towards a transcriptionally active state. On a structural basis, the histone N-terminal tails only briefly interact with the DNA superhelix, and so it is unlikely that neutralization of the positively charged residues of the histone tails will drastically affect DNA-nucleosome binding [32]. Rather, the hypothesis that there is a "histone code," in which many histone N-terminal modifications are working in concert to open the chromatin for the transcriptional machinery, is more likely [33].

Here, we describe acetylation and methylation of histone N-termini, key components of the histone code. Acetyl groups are covalently bound to specific lysine residues within the histone N-terminal tails through the action of histone acetyltransferases (HATs) (Fig. 10.8) [34–36]. HATs are categorized according to location within the cell and sequence homology. Type-A HATs acetylate deposited histones within the nucleus and may belong to the GNAT, MYST, or CBP/p300 family of acetyltransferases [34–36]. Type-B HATs acetylate histones within the cytoplasm prior to their deposition upon the chromatin [34, 36]. These HATs are closely related to the first characterized type-B HAT: yeast HAT1 (yHAT1) [35, 37]. Yeast GCN5 bears considerable sequence similarity to human GCN5 within several domains, and so a discussion of the catalytic mechanism of yGCN5 is not entirely irrelevant [34]. The mechanism GCN5-mediated acetyl group transfer from acetyl-CoA to the

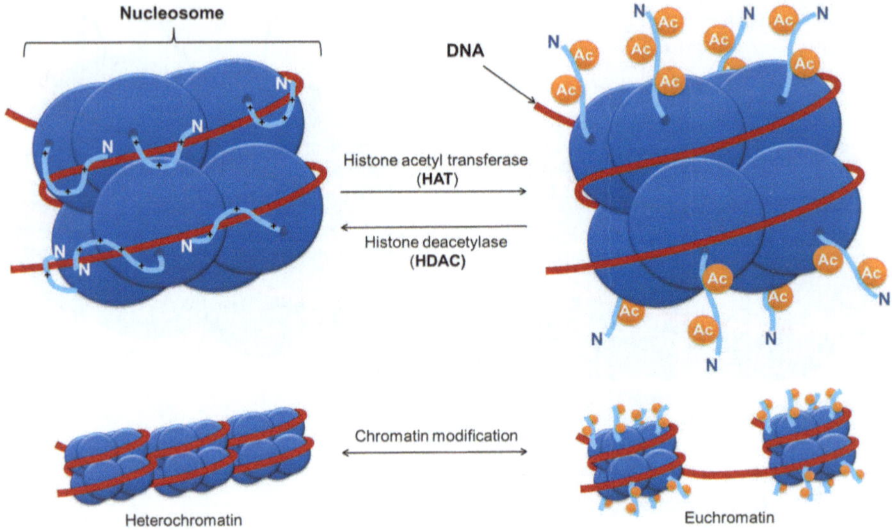

Fig. 10.8 Representation of the deacetylated (*left*) and acetylated nucleosome (*right*). When the N-terminal regions of the histones are acetylated, the tails become neutral, losing affinity to the DNA and contributing for a less condensed chromatin. The open chromatin (euchromatin) allows transcription factors to bind to the DNA, therefore allows transcription

positively charged *N*-ε-amino group of the lysine residue has been elucidated, and occurs according to a mechanism in which the enzyme forms a ternary complex with both substrates. The glutamate active-site residue (E173) abstracts a proton from the *N*-ε-amino group. The activated nitrogen of the ε-amino group then attacks the chiral carbon of the acetyl group through a nucleophilic addition reaction, forming a tetrahedral intermediate. The S-CoA group leaves giving the final product [38]. The acetylated lysine is uncharged [38], and so this modification contributes to a less condensed chromatin state [33]. GCN5 is capable of acetylating nucleosomal H3 and H4 when a member of the yeast Ada and SAGA co-activator complexes [39].

Methylation of the histone N-terminal tails occurs at specific lysine and arginine residues [35] (Fig. 10.9). The addition of methyl groups does not modify charge, and so the chromatin structure is not disrupted. Rather, the nature and degree of methylation determines the transcriptional effect of the modification. Lysine residues have been found to be mono-, di-, or tri-methylated [40]. Arginine residues have been found to be mono-methylated, symmetrically or asymmetrically di-methylated [41]. Histone N-terminal methylation is catalyzed by histone methyltransferases (HMTs). Here, we shall be concerned with the SET domain containing family of histone lysine methyltransferases (HKMTs) and the protein arginine *N*-methyltransferase (PRMT) family of histone arginine methytransferases [40, 41]. The mechanism of SET domain and PRMT methyltransfer is discussed elsewhere [42].

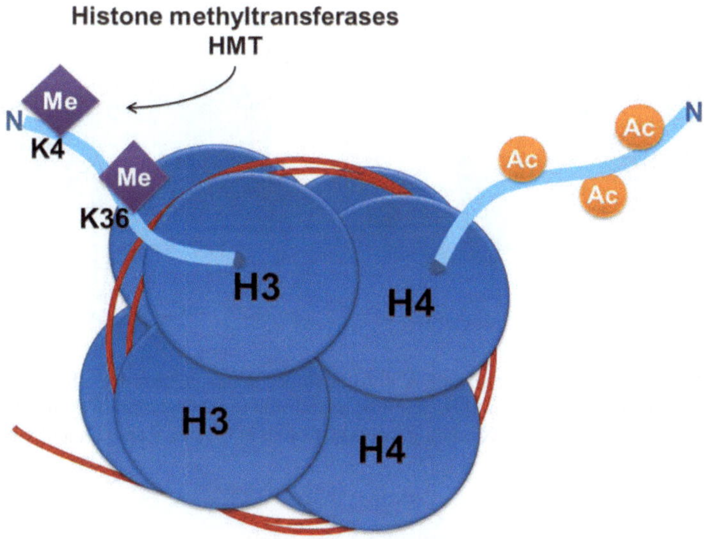

Fig. 10.9 Representation of histone methylation. The methylation of the histone N-terminal tails occurs at specific lysine (K) and arginine residues. The methylation of the histones does not alter the charge; therefore, the chromatin structure is not modified

Barski et al. in a comprehensive survey of histone methylation within the human genome found that H3K4me3, H3K27me, H3K36me3, H3K9me, H4K20me, H3K79me, and H2BK5me were all found at active promoters. H3K27me3 was enriched at silent genes [43]. In the flowering plant *Arabidopsis thaliana*, plants with nonfunctional SGD8, a methyltransferase specific for H3K36me2 and H3K36me3 marks in this system, had a differential expression profile in comparison with the wild type. A greater portion of the genes investigated were differentially downregulated, suggesting that H3K36me2 and H3K36me3 function towards gene activation [44]. Methylation of arginine 3 on the H4 N-terminus has been observed; however, its exact effect upon transcription remains unclear [43]. The H4Rme2a mark has been suggested to block H3K4 trimethylation, and thus has a negative effect upon gene expression [45]. Indeed, methylation of the histone N-termini is complex. Not all histone N-terminal methylations mediate gene transcription, and those that do vary in their location within active genes. Additionally, the degree of methylation determines the effect of modification (e.g., H3K27me is associated with active genes, whereas H327me3 is associated with repression) [43].

While a single covalent histone modification may be attributed to the transcriptional state of a gene, it is likely that multiple modifications behave cooperatively to achieve a specific effect. Compelling evidence for a histone code has been provided by Wang et al. [46], who determined the distribution of different histone acetylations and methylations across the human genome. Out of 12,541 gene promoters surveyed, 4,339 contained unique patterns of methylation and acetylation. A particular set of 16 modifications, along with H2A variant H2A.Z, was found at 3,286

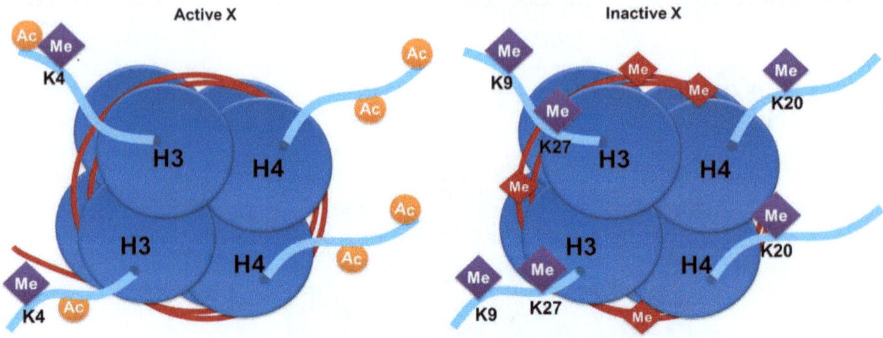

Fig. 10.10 Chromatin modification of X chromosome. When active, the chromatin of the X chromosome contains acetylation in H3 and H4, in addition to methylation of lysine 4 (K4) of H3. When inactive, the chromosome X contains methylation of H3 lysine 9 and 27, and methylation of H4, lysine 20

promoters. Genes with these histone modifications, termed the "backbone" by the authors, were more expressed than those lacking. Addition of other marks, such as H4K20me1, to the backbone increased expression above that of the backbone itself [46]. Such a unique set of histone modifications work in concert make the chromatin accessible to transcription factors and communicate that particular genes are ready for transactivation. The favorable backbone of histone modifications included a majority of acetylated lysine residues, as well as H3K4 mono-, di-, and tri-methylation [46]. Increased acetylation may open up the chromatin through neutralization of the positively charged lysine residues and H3K4 methylation may signal to transcription factors. In the same sense, transcriptionally inactive chromatin is lacking in histone acetylation and bears the repressive H3K27me3 mark, as seen in X-chromosome inactivation [47, 48] (Fig. 10.10).

10.4 The DNA Sequencing and Epigenomic Events

Methods to evaluate the transcriptome has become widely available, and as better understanding of how the epigenetics events change the transcriptome becomes necessary, techniques to understand the gene expression regulation have emerged. Methods to evaluate transcription factors and their binding sites, chromatin remodeling, DNA methylation, microRNAs, and histone modifications have rapidly developed [49]. In addition, the need for high-throughput techniques made the microarray industry and more recently the next-generation sequencing to grow exponentially. Due to the advantages of the next-generation sequencing over the microarray technology, such as more in-depth information of the studied DNA, no need of previous knowledge about the genome being studied, the fast decrease of costs and increase in the high-throughput ability, this technology has gained popularity.

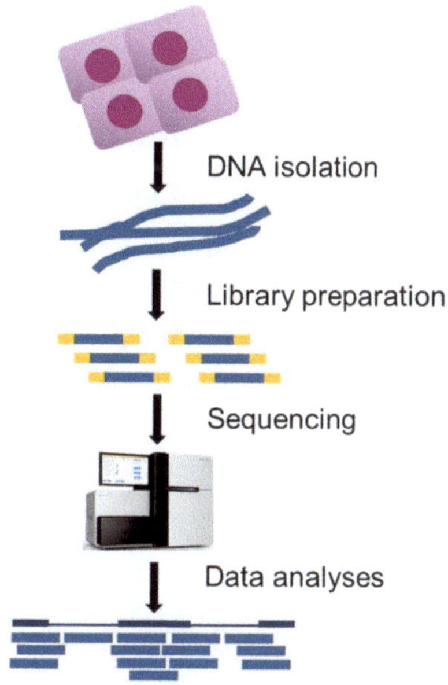

Fig. 10.11 Steps for DNA sequencing. First, the DNA is isolated from the cells of interest. Then the DNA is fragmented and prepared for the sequencing (library preparation). Once the library is prepared and reaches the quality and quantities required, the library is introduced to the sequencer. Lastly, the output data is analyzed

The normal human breast, as well as breast cancers, is very heterogeneous and complex [50], the information that DNA sequencing can provide us will help in deciphering the molecular complexity of the human breast [51], as well as the molecular mechanisms in which some women are protected while others are more prone to develop breast tumors.

The basic steps to perform the sequencing are the same for RNA-seq or DNA-seq, which are described in detail in Chap. 9 of this book. The conventional approach consists in isolate the DNA cells of interest, prepare the library of DNA fragments, and then perform the sequencing (Fig. 10.11). The DNA can be extracted from whole or sections of a tissue, if a more cell-specific analysis is desired, it is possible to perform flow cytometry of cell culture or yet, laser-capture microdissection of the tissue [52].

In addition, one can target the subset of DNA of interest (e.g., epigenetic events). Epigenetics refers to the biochemical changes in proximities of the DNA, including DNA methylation and modifications of the histone proteins, these modifications in addition to chromatin accessibility and transcription factor binding, can all be detected using sequencing [52].

The techniques to target the epigenetic events will be discussed later in this chapter (Sects. 10.4.1 and 10.4.2). During the targeting process or afterwards, the DNA library is constructed, in which fragmented DNAs are linked to adaptors specific for the platform in use, and they might be PCR-amplified or not. In addition, samples can have indexes, also known as barcodes, added to the DNA library (Fig. 10.12). The indexes tag different samples that are combined in one sequencing lane, allowing increased number of samples studied per run. Once this process has been completed, the samples are added to the sequencer (Fig. 10.12).

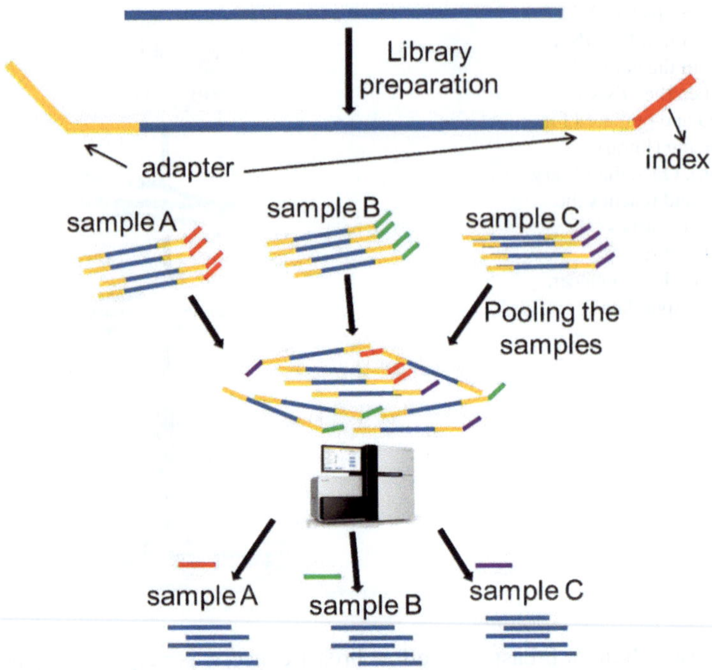

Fig. 10.12 During the library preparation, indexes or barcodes may be added to the samples. The different indexes have different nucleotide sequences. The indexes tag the different samples, which can now be combined in one reaction. At the end of the sequencing, the different samples are identified accordingly with the index sequence used

Some common challenges faced when using sequencing to study the epigenomic events are limitation of specific antibodies (e.g., antibodies that recognize methylated DNA, or histone modification), adapted protocols for low amounts of starting material and methodologies that measure dynamics of epigenetic regulation instead of the static snapshots of the events [52].

10.4.1 Targeting Methylated DNA

To separate the methylated DNA from the unmethylated DNA in order to perform the sequencing (Methyl-seq), there are three main strategies (Fig. 10.13), bisulfite conversion, use of restriction endonucleases, and use of affinity properties [53]. The most commonly used is the bisulfite conversion (BS-seq). In this technique, the genomic DNA is treated with sodium bisulfite, which converts unmethylated cytosines into uracil residues. The sodium bisulfite does not affect methylated cytosines, which remain unchanged in the sequence. In the second technique, the genomic DNA is treated with methylation-sensitive restriction endonucleases, which discriminate sequences based on methylation status. In principle, the methylated DNAs

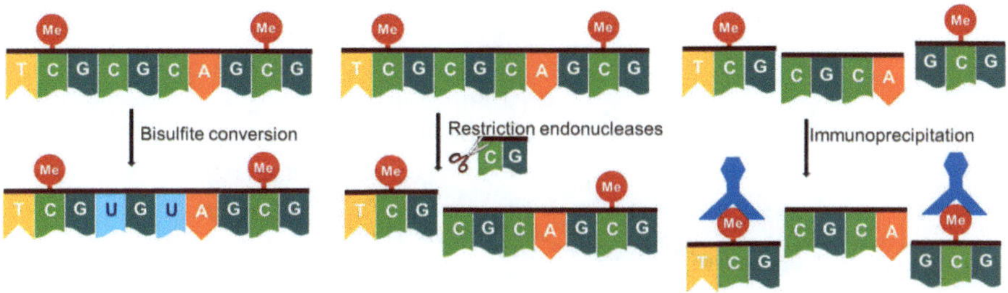

Fig. 10.13 Methods of separation of the methylated DNA. There are three main strategies to separate the methylated DNA for later analysis. (1) Bisulfite conversion, in which the genomic DNA is treated with sodium bisulfite. The bisulfite converts the unmethylated cytosines into uracil residues. (2) Use of methylation-sensitive restriction endonucleases, which discriminate the DNA sequences based on methylation status. (3) Use of affinity properties, which antibodies that recognize methylated areas are used. Once the antibody is bound to the methylated DNA, it is separated from the unmethylated DNA, using, e.g., immunoprecipitation

resist the restriction digestions, and can be amplified, while the unmethylated DNA is fragmented and it fails to be amplified [53]. The third method is based on immunoprecipitation (MeDIP-seq). In the MeDIP, the genomic DNA is fragmented and the incubated with anti-5-methylcytosine antibody, the product resultant of the immunoprecipation is the hypermethylated DNA. In this technique a limiting factor is the sensitivity of the antibody, which usually only densely methylated sequences [53]. A modification of the MeDIP is the MBDCap-seq, which is based on affinity chromatography over a methyl-binding domain (MBD) [53]. The MBD is linked to beads and specifically binds to CpG; therefore, the DNA fragments interacting with the MBD are hypermethylated regions of the DNA [49], more details will be discussed later in this chapter (Sect. 10.5.1).

BS-seq is considered the "gold standard" method, and although it provides the most comprehensive and accurate picture of the methylome, extensive sequencing of the genome is required to obtain a good coverage. Both MeDIP-seq and MBDCap-seq identify methylation-enriched regions; MeDIP-seq is more sensitive in regions of high CpG density and MBDCap-seq more sensitive in regions of moderate CpG density [49]. In order to study different methyl-CpG densities, a step-wise elution of the MBD-captured DNA, using crescent salt concentrations, can separate the genome in different fractions containing different methyl-CpG densities, because the affinity between the MBD and the DNA depends on the density of CpGs [54].

10.4.2 Targeting Other Epigenetic Events

Chromatin immunoprecipitation (ChIP) is one of the major techniques to target DNA–protein interactions. In the ChIP, the cells are fixed to preserve the interactions between the DNA and proteins and cross-linked (Fig. 10.14). The chromatin

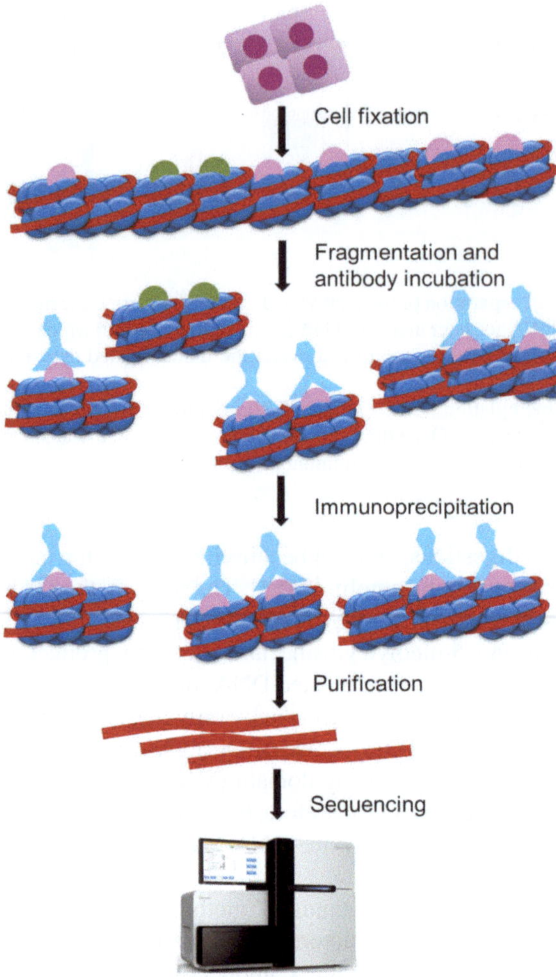

Fig. 10.14 Steps of chromatin immunoprecipitation (ChIP). First, the cells are fixed to preserve the interactions between the DNA and proteins, such as transcription factors. Next, the chromatin is fragmented and incubated with specific antibody against the protein of interest. Immunoprecipitation is used to separate the protein of interest bonded to the DNA from the rest of the unwanted fragments of DNA. Once these fragments are separated, they are then purified, releasing the bound proteins. These DNA fragments are later used for library preparation and sequencing (ChIP-Seq)

is then fragmented and incubated with specific antibody against the protein of interest. Immunoprecipitation is used to separate the protein of interest bonded to the DNA. In the following steps, the proteins are released from the DNA, which is purified, used for library preparation and then sequenced (ChIP-seq) [49]. To analyze histone association with the DNA, it is not necessary to perform the fixation and cross-linking steps since the histones are so intimately associated with the

DNA, this process is known as Native-ChIP (N-ChIP) [49]. In order to determine the relationship between genes and chromatin, additional methods are being developed [49, 52], such as the nucleases accessibility (NA-seq). In the NA-seq, exogenous nucleases are used and in the areas of the chromatin that the DNA is exposed, the DNA will be digested. The fragmented DNA is then sequenced, and the areas of the chromatin in which regulatory process can take place are identified [49, 55]. The other method is called FAIRE (formaldehyde-assisted isolation of regulatory elements) [52], in which cells are cross-linked briefly with formaldehyde. The sheared chromatin is subjected to phenol/chloroform extraction and because DNA regions without nucleosomes are less efficiently cross-linked to protein, they will segregate to the aqueous phase during phenol/chloroform extraction. In contrast, the DNA linked to proteins will be trapped between the organic and aqueous phase. The DNA in the aqueous phase, which contains the open chromatin, is then purified [56, 57].

10.5 Epigenome of the Human Breast

An understanding of gene expression and its regulation within the normal breast is necessary in the understanding of the origins of breast cancer and how key developmental events during the life of the female, such as pregnancy, confer protection from this disease [3]. The breast structure and development have been described in detail in Chaps. 1 and 2 of this book.

The parous breast possesses a unique pattern of gene expression that differs from the nulliparous breast. Russo and Russo [58] investigated the extent to which 1,176 genes were differentially expressed between the Lob1 of the nulliparous and the Lob3 of the parous premenopausal breast. Eighty-two genes were found to be differentially expressed between the two glands. These genes were up-regulated in the Lob3 and covered a broad range of functions, including cell cycle control, transcriptional regulation, tumor suppression, and cell growth [58]. Asztalos et al. [5] found differential expression of genes between parous, premenopausal women (0–2 and 5–10 years and after pregnancy) and nulliparous, premenopausal women. Differentially expressed genes were involved in involution by inflammation, ECM remodeling, angiogenesis, and estrogen signaling [5].

The genomic signature of pregnancy persists through the involution of the lobules and into postmenopause. Peri et al. [6] reported 208 genes differentially expressed between parous and nulliparous postmenopausal breast tissue. Again, these genes covered a wide array of functions including apoptosis, DNA repair, cell adhesion, cell cycle control, cell differentiation, cell proliferation, and cell motility [6, 7]. Here, we report the unique methylation pattern of the parous, postmenopausal breast, which may explain the genomic signature as it exists well after the events of pregnancy. All somatic cells possess the same genetic code, however as we have seen, they do not possess the same epigenetic code.

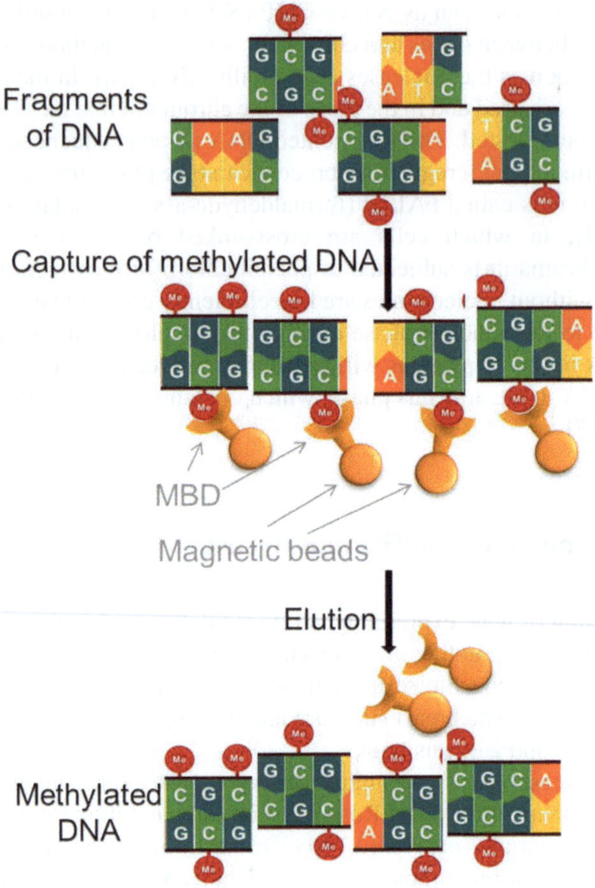

Fig. 10.15 Scheme of MDB-Cap assay. The fragmented DNA is incubated with the MBD protein, which is coupled to magnetic beads via a biotin linker. The CpG-methylated DNA is captured by the MBD protein. Then the methylated DNA is eluded using 2000 mM NaCl

10.5.1 Identification of Hypermethylated Genes Through MBDCap-seq Assay

In order to gain a better understanding of gene methylation in the postmenopausal breast, core needle biopsies (CNBs) were taken from eligible parous and nulliparous postmenopausal women. All women who donated breast tissue in this study were determined eligible according to the criteria described by Belitskaya-Levy et al. [8]. A total of five CNBs were used from each group (parous, postmenopausal and nulliparous, postmenopausal). DNA extracted from the CNBs of the parous and nulliparous samples were studied using the MBDCap-seq. The MBDCap-seq technique makes use of the methyl-CpG-binding domain of human MBD2 protein, which is coupled to magnetic beads. The MBD has an affinity for methylated cytosine residues and is used to capture the DNA high in CpG content (Fig. 10.15).

The greater the methyl-CpG density of a given DNA sequence, the more strongly it will be bound to the MBD, and a higher salt concentration will be needed for elution. The captured methylated DNA was eluted using a high concentration of salt, which select the highly methylated fragments of the DNA. The eluted DNA was used to generate libraries for sequencing. MBDCap-seq libraries were sequenced using Illumina system. Image analysis and base calling were performed with the standard Illumina pipeline, and the reads were aligned to human reference genome (hg18) (Fig. 10.15). DMRs were identified by comparing the difference of averaged methylation values between parous and nulliparous samples using by Student's t-test ($P<0.01$).

The MBDCap-seq identified a unique set of hypermethylated genes for each studied group (parous and nulliparous). In the parous breast, 455 genes were significantly hypermethylated (Table 10.1), whereas 128 genes were identified as significantly hypermethylated genes in the nulliparous breast (Table 10.2). To evaluate these lists of genes, we used complementary approaches. We performed the visualization of the sequenced data using Integrative Genomics Viewer (IGV) [59, 60], gene ontology (GO) classification through DAVID [61], canonical pathways, and networks were evaluated through the use of IPA (Ingenuity® Systems, www.ingenuity.com), and literature search.

10.5.2 Visualization of the Hypermethylated Genes

In order to better understand the methylation profile of the 583 genes found differentially methylated and to identify the areas where methylation levels differed between the groups we used the Integrative Genomics Viewer, IGV for short (www.broadinstitute.org/igv/) [59, 60]. IGV is a visualization tool which allows for the exploration and examination of large genomic datasets compared to a reference sequence. The program itself can be used for almost any data type that can be mapped to DNA and supports anything from the next-generation sequencing to array-based platforms [59]. It supports data files that are hundreds of gigabytes in size; in addition, IGV allows the user to examine their data at multiple different view levels, from whole genome to individual base pairs. It also gives users the opportunity to look at separate pieces of data, whether from the same sample genome or not, in adjacent windows to allow for easier examination [60]. For this experiment, IGV was used to compare the ten volunteers' methylome against the "Human hg 18" reference genome and against each other.

The ".bam" files resulted from the alignment were sorted so they would be correctly read by the IGV program. Once loaded by the program, each file contains the results of each read done on the genome and the overall coverage at each base pair. For this methylation study, we examined the overall coverage, once that the existence of a read at any point indicates methylation in that area, and the more reads completed, the more likely that area is to be methylated within the genome. To be able to properly see any differences in methylation between the groups, all the files were loaded simultaneously and were judged against each other.

Table 10.1 List of genes found hypermethylated genes in the breast of parous postmenopausal women

ACACA	C6orf124	DMTF1	GLRA3	KY	NMB	PRKAB2	SFRS18	TRIM29
ACAD11	C6orf130	DNAJC13	GOSR2	LANCL1	NOL4	PRKAR1A	SGOL2	TRIM68
ACTR1B	C6orf138	DNAJC6	GPR63	LARP6	NOVA1	PRKAR2B	SIDT1	TTC29
ACVR1C	C9orf150	DNER	GRIK5	LATS2	NPY5R	PRKD1	SKAP1	TTC30B
ADAD1	CA10	DNM3	GRP	LOC100128164	NR2F2	PRMT6	SLAIN2	TTLL7
ADCYAP1	CACNA1D	DOCK2	GSK3B	LOC100133469	NR3C1	PROK2	SLC1A2	TWISTNB
ADD3	CACNA2D1	DPH3	GUCY1A3	LOC145783	NR3C2	PROX1	SLC25A46	UBA5
AGBL4	CACNG3	DPP10	GULP1	LOC389033	NSD1	PRPF40A	SLC27A1	UBC
AK5	CADPS2	DPYD	HBM	LOC554202	NT5E	PSAT1	SLC35D1	UBE2E3
AKIRIN2	CAPS	DUSP1	HCCS	LOX	NUDT9P1	PTPN13	SLC43A3	UBE3A
AKR1B1	CAV1	DUSP6	HCRTR2	LRFN2	NUF2	PTPRK	SLC46A2	UBR5
ALG10B	CBR4	DYNC1I2	HDAC2	LRRFIP2	OBFC1	PTTG1	SLCO6A1	UFM1
ALOXE3	CCDC23	DYRK2	HDGFRP3	LRRIQ3	ORMDL1	PXN	SLIT2	UNC5C
ANKMY1	CCDC50	EDIL3	HERC4	MAF	OSBP	RAB32	SMAD9	UROS
ANKRD12	CD34	EFNA3	HIST1H2AB	MAL2	PAG1	RAB3C	SMG7	USP15
ANKRD50	CD55	EIF3H	HMBOX1	MALL	PAPPA	RAI2	SNHG1	USP31
AOX1	CD59	ELAVL4	HMGCLL1	MAN1A1	PARP15	RALYL	SNRK	USP9X
APOLD1	CD99L2	ENPP1	HOPX	MAN1A2	PC	RBKS	SNX2	VEZF1
APPL1	CDC42BPA	EPC2	HOXB13	MBTD1	PCDH19	RCBTB1	SORCS3	VGLL3
ARGLU1	CDC42EP3	EPHA5	HOXD13	MC3R	PCF11	RECQL	SPATA5L1	WASL
ARID5B	CDK13	EPHA6	HSF2	MDH1B	PCSK1	REV3L	SPATA6	WBSCR17
ARNTL	CEP350	ESR2	HSPA13	MIER1	PDE3A	RFX3	SPHKAP	WDR72
ASAP2	CHD6	ETAA1	HSPA4	MIR663	PDE7A	RGMB	STAMBPL1	WDR75
ASCL1	CHIC2	ETV3	HTR1A	MIR886	PDGFC	RGS2	STARD4	WDR78
ASPM	CHRDL1	EXTL2	IFITM4P	MLLT10	PDIA3P	RGS7BP	STIM2	WTAP

ATF1	CHSY3	FABP5	IFT74	MMP1	PDIA5	RIPK2	STK38	YIPF5
AZI2	CMAH	FAM110C	IL6ST	MOBKL2B	PELI2	RLN2	STL	YTHDC2
B4GALT5	CNOT2	FAM120C	INPP4B	MOCS1	PERP	RNF103	STRAP	ZAK
BAT2L2	CNTLN	FAM13C	INSIG2	MOV10	PF4	RNF13	STRN3	ZBTB41
BCL2L10	CNTN4	FAM159A	INTS6	MOXD1	PF4V1	RNF175	SUPT16H	ZCCHC12
BEND5	COL27A1	FAM160A2	ISL1	MRPS18A	PGAP3	RNLS	SVEP1	ZCCHC6
BHLHE41	COL2A1	FAM59A	ITGA2	MSX2	PGRMC2	RNMT	SYPL1	ZCRB1
BTBD11	COX18	FANCL	ITGA4	MT2A	PHACTR2	ROBO1	SYT14	ZFP36L1
BTG3	CP110	FBXL5	ITGBL1	MTMR7	PHF12	ROPN1	TAF1D	ZFPM2
C11orf88	CPEB2	FBXO38	KARS	MTX1	PHF15	RORB	TAOK2	ZNF219
C12orf61	CRISPLD1	FGFR1OP	KCNA4	MUT	PHF16	RPIA	TBC1D9	ZNF230
C14orf135	CRLF1	FKBP14	KCNJ2	MYCBP2	PHF20L1	RRBP1	TFAP2C	ZNF24
C15orf33	CTGF	FKBP3	KCNJ3	MYL12B	PIGV	RXFP3	TFAP2D	ZNF292
C15orf60	CTNNAL1	FLRT2	KCNK3	NAP1L4	PLA2G7	SART3	TFIP11	ZNF300
C1orf146	CWC22	FNDC3A	KCNV1	NCKIPSD	PLCL1	SAV1	TGFBI	ZNF492
C1orf88	CXXC5	FNDC4	KCTD12	NCOA7	PLD6	SCARB2	TGFBR2	ZNF518A
C1orf9	CYCS	FNIP2	KIAA0564	NCRNA00081	PLEKHA3	SCN9A	THAP9	ZNF542
C1orf94	CYP2J2	FRAT2	KIAA1009	NDUFC2	PLSCR4	SDC2	THYN1	ZNF620
C20orf103	CYP2R1	FSD1L	KIAA1257	NECAB1	POU4F1	SDK1	TICAM2	ZNF622
C20orf194	DACT1	FTHL17	KIAA1383	NEGR1	PPIC	SEC24D	TIFA	ZNF642
C21orf66	DCAF5	GABRA2	KIAA1586	NFIB	PPIL4	SEC61A1	TLE4	ZNF793
C4orf32	DCLK1	GABRB3	KIFAP3	NIPBL	PPP1R2P3	SECISBP2	TLL2	ZNF804B
C5orf13	DCUN1D5	GADD45A	KLF8	NKIRAS1	PPP2CA	SEPT4	TMEM196	
C5orf30	DDX59	GFI1	KLHL20	NKX2-4	PRDM12	SERP1	TMEM59L	
C5orf51	DIRC2	GHR	KLHL7	NKX2-6	PREP	SFRP2	TNFAIP8	
C5orf58	DMRTA1	GIGYF2	KRT72	NLGN1	PRKAA2	SFRS12IP1	TOX3	

Table 10.2 List of genes found hypermethylated genes in the breast of nulliparous postmenopausal women

AIF1L	CELSR1	GDF15	LONRF3	PMM1	SLC19A1
ASB6	CITED1	GNAS	LRRC37A3	PPP1R1B	SLC45A4
ATP8B3	COBRA1	GNG13	MAFA	PQBP1	SOHLH1
C16orf11	COL5A1	GP1BB	MAFB	PRIC285	SOX10
C16orf38	CYBA	GPR123	MIR1225	PRPH	SOX3
C16orf81	DGKK	GPR64	MIR127	PRR25	SRMS
C16orf91	DMRT2	GRAMD4	MIR1909	PTGDS	STAG2
C17orf55	DOK7	GSTT1	MIR200B	PTK6	TBX3
C19orf20	DPP7	H19	MIR675	RAX2	TELO2
C19orf34	EFNA2	HCFC1	MUC6	RFNG	THAP11
C19orf77	EPPK1	HDDC3	MUPCDH	RHOT2	TNFRSF18
C20orf166	ESX1	IGFALS	NAA10	RPL10	TPPP
C20orf200	ETV2	IRF7	NCAPH2	RPS6KA3	TTLL5
C21orf81	FAM125A	IRX1	NCLN	RRP1	TULP3
C22orf45	FAM3A	IRX2	NCOA3	SAMD10	WDR44
C6orf195	FHL1	IRX5	NELF	SCRIB	ZAP70
C9orf86	FLJ90757	KCNT1	NHSL2	SDF4	ZIM2
CACNA1H	FOXJ1	LMF1	PCYT2	SELO	ZNF414
CBLN1	FZD1	LMF2	PDHA1	SEPX1	
CCDC154	FZR1	LOC100129637	PEG3	SH3KBP1	
CDKN3	GAA	LOC100130522	PITX2	SHROOM4	
CEBPB	GALR1	LOC391322	PKP3	SLC16A2	

For example, COBRA1 had a DMR near the end of the gene, while PRKAR2B, gene found hypermethylated in the parous women, had two DMRs near the end of the gene. Every gene within the 583 genes was closely examined for DMRs and we found 50 genes with DMRs. Of the 455 parous hypermethylated genes, we identified 39 DMRs (Table 10.3) and among 128 nulliparous hypermethylated genes, 11 DMRs were identified (Table 10.4).

10.5.3 Analysis of Hypermethylated Genes in Parous and Nulliparous Postmenopausal Breast

In order to understand the biological processes represented among the genes found hypermethylated, we performed gene ontology (GO) functional clustering through DAVID bioinformatics resources [61]. DAVID bioinformatics resource is an integrated biological database combined with tools to extract biological meaning from large lists of genes [61]. We uploaded the lists of hypermethylated genes and performed functional cluster analysis, only those clusters statistically significant ($P<0.05$) according to a two-tailed Fisher's exact test were considered for later analysis.

GO functional clustering placed these genes within biological processes such as regulation of transcription, cell motility, cell adhesion, cell cycle control, DNA

Table 10.3 Differentially methylated regions (DMRs) within parous hypermethylated genes

Gene	Chromosome	DMR location
NEGR1	chr1	71702567–71703327
		72142369–72142934
NUF2	chr1	161576182–161576653
SYT14	chr1	208309959–208310406
		208206495–208206910
POU4F1	chr13	78072725–78073146
FLRT2	chr14	85155301–85155789
ASAP2	chr2	9266977–9267464
		9432659–9433115
IFITM4P	Chr6	29826792–29827266
ZNF292	Chr6	88022117–88022631
SDK1	Chr7	4121961–4122279
		4230104–4230384
ELAVL4	Chr1	50387715–50388146
DACT1	Chr14	58182547–58182717
SPATA5L1	Chr15	43494615–43495210
DYNC1I2	Chr2	172279940–172280462
NLGN1	Chr3	175147546–175148159
		175156296–175156626
		175277928–175278476
MAN1A1	Chr6	119623891–119624320
AK5	Chr1	77616541–77616886
		77655265–77655548
DPYD	Chr1	98153997–98154252
PROX1	Chr1	212267523–212267905
PDE3A	Chr12	20432463–20432808
NOVA1	Chr14	26015695–26016215
SKAP1	Chr17	43591761–43592022
ANKRD12	Chr18	9168269–9168654
B4GALT5	Chr20	47704095–47704520
CNTN4	Chr3	2572819–2573349
ROBO1	Chr3	79026030–79023709
INPP4B	Chr4	143292977–143293319
		143347212–143347585
		143966478–143966985
FNIP2	Chr4	159911129–159911596
		160015288–160015809
IL6ST	Chr5	55271135–55271466
TICAM2	Chr5	114955685–114955992
		114956473–114956938
PPP2CA	Chr5	133567556–133567871
C6orf138	Chr6	48025616–48025836
		48067151–48067418
PRKAR2B	Chr7	106573431–106573642
		106574760–106574889

(continued)

Table 10.3 (continued)

Gene	Chromosome	DMR location
TTLL7	Chr1	84185339–84185660
MAN1A2	Chr1	117816180–117816444
CDC42BPA	Chr1	225520202–225520399
OSBP	Chr11	59121100–59121437
		59121927–59122155
STIM2	Chr4	26572404–26572775
NR3C2	Chr4	149367631–149368052
REV3L	Chr6	111804054–111804285

Table 10.4 DMRs within nulliparous hypermethylated genes

Gene	Chromosome	DMR location
NHSL2	chrX	71270541–71271527
C16orf38 (PTX4)	Chr16	1476600–1476773
LRRC37A3	Chr17	60311872–60311982
TPPP	Chr5	742334–742618
NELF	Chr9	139471353–139471653 (hypo)
		139471653–139471895
SAMD10	Chr20	62077471–62077661
CELSR1	Chr22	45272965–45273071
FZD1	Chr7	90733372–90733621
TNFRSF18	Chr1	1130349–1130634
SRMS	Chr20	61646714–61647041
COBRA1	Chr9	139285424–139285977

repair, and apoptosis. Of interest was the cluster enriched by transcriptional regulation factors, as the methylation of transcription factors, transcriptional co-activators, and histone-modifying enzymes could have a widespread effect upon many cellular processes that could be correlated with the reduced occurrence of breast neoplasia within parous women. Seventy six (18 %) of the hypermethylated genes in the parous breast were associated with regulation of transcription (Table 10.5). In contrast, in the nulliparous breast, 29 (11 %) of the hypermethylated genes were related to regulation of transcription (Table 10.6). These hypermethylated transcriptional regulators were combined with the genes found to be differentially expressed between parous and nulliparous postmenopausal breast [6] and were analyzed using IPA software (Ingenuity® Systems, www.ingenuity.com), in order to investigate the relationships between the methylation and expression landscapes.

Through analysis of the interactions, we identified several methylated genes encoding histone-modifying enzymes, interacting with TFF1, which was up-regulated in the parous women. TFF1 (trefoil factor 1), a gene up-regulated within the parous postmenopausal breast, may be regulated by several genes, of which include histone deacetylase 2 (HDAC2), nuclear receptor-binding SET domain protein 1 (NSD1),

Table 10.5 Parous hypermethylated genes that enriched the functional cluster related to regulation of transcription ($p = 0.04$)

Symbol	Gene name
AKIRIN2	Akirin 2
ARID5B	AT-rich interactive domain 5B (MRF1-like)
ARNTL	Aryl hydrocarbon receptor nuclear translocator-like
ASCL1	Achaete-scute complex homolog 1
ATF1	Activating transcription factor 1
BHLHE41	Basic helix-loop-helix family, member e41
C21orf66	PAX3- and PAX7-binding protein 1
CHD6	Chromodomain helicase DNA-binding protein 6
CNOT2	CCR4-NOT transcription complex, subunit 2
DMRTA1	DMRT-like family A1
DMTF1	Cyclin D-binding myb-like transcription factor 1
EPC2	Enhancer of polycomb homolog 2 (*Drosophila*)
ESR2	Estrogen receptor 2 (ER beta)
ETV3	Ets variant 3
GFI1	Growth factor independent 1 transcription repressor
HDAC2	Histone deacetylase 2
HMBOX1	Homeobox containing 1
HOPX	HOP homeobox
HOXB13	Homeobox B13
HOXD13	Homeobox D13
HSF2	Heat shock transcription factor 2
ISL1	ISL LIM homeobox 1
KLF8	Kruppel-like factor 8
MAF	V-maf musculoaponeurotic fibrosarcoma oncogene homolog (avian)
MBTD1	Mbt domain containing 1
MIER1	Mesoderm induction early response 1 homolog (Xenopus laevis)
MOV10	Mov10, Moloney leukemia virus 10, homolog (mouse)
MSX2	Msh homeobox 2
MYCBP2	MYC-binding protein 2, E3 ubiquitin protein ligase
NCOA7	Nuclear receptor co-activator 7
NFIB	Nuclear factor I/B
NR2F2	Nuclear receptor subfamily 2, group F, member 2
NR3C1	Nuclear receptor subfamily 3, group C, member 1 (glucocorticoid receptor)
NR3C2	Nuclear receptor subfamily 3, group C, member 2
NSD1	Nuclear receptor-binding SET domain protein 1
PARP15	Poly (ADP-ribose) polymerase family, member 15
PHF12	PHD finger protein 12
POU4F1	POU class 4 homeobox 1
PPP2CA	Protein phosphatase 2, catalytic subunit, alpha isozyme
PRDM12	PR domain containing 12
PRKAR1A	Protein kinase, cAMP-dependent, regulatory, type I, alpha
PRMT6	Protein arginine methyltransferase 6
PROX1	Prospero homeobox 1
PTPRK	Protein tyrosine phosphatase, receptor type, K

(continued)

Table 10.5 (continued)

Symbol	Gene name
RCBTB1	Regulator of chromosome condensation (RCC1) and BTB (POZ) domain containing protein 1
RFX3	Regulatory factor X, 3 (influences HLA class II expression)
RGMB	RGM domain family, member B
RIPK2	Receptor-interacting serine-threonine kinase 2
RORB	RAR-related orphan receptor B
SKAP1	Src kinase-associated phosphoprotein 1
SMAD9	SMAD family member 9
STRAP	Serine/threonine kinase receptor-associated protein
STRN3	Striatin, calmodulin-binding protein 3
SUPT16H	Suppressor of Ty 16 homolog (*S. cerevisiae*)
TAF1D	TATA box-binding protein (TBP)-associated factor, RNA polymerase I, D, 41 kDa
TFAP2C	Transcription factor AP-2 gamma (activating enhancer-binding protein 2 gamma)
TFAP2D	Transcription factor AP-2 delta (activating enhancer-binding protein 2 delta)
TLE4	Transducin-like enhancer of split 4 (E(sp1) homolog, *Drosophila*)
UBC	Ubiquitin C
VEZF1	Vascular endothelial zinc finger 1
VGLL3	Vestigial like 3 (*Drosophila*)
WASL	Wiskott–Aldrich syndrome-like
ZBTB41	Zinc finger and BTB domain containing 41
ZCCHC12	Zinc finger, CCHC domain containing 12
ZFPM2	Zinc finger protein, FOG family member 2
ZNF219	Zinc finger protein 219
ZNF230	Zinc finger protein 230
ZNF24	Zinc finger protein 24
ZNF292	Zinc finger protein 292
ZNF300	Zinc finger protein 300
ZNF492	Zinc finger protein 492
ZNF518A	Zinc finger protein 518A
ZNF542	Zinc finger protein 542
ZNF620	Zinc finger protein 620
ZNF642	ZFP69 zinc finger protein
ZNF793	Zinc finger protein 793

protein arginine *N*-methyltransferase 6 (PRMT6), and nuclear receptor co-activator 3 (NCOA3). HDAC2, NSD1, and PRMT6 were found hypermethylated within the parous breast, and NCOA3 was hypermethylated in the nulliparous woman. It is possible that methylation of HDAC2, PRMT6, NSD1, and NCOA3 resulted in a specific histone code that allowed for expression of a variety of genes including TFF1 in the parous breast. The gene network formed in IPA indicated that HDAC2 directly interacted with TFF1, which we could show an association with the TFF1 promoter. PRMT6 and NSD1 indirectly affected TFF1 expression, which could be through modification of the H3 N-termini. NCOA3, hypermethylated in the nulliparous women, and known co-activator of ERα could be also playing a role in the expression of TFF1 [62].

Table 10.6 Nulliparous hypermethylated genes that enriched the functional cluster related to regulation of transcription ($p=0.005$)

Symbol	Gene name
CEBPB	CCAAT/enhancer-binding protein (C/EBP), beta
DMRT2	DMRT2 doublesex and mab-3-related transcription factor 2
ESX1	ESX homeobox 1
ETV2	Ets variant 2
FOXJ1	Forkhead box J1
FZD1	Frizzled family receptor 1
GPR123	G protein-coupled receptor 123
GRAMD4	GRAM domain containing 4
PRIC285	Helicase with zinc finger 2
HCFC1	Host cell factor C1 (VP16-accessory protein)
IRF7	Interferon regulatory factor 7
IRX1	Iroquois homeobox 1
IRX2	Iroquois homeobox 2
IRX5	Iroquois homeobox 5
COBRA1	Negative elongation factor complex member B
NCOA3	Nuclear receptor co-activator 3
PITX2	Paired-like homeodomain 2
PEG3	Paternally expressed 3
PQBP1	Polyglutamine-binding protein 1
RAX2	Retina and anterior neural fold homeobox 2
SOHLH1	Spermatogenesis and oogenesis-specific basic helix-loop-helix 1
SOX10	SRY (sex determining region Y)-box 10
SOX3	SRY (sex determining region Y)-box 3
TBX3	T-box 3
THAP11	THAP domain containing 11
TULP3	Tubby-like protein 3
MAFB	V-maf musculoaponeurotic fibrosarcoma oncogene homolog B
ZNF414	Zinc finger protein 414
ZIM2	Zinc finger, imprinted 2

10.6 In Search of an Epigenetic Pathway

10.6.1 The Role of TFF1

TFF1 increased expression in the parous postmenopausal breast may be the result of the underlying epigenetic landscape. Accordingly, the methylated states of HDAC2, NSD1, PRMT6, and NCOA3 may allow for a unique histone code that induces TFF1 expression. TFF1 expression is of particular significance in ER-positive breast cancer, where it may exert control over the cell cycle and promote both proliferation and invasiveness [63]. If TFF1 serves as a mitogen, we question its expression within the differentiated parous postmenopausal breast. Interestingly, a recent study that included TFF1 gain-of-function and loss-of-function within normal and malignant breast cells found that TFF1 does not exert control over the cell cycle; however, it does enhance motility and invasiveness [64]. TFF1 has been found to be secreted as a component of breast milk in human [65]. While TFF1 expression may appear dependent upon ERα for transcription [66], Peri et al. have not found a significant difference in ERα expression between the parous and nulliparous postmenopausal breast [6]. Certain ERα-dependent genes in the parous postmenopausal breast were up-regulated, suggesting that TFF1 expression in the parous postmenopausal breast may be mediated by an alternate pathway [6]. Interestingly, NCOA3 serves as a co-activator of ERα [62], and this gene was hypermethylated in the nulliparous breast. If members of the nuclear co-activator family are expressed in the parous breast, specifically a known co-activator of ERα, it is possible that they are functioning as members of other co-activator complexes with other receptors to stimulate transcription. A signaling cascade that does not require estrogen and ERα may be activated and may lead to expression of such genes as TFF1.

10.6.2 The Role of HDAC2

HDAC2, which acts contrary to the previously described HATs [67], occupies the TFF1 promoter in E2-deplete MCF-7 cells. Addition of E2 resulted in the eventual reduction in HDAC2 at the TFF1 promoter, increased association of ER and cofactors at the TFF1 promoter, and increased acetylation of core histones 3 and 4 [68]. HDAC2 clearly modulates chromatin acylation, favoring increased acetylation in the presence of E2. Baek et al. [69] found that HDAC2 was bound to the TFF1 promoter in tamoxifen-treated MCF-7 cells. N-CoR and mSin3 were also found to occupy the TFF1 promoter in and ER-antagonist bound environment, suggesting that HDAC2 is a member of this co-repressor complex [69]. The absence of HDAC2 in the parous breast may abrogate the repressive function of its member complexes. In this way, ERα regulated genes that would be repressed by occupancy of an HDAC2-containing co-repressor complex may instead exist in a more transcriptionally favorable state.

10.6.3 The Role of NSD1 and PRMT6

NSD1 is a member of the SET domain containing family of methyltransferases. This methyltransferase di-methylates lysine 36 of H3 [44, 70]. PRTM6 asymmetrically di-methylates arginine 3 of H3, a modification that may block H3K4me3 [45]. Taken together, the silencing of HDAC2, NSD1, and PRTM6 may create a unique histone code of increased acetylation at histone N-terminal lysine residues, and perhaps transcriptionally favorable methylation marks.

10.6.4 The Role of NCOA3

NCOA3 is a member of the p160 steroid receptor co-activator family and has been found to play a pivotal role in important biological processes such as somatic growth, reproductive function, uterine growth, blastocyst implantation, and mammary gland differentiation as well as many types of cancers where the deletion or overexpression results in formation of neoplasms. NCOA3 initiates chromatin remodeling and activates gene expression by recruiting HATs as well as methyltransferases when interacting with nuclear receptors or transcription factors. Estrogen receptor transcriptional activity is enhanced by the binding of NCOA3. NCOA3 then recruits the HATs CBP/p300 and PCAF. NCOA3 is methylated the nulliparous postmenopausal breast tissue, which can impair the expression of a number of different genes such as Nanog, Oct4, and Sox2 [71, 72].

10.6.5 Other Epigenomic Pathways That Can Regulate Differentiation of the Human Breast

We have also observed the hypermethylation of key players in the TGF-β signaling pathway, notably TGF-β receptor 2 (TGFBR2). Ligand-bound TGFBR2 recruits and phosphorylates TGF-β receptor 1 (TGFBR1) [73]. TGFBR1 then activates the downstream effector proteins. In this case, the effector proteins are members of the SMAD family of transcription factors. Phosphorylated SMADS (RSMADS) associate with other SMADs, usually SMAD4, and promote transcription of target genes [74]. We have also identified a potential downstream target of the TGF-β signaling cascade, connective tissue growth factor (CTGF). Expression of CTGF is substantially increased within human skin fibroblasts after the addition of TGF-β [75], and being that TGF-β is a specific ligand of TGFBR2 [73], it can be inferred that this receptor would be necessary for CTGF transactivation. That a growth factor is methylated in the parous postmenopausal breast is a step towards an explanation of the protective effect of this epigenomic signature against breast cancer. Growth factors are mitogenic by nature, and the selective methylation of these genes would account for the observed difference in proliferation between the Lob1 of the nulliparous breast and the Lob1 of the parous breast [3, 58]. Further analysis of the significance of CTGF methylation, as well as a more complete understanding of TGF-β signaling and its apparent suppression in the parous postmenopausal breast is necessary.

In the nulliparous postmenopausal breast, we also found CCAAT/enhancer-binding protein beta (CEBPB) to be hypermethylated. CEBPB is a bZIP transcription factor that controls the differentiation and development of several different types of tissues. This gene may play a significant role in the increased risk for developing breast cancer in the nulliparous postmenopausal breast. Deletion studies of CEBPB have shown to greatly inhibit lobulo-alveolar development and differentiation, as well as induce alterations in ductal morphogenesis [76]. The deletion of CEBPB in germline alters progenitor cell fate, as well as altered expression levels of ERα, progesterone receptor (PGR), and prolactin receptor (PRLR). CEBPB has two different protein isoforms: liver-enriched activating protein (LAP) and liver-enriched inhibitory protein (LIP). Tumorigenesis in the mammary gland is caused by the dysregulated expression of these isoforms. The methylation of a transcription factor involved in the development and differentiation of several different types of tissue can clearly result in dysregulated pathways. Methylation of CEBPB has been shown to impair the development and differentiation of the mammary gland [77].

Literature Cited

1. Ferlay J, Shin HR, Bray F, Forman D, Mathers C, Parkin DM (2010) Estimates of worldwide burden of cancer in 2008: GLOBOCAN 2008. Int J Cancer 127:2893–2917
2. Russo IH, Koszalka M, Russo J (1991) Comparative study of the influence of pregnancy and hormonal treatment on mammary carcinogenesis. Br J Cancer 64:481–484
3. Russo J, Russo IH (2004) Molecular basis of breast cancer: prevention and treatment. Springer, Berlin
4. Russo J, Rivera R, Russo IH (1992) Influence of age and parity on the development of the human breast. Breast Cancer Res Treat 23:211–218
5. Asztalos S, Gann PH, Hayes MK, Nonn L, Beam CA, Dai Y, Wiley EL, Tonetti DA (2010) Gene expression patterns in the human breast after pregnancy. Cancer Prev Res (Phila) 3:301–311
6. Peri S, de Cicco RL, Santucci-Pereira J, Slifker M, Ross EA, Russo IH, Russo PA, Arslan AA, Belitskaya-Levy I, Zeleniuch-Jacquotte A, Bordas P, Lenner P, Ahman J, Afanasyeva Y, Johansson R, Sheriff F, Hallmans G, Toniolo P, Russo J (2012) Defining the genomic signature of the parous breast. BMC Med Genomics 5:46
7. Russo J, Santucci-Pereira J, de Cicco RL, Sheriff F, Russo PA, Peri S, Slifker M, Ross E, Mello ML, Vidal BC, Belitskaya-Levy I, Arslan A, Zeleniuch-Jacquotte A, Bordas P, Lenner P, Ahman J, Afanasyeva Y, Hallmans G, Toniolo P, Russo IH (2012) Pregnancy-induced chromatin remodeling in the breast of postmenopausal women. Int J Cancer 131:1059–1070
8. Belitskaya-Levy I, Zeleniuch-Jacquotte A, Russo J, Russo IH, Bordas P, Ahman J, Afanasyeva Y, Johansson R, Lenner P, Li X, de Cicco RL, Peri S, Ross E, Russo PA, Santucci-Pereira J, Sheriff FS, Slifker M, Hallmans G, Toniolo P, Arslan AA (2011) Characterization of a genomic signature of pregnancy identified in the breast. Cancer Prev Res (Phila) 4:1457–1464
9. Jablonka E, Lamb MJ (2002) The changing concept of epigenetics. Ann N Y Acad Sci 981:82–96
10. Holliday R (1987) The inheritance of epigenetic defects. Science 238:163–170
11. Holliday R (2005) DNA methylation and epigenotypes. Biochemistry (Mosc) 70:500–504
12. Cohn WE (1951) The isolation and identification of desoxy-5-methylcytidylic acid from thymus nucleic acid1. J Am Chem Soc 73:1539–1541
13. Holliday R (2006) Epigenetics: a historical overview. Epigenetics 1:76–80
14. Doskocil J, Sorm F (1962) Distribution of 5-methylcytosine in pyrimidine sequences of deoxyribonucleic acids. Biochim Biophys Acta 55:953–959
15. Doerfler W (1981) DNA methylation–a regulatory signal in eukaryotic gene expression. J Gen Virol 57:1–20
16. Gardiner-Garden M, Frommer M (1987) CpG islands in vertebrate genomes. J Mol Biol 196:261–282
17. Holliday R, Pugh JE (1975) DNA modification mechanisms and gene activity during development. Science 187:226–232

18. Futscher BW, Oshiro MM, Wozniak RJ, Holtan N, Hanigan CL, Duan H, Domann FE (2002) Role for DNA methylation in the control of cell type specific maspin expression. Nat Genet 31:175–179
19. Bird A (2002) DNA methylation patterns and epigenetic memory. Genes Dev 16:6–21
20. Deaton AM, Bird A (2011) CpG islands and the regulation of transcription. Genes Dev 25:1010–1022
21. Jaenisch R (1997) DNA methylation and imprinting: why bother? Trends Genet 13:323–329
22. Smith ZD, Chan MM, Mikkelsen TS, Gu H, Gnirke A, Regev A, Meissner A (2012) A unique regulatory phase of DNA methylation in the early mammalian embryo. Nature 484:339–344
23. Wolffe AP, Matzke MA (1999) Epigenetics: regulation through repression. Science 286:481–486
24. Bestor TH (2000) The DNA methyltransferases of mammals. Hum Mol Genet 9:2395–2402
25. Chuang LS, Ian HI, Koh TW, Ng HH, Xu G, Li BF (1997) Human DNA-(cytosine-5) methyltransferase-PCNA complex as a target for p21WAF1. Science 277:1996–2000
26. Balmain A, Gray J, Ponder B (2003) The genetics and genomics of cancer. Nat Genet 33(suppl):238–244
27. Okano M, Bell DW, Haber DA, Li E (1999) DNA methyltransferases Dnmt3a and Dnmt3b are essential for de novo methylation and mammalian development. Cell 99:247–257
28. Hermann A, Gowher H, Jeltsch A (2004) Biochemistry and biology of mammalian DNA methyltransferases. Cell Mol Life Sci 61:2571–2587
29. Klimasauskas S, Kumar S, Roberts RJ, Cheng X (1994) HhaI methyltransferase flips its target base out of the DNA helix. Cell 76:357–369
30. Klug A, Rhodes D, Smith J, Finch JT, Thomas JO (1980) A low resolution structure for the histone core of the nucleosome. Nature 287:509–516
31. Luger K, Mader AW, Richmond RK, Sargent DF, Richmond TJ (1997) Crystal structure of the nucleosome core particle at 2.8 A resolution. Nature 389:251–260
32. Luger K, Richmond TJ (1998) The histone tails of the nucleosome. Curr Opin Genet Dev 8:140–146
33. Strahl BD, Allis CD (2000) The language of covalent histone modifications. Nature 403:41–45
34. Sterner DE, Berger SL (2000) Acetylation of histones and transcription-related factors. Microbiol Mol Biol Rev 64:435–459
35. Bannister AJ, Kouzarides T (2011) Regulation of chromatin by histone modifications. Cell Res 21:381–395
36. Brownell JE, Allis CD (1996) Special HATs for special occasions: linking histone acetylation to chromatin assembly and gene activation. Curr Opin Genet Dev 6:176–184
37. Parthun MR (2007) Hat1: the emerging cellular roles of a type B histone acetyltransferase. Oncogene 26:5319–5328
38. Tanner KG, Trievel RC, Kuo MH, Howard RM, Berger SL, Allis CD, Marmorstein R, Denu JM (1999) Catalytic mechanism and function of invariant glutamic acid 173 from the histone acetyltransferase GCN5 transcriptional coactivator. J Biol Chem 274:18157–18160
39. Grant PA, Eberharter A, John S, Cook RG, Turner BM, Workman JL (1999) Expanded lysine acetylation specificity of Gcn5 in native complexes. J Biol Chem 274:5895–5900
40. Lan F, Shi Y (2009) Epigenetic regulation: methylation of histone and non-histone proteins. Sci China C Life Sci 52:311–322
41. Bedford MT (2007) Arginine methylation at a glance. J Cell Sci 120:4243–4246
42. Smith BC, Denu JM (2009) Chemical mechanisms of histone lysine and arginine modifications. Biochim Biophys Acta 1789:45–57
43. Barski A, Cuddapah S, Cui K, Roh TY, Schones DE, Wang Z, Wei G, Chepelev I, Zhao K (2007) High-resolution profiling of histone methylations in the human genome. Cell 129:823–837
44. Xu L, Zhao Z, Dong A, Soubigou-Taconnat L, Renou JP, Steinmetz A, Shen WH (2008) Di- and tri- but not monomethylation on histone H3 lysine 36 marks active transcription of genes involved in flowering time regulation and other processes in Arabidopsis thaliana. Mol Cell Biol 28:1348–1360

45. Hyllus D, Stein C, Schnabel K, Schiltz E, Imhof A, Dou Y, Hsieh J, Bauer UM (2007) PRMT6-mediated methylation of R2 in histone H3 antagonizes H3 K4 trimethylation. Genes Dev 21:3369–3380
46. Wang Z, Zang C, Rosenfeld JA, Schones DE, Barski A, Cuddapah S, Cui K, Roh TY, Peng W, Zhang MQ, Zhao K (2008) Combinatorial patterns of histone acetylations and methylations in the human genome. Nat Genet 40:897–903
47. Jeppesen P, Turner BM (1993) The inactive X chromosome in female mammals is distinguished by a lack of histone H4 acetylation, a cytogenetic marker for gene expression. Cell 74: 281–289
48. Brinkman AB, Roelofsen T, Pennings SW, Martens JH, Jenuwein T, Stunnenberg HG (2006) Histone modification patterns associated with the human X chromosome. EMBO Rep 7:628–634
49. Cullum R, Alder O, Hoodless PA (2011) The next generation: using new sequencing technologies to analyse gene regulation. Respirology 16:210–222
50. Perou CM, Sorlie T, Eisen MB, van de Rijn M, Jeffrey SS, Rees CA, Pollack JR, Ross DT, Johnsen H, Akslen LA, Fluge O, Pergamenschikov A, Williams C, Zhu SX, Lonning PE, Borresen-Dale AL, Brown PO, Botstein D (2000) Molecular portraits of human breast tumours. Nature 406:747–752
51. Cavallaro S, Paratore S, de Snoo F, Salomone E, Villari L, Buscarino C, Ferrau F, Banna G, Furci M, Strazzanti A, Cunsolo R, Pezzino S, Gangi S, Basile F (2012) Genomic analysis: toward a new approach in breast cancer management. Crit Rev Oncol Hematol 81:207–223
52. Shendure J, Lieberman Aiden E (2012) The expanding scope of DNA sequencing. Nat Biotechnol 30:1084–1094
53. Zuo T, Tycko B, Liu TM, Lin HJ, Huang TH (2009) Methods in DNA methylation profiling. Epigenomics 1:331–345
54. Brinkman AB, Simmer F, Ma K, Kaan A, Zhu J, Stunnenberg HG (2010) Whole-genome DNA methylation profiling using MethylCap-seq. Methods 52:232–236
55. Gargiulo G, Levy S, Bucci G, Romanenghi M, Fornasari L, Beeson KY, Goldberg SM, Cesaroni M, Ballarini M, Santoro F, Bezman N, Frige G, Gregory PD, Holmes MC, Strausberg RL, Pelicci PG, Urnov FD, Minucci S (2009) NA-Seq: a discovery tool for the analysis of chromatin structure and dynamics during differentiation. Dev Cell 16:466–481
56. Azvolinsky A, Giresi PG, Lieb JD, Zakian VA (2009) Highly transcribed RNA polymerase II genes are impediments to replication fork progression in Saccharomyces cerevisiae. Mol Cell 34:722–734
57. Simon JM, Giresi PG, Davis IJ, Lieb JD (2012) Using formaldehyde-assisted isolation of regulatory elements (FAIRE) to isolate active regulatory DNA. Nat Protoc 7:256–267
58. Russo J, Russo IH (2004) Development of the human breast. Maturitas 49:2–15
59. Robinson JT, Thorvaldsdottir H, Winckler W, Guttman M, Lander ES, Getz G, Mesirov JP (2011) Integrative genomics viewer. Nat Biotechnol 29:24–26
60. Thorvaldsdottir H, Robinson JT, Mesirov JP (2013) Integrative Genomics Viewer (IGV): high-performance genomics data visualization and exploration. Brief Bioinform 14:178–192
61. da Huang W, Sherman BT, Lempicki RA (2009) Systematic and integrative analysis of large gene lists using DAVID bioinformatics resources. Nat Protoc 4:44–57
62. Robyr D, Wolffe AP, Wahli W (2000) Nuclear hormone receptor coregulators in action: diversity for shared tasks. Mol Endocrinol 14:329–347
63. Perry JK, Kannan N, Grandison PM, Mitchell MD, Lobie PE (2008) Are trefoil factors oncogenic? Trends Endocrinol Metab 19:74–81
64. Buache E, Etique N, Alpy F, Stoll I, Muckensturm M, Reina-San-Martin B, Chenard MP, Tomasetto C, Rio MC (2011) Deficiency in trefoil factor 1 (TFF1) increases tumorigenicity of human breast cancer cells and mammary tumor development in TFF1-knockout mice. Oncogene 30:3261–3273
65. Vestergaard EM, Nexo E, Wendt A, Guthmann F (2008) Trefoil factors in human milk. Early Hum Dev 84:631–635

66. Sun JM, Spencer VA, Li L, Yu Chen H, Yu J, Davie JR (2005) Estrogen regulation of trefoil factor 1 expression by estrogen receptor alpha and Sp proteins. Exp Cell Res 302:96–107
67. Cress WD, Seto E (2000) Histone deacetylases, transcriptional control, and cancer. J Cell Physiol 184:1–16
68. Sun JM, Chen HY, Davie JR (2001) Effect of estradiol on histone acetylation dynamics in human breast cancer cells. J Biol Chem 276:49435–49442
69. Baek SH, Ohgi KA, Rose DW, Koo EH, Glass CK, Rosenfeld MG (2002) Exchange of N-CoR corepressor and Tip60 coactivator complexes links gene expression by NF-kappaB and beta-amyloid precursor protein. Cell 110:55–67
70. Rayasam GV, Wendling O, Angrand PO, Mark M, Niederreither K, Song L, Lerouge T, Hager GL, Chambon P, Losson R (2003) NSD1 is essential for early post-implantation development and has a catalytically active SET domain. EMBO J 22:3153–3163
71. Gojis O, Rudraraju B, Gudi M, Hogben K, Sousha S, Coombes RC, Cleator S, Palmieri C (2010) The role of SRC-3 in human breast cancer. Nat Rev Clin Oncol 7:83–89
72. Wu Z, Yang M, Liu H, Guo H, Wang Y, Cheng H, Chen L (2012) Role of nuclear receptor coactivator 3 (Ncoa3) in pluripotency maintenance. J Biol Chem 287:38295–38304
73. Wrana JL, Attisano L, Wieser R, Ventura F, Massague J (1994) Mechanism of activation of the TGF-beta receptor. Nature 370:341–347
74. Massague J (2000) How cells read TGF-beta signals. Nat Rev Mol Cell Biol 1:169–178
75. Grotendorst GR, Okochi H, Hayashi N (1996) A novel transforming growth factor beta response element controls the expression of the connective tissue growth factor gene. Cell Growth Differ 7:469–480
76. Grimm SL, Rosen JM (2003) The role of C/EBPbeta in mammary gland development and breast cancer. J Mammary Gland Biol Neoplasia 8:191–204
77. Russell A, Boone B, Jiang A, Sealy L (2010) Genomic profiling of C/EBPbeta2 transformed mammary epithelial cells: a role for nuclear interleukin-1beta. Cancer Biol Ther 10:509–519

Index

J. Russo and I.H. Russo, *Techniques and Methodological Approaches in Breast Cancer Research*, DOI 10.1007/978-1-4939-0718-2,

MIX
Papier aus verantwortungsvollen Quellen
Paper from responsible sources
FSC® C105338

If you have any concerns about our products,
you can contact us on
ProductSafety@springernature.com

In case Publisher is established outside the EU,
the EU authorized representative is:
Springer Nature Customer Service Center GmbH
Europaplatz 3, 69115 Heidelberg, Germany

Printed by Libri Plureos GmbH
in Hamburg, Germany